Environmental Science and Engineering
Subseries: Environmental Science

Series Editors: R. Allan • U. Förstner • W. Salomons

Susanne Stoll-Kleemann
Martin Welp (Eds.)

Stakeholder Dialogues in Natural Resources Management

Theory and Practice

With 20 Figures

Springer

EDITORS:

PD DR. SUSANNE STOLL-KLEEMANN
HUMBOLDT-UNIVERSITÄT ZU BERLIN, DEPARTMENT OF AGRICULTURAL ECONOMICS AND SOCIAL SCIENCES, RESEARCH GROUP GOBI,
LUISENSTR. 53
10099 BERLIN, GERMANY

E-mail: susanne.stoll-kleemann@agrar.hu-berlin.de

PROF. DR. MARTIN WELP
SOCIOECONOMICS AND COMMUNICATION
UNIVERSITY OF APPLIED SCIENCES EBERSWALDE, FACULTY OF FORESTRY
ALFRED-MOELLER-STR. 1
16225 EBERSWALDE, GERMANY

E-mail: martin.welp@fh-eberswalde.de

ISSN- 1863-5520
ISBN 10 3-540-36916-3 **Springer Berlin Heidelberg New York**
ISBN 13 978-3-540-36916-5 **Springer Berlin Heidelberg New York**

Library of Congress Control Number: 2006934202

Springer is a part of Springer Science+Business Media
springeronline.com

Cover design: E. Kirchner, Heidelberg
Production: A. Oelschläger
Typesetting: Camera-ready by the Editors
Printed on acid-free paper 30/2132/AO 543210

To Luisa and Elina

Contents

Part II Theories and Tools

Part III Case Studies in Environmental Policy, Management and Science

Part IV Perspectives

List of Figures

List of Tables

List of Contributors

Dr. Christiane Averbeck is executive director of Transfer-21, a Germany-wide program on education for sustainable development. She has worked as a consultant for the German Council for Sustainable Development, conducted a research project on sustainable resource use in Uganda, worked as a senior advisor for a German NGO in Uganda, and as senior researcher in a project funded by the Federal Environmental Agency on marine pollution. She holds a Master's degree in biology and a Ph.D. in natural science.

Augustin Berghöfer works for the Omora Foundation in the recently established Cape Horn Biosphere Reserve, coordinating the outreach activities of this small Chilean NGO dedicated to bio-cultural research and conservation. With a background in economics and political science, he has been studying the obstacles to participatory endeavours in post-conflict development assistance. As member of Susanne Stoll-Kleemann's research group on biodiversity governance, he investigated experiences of participation in protected area management.

Uta Berghöfer works at the UFZ Centre for Environmental Research in Leipzig. Trained in geography, she is currently completing her Ph.D., exploring the diverse ways that humans relate to nature and the significance of these relationships for biodiversity conservation. As a member of BIOKONCHIL - a German-Chilean research project studying possibilities for implementing the Ecosystem Approach of the Convention on Biological Diversity - she has been conducting qualitative social research in southern Chile since 1998. In this context she has accompanied the initiative to establish the Cape Horn Biosphere Reserve, advising in the design and organisation of the participatory processes and the associated local information campaign.

Melanie Eben studied ecology at the University of East Anglia (UK) and obtained a Master´s degree in conservation from University College London. Her interest in conservation, environmental education, monitoring systems, and participatory methods and instruments has led her to undertake research in various Latin American countries. She is currently living in Ecuador, where she is working as a freelance consultant.

Cornelia Fürstenau is a junior research scientist at the Potsdam Institute for Climate Impact Research (Department of Global Change and Natural Systems). She has a university degree (Diploma) in forest science. Her research focuses on the impact of forest management and climate change on different forest functions such as carbon sequestration, ground water recharge, income from timber production, and biodiversity in temperate forest ecosystems. In a science-based stakeholder dialogue, she has looked into the interests of forest user groups in the management of forest ecosystems, their goods and services, and their awareness of the future impacts of climate change on forest ecosystems.

Dr. Eeva Hellström is director of the Forest Academy for Decision-Makers, which is a discussion forum on forest issues directed at top-level decision makers throughout Finnish society. It is organised by the Finnish Forest Association, which is a co-operation and communication body that links the Finnish forest sector and related fields. She has also held numerous positions of trust related to forest policy, natural resource businesses, professional unions, rural employment, and forest science. For example, Eeva was a member of the steering group of Finland's National Forest Program during its formative period (1998-99). As a member of the Committee for Forest Protection in Southern Finland (2000-02), she chaired a working group assigned to develop new policy means of forest protection.

Dr. Jörg Linke is working for the German technical cooperation agency GTZ, where he is currently serving as an advisor in the Tunisian Ministry of Environment. In this role he coordinates two projects: i) Implementation of the United Nations Convention on Combat of Desertification (UNCCD), ii) Implementation of the UN Framework Convention on Climate Change (UNFCCC). Before this, he was coordinator of the GTZ programme component: "Sustainable management of forests, protected areas and forestry politics" in the Ministry of Environment of Ecuador. He has a Master's Degree in Forestry Science in the temperate zones and a Master's Degree in Forestry in the tropics and subtropics, as well as a Ph.D. in Forestry. Jörg also has gained broad knowledge working with pluralistic stakeholder forums (including indigenous communities and the private sector) and from experience in the conceptualisation and implementation of financing instruments for environmental management and protection, including payment systems for environmental services.

Dr. Marleen Maarleveld studied social and organizational psychology at Leyden University, the Netherlands. She worked as a researcher at the Department of Communication and Innovation Studies, Wageningen University and Research. In addition to her Ph.D. research on social-environmental learning for sustainable natural research management, she has facilitated participatory projects in water management and management of change. She worked as a knowledge and innovation manager in the field of planning and governance and as a program manager leading an organizational development program that aimed to improve the "customer-orientation" of the organization. Currently she is working as a consultant in the field of water management and spatial development for Arcadis, an engineering and consultancy firm.

Dr. Angela Oels is Assistant Professor in International Relations at the Institute of Political Science at the University of Hamburg. She teaches the politics of international trade, environmental politics, globalisation/global governance and political theory. Dr. Oels was trained in environmental engineering and has a Ph.D. in environmental sciences. Her Ph.D. was published by LIT-Verlag under the title 'Evaluating Stakeholder Participation in the Transition to Sustainable Development: methodology, case studies, and policy implications'. She is currently working on a book project toward her German post-doctoral Habilitation degree that draws on Foucault's concept of governmentality to discuss current changes in the role and power of the state in multi-level, multi-actor governance. For a detailed CV see www.angelaoels.de.

Prof. Dr. Tim O'Riordan is Professor of Environmental Sciences at the University of East Anglia and closely associated with the Tyndall Centre for Climate Change Research, the Leverhulme Programme on Understanding Risk, and the Centre for Social and Economic Research on the Global Environment. Tim began his studies in geography at the Edinburgh University, in his home City of Edinburgh, and received his Ph.D. in Geography at Cambridge University. After a spell of teaching geography at Simon Fraser University, Vancouver, British Columbia, Canada, he took up a visiting lectureship in geography at the University of Canterbury in 1971. Returning to East Anglia in 1974, he embarked on a course in conventional politics and wrote a book on the meaning and role of environmentalism. This was the beginning of a series of publications on environmental politics and law, citizen science, the precautionary principle, and the tortuous transition to sustainable development. His research spans countryside management, biodiversity politics, risk and nuclear power, community involvement in sustainable futures, and the deeper politics of sustainability in the contemporary age. Tim is a member of the UK Sustainable Development Commission and an academic advisor to a number of integrated scientific bodies. He also works with business on the transition to sustainability.

Irene Pleizier studied earth sciences at the Vrije Universiteit Amsterdam. At the end of her studies, she undertook an internship at the company Geodan, working on 3D visualisation and 3D GIS. After finishing her studies, she started working both for Spinlab as a researcher on 3D visualisation and for the research and development department of Geodan. Her main focuses at Geodan are 3D visualisation of large spatial areas and GIS for Secondary education. At the Vrije Universiteit, Irene is currently working on her Ph.D. on the influence of GIS on the increase of knowledge of secondary school geography students. This is being done within the EduGIS project (www.edugis.nl).

Prof. Dr. Ortwin Renn serves as full professor and chair of environmental sociology at Stuttgart University. He directs the Interdisciplinary Research Unit for Risk Governance and Sustainable Technology Development (ZIRN) at the University of Stuttgart and the non-profit company DIALOGIK, a research institute for the investigation of communication and participation processes in environmental policy making. Ortwin Renn has a doctoral degree in sociology and social psychology from the University of Cologne. He is a member of the panel on "Public Participation in Environmental Assessment and Decision Making" of the U.S. National Academy of Sciences in Washington, D.C., a member of the Berlin-Brandenburg Academy of Sciences, the German Academy for Technology and Engineering, and the European Academy of Science and Arts (Vienna and Salzburg). His honours include the "Distinguished Achievement Award" of the Society for Risk Analysis (SRA) and the Outstanding Publication Award from the Environment and Technology Section of the American Sociological Association for the book "Risk, Uncertainty and Rational Action“ co-authored with C. Jaeger, G. Rosa und Th. Webler. Among his political activities is the chairmanship of the State Commission for Sustainable Development (German State of Baden-Württemberg). Renn is primarily interested in risk governance, political participation, and technology assessment. He has published more than 30 books and 200 articles.

Jorge Samaniego Rivera is a marine biologist. He has worked in environmental education as a staff member of the Machalilla National Park and the local authority of Puerto López, Ecuador. In this function he has worked as a counterpart of the German Development Service DED. He is currently an advisor in the Corpocación de Manejo Forestal Sustentable (COMAFORS) in the province of Manabí, Ecuador.

Prof. Dr. Jürgen Scheffran is a senior research scientist with ACDIS at the University of Illinois at Urbana-Champaign, and has adjunct faculty positions at the Departments of Political Science and Atmospheric Sciences. After his Ph.D. in Physics at the University of Marburg in Germany, he worked as a researcher and assistant professor in the interdisciplinary research group IANUS and the mathematics department at the Technical University of Darmstadt. After a research project at Hamburg University, he joined the Potsdam Institute for Climate Impact Research in 2001, and in 2003 he was temporary Visiting Professor at the University of Paris (Pantheon/Sorbonne). His research and teaching interests include energy, environment and climate change; complex systems analysis and computer modeling; technology assessment, arms control and international security. Currently he is coordinating a research project on renewable energy and land use.

PD Dr. Susanne Stoll-Kleemann is an Associate Professor at the Humboldt University of Berlin and is trained in geography and social sciences. She leads the Research Group GoBi (Assessing Biodiversity Governance and Management Approaches). In this research project, together with her five Ph.D. and six Master's students, she investigates success and failure factors of protected area management and governance. Before this recent appointment, she was a senior researcher at the Potsdam Institute for Climate Impact Research (Department of Global Change and Social Systems). Susanne is an interdisciplinary social scientist focusing on human-environment relations, especially stakeholder dialogues. Her research interests are in the human dimensions of global environmental change and sustainability science. Susanne is the Vice-President of the German Society of Human Ecology.

Dr. Michael Sturm started his doctor's degree in Geography at the Technical University of Berlin on sustainable land-use in Monteverde, Costa Rica, finishing it at the Institute for Geography at the Humboldt University Berlin in 1995. He has worked as a National Park Manager for the German Development Service DED in the Machalilla National Park, Ecuador (1995 - 1997) and as a Project Manager of EXPO 2000, a world exhibition project in north Germany, on Implementing Low and Renewable Energy Concepts (1999 - 2001). From the end of 2001 until the end of 2004 he worked as co-ordinator in a project in Flensburg, Germany on strategies of energy reduction in schools (based on the Eco Management and Audit Scheme EMAS). Currently he leads his own data management company, “Agentur Sturm“, in Flensburg. He plans and realizes projects with environmental and infrastructural subject matters, among other things for people with activity limitations.

Joost van Uum graduated with a Master of Science degree in Tropical Civil Engineering and Water Management from the Wageningen University and Research Centre, the Netherlands. As a geo-adviser he has coordinated GIS projects at several governmental institutes. His expertise is in programming, implementing, and the use of GIS. As a teacher at Bureau Nieuwland, he coordinated educational GIS programs and gave GIS courses to civil engineering companies and government organizations. He is currently working at the Government Service for Land and Water Management as geo-adviser and program manager to coordinate the development of new initiatives in GIS.

Dr. Anne Cristina de la Vega-Leinert is an independent researcher on societal perception of climate change, environmental conflicts, the communication of scientific knowledge, social learning, and participatory research. She has an M.Sc. in Quaternary Sedimentology and Geomorphology (University College Dublin) and a Ph.D. in Holocene Coastal Environmental changes (Coventry University). At the Flood Hazard Research Center (London), her research focused on the impacts of accelerated sea-level rise and integrated coastal zone management. At the Potsdam Institute for Climate Impact Research (Germany), she worked as a scientific and stakeholder dialogue coordinator within the DINAS-COAST (http://www.dinas-coast.net/) and the ATEAM projects (http://www.pik-potsdam.de/ateam/). She is currently working in the GoBi project (http://www.biodiversitygovernance.de/), assessing success and failure factors in management of tropical biosphere reserves.

Rob van de Velde studied human geography at the Vrije Universiteit Amsterdam. He has worked as GIS-program manager at the National Physical Planning Agency and the National Institute of Public Health and Environment. He has undertaken applied research on various geo-information issues, such as spatial decision support systems, environmental assessments, and land-use policy scenarios. Currently, he leads the GIS Competence Centre at the Government Service for Land and

Water Management, an executive body of the Ministry of Agriculture, Nature and Food Quality. Since May 2003 he has been part of the Spinlab team of the Vrije Universiteit Amsterdam, where he focuses on research about spatial virtual environments and decision making.

Prof. Dr. Martin Welp serves as professor at the University of Applied Sciences Eberswalde (near Berlin) at the Faculty of Forestry. Before this recent appointment, he was a senior researcher at the Potsdam Institute for Climate Impact Research (Department of Global Change and Social Systems). He holds a Master's Degree in Forestry and a Ph.D. Degree in Agriculture. Martin has researched public participation and stakeholder dialogues in various fields of environmental management, including forest management, coastal management, river basin management, and protected area management. His current research activities focus on linking stakeholder involvement and computer-based modelling in global-change mitigation and adaptation. The European Climate Forum (ECF), which engages researchers, companies, NGOs and other actors in science-based stakeholder dialogues, provides a context and basis for his research on the practice and theory of dialogues.

Acknowledgements

The idea for this book was born after an ambitious workshop organized by the Centre for Environmental Research (UFZ) Halle-Leipzig. The session focussed on participatory methods and multicriteria analysis. During a discussion on the way back home, we concurred that the present literature on natural resources management does not adequately connect participatory approaches with some theories and new tools that in our view are highly relevant. This relates in particular to the new field of stakeholder dialogues and its theoretical underpinning.

Fortunately we were able to convince a few of our colleagues, some of whom were practitioners and some academics, to reflect on the art and practice of stakeholder dialogues. We thank all the authors of the present volume for sharing their specific knowledge, insights, and experiences in the articles. In particular, we are grateful to Prof. Ortwin Renn and Prof. Tim O'Riordan for their valuable observations in the introduction and epilogue.

The stakeholder task force at our former affiliation, the Potsdam Institute for Climate Impact Research, Department of Global Change and Social Systems, provided a valuable platform for new ideas. We would like to thank Prof. Carlo C. Jaeger (Head of Department), Dr. Anne C. de la Vega Leinert, and Antonella Battaglini for many intellectually challenging discussions. Exchanges with many other colleagues in Germany and abroad are also deeply appreciated, in particular those with Dr. Fritz Reusswig, Dr. Hermann Lotze-Campen, Dr. Jürgen Kropp, Prof. Bernhard Glaeser, Prof. Ludwig Ellenberg, Prof. Konrad Ott, Prof. Lenelis Kruse-Graumann, Dr. Marc Hockings, Dr. Irene Ring, Dr. Frank Wätzold, Prof. Uwe Jens Nagel, Prof. Eckart Ehlers, Prof. Craig ZumBrunnen, Monika Bertzky, and Prof. Klaus Hasselmann.

The European Climate Forum provided a platform for dialogue and the interchange of ideas with stakeholders representing companies, NGOs, and policymakers. The German Society for Human Ecology was a valuable academic forum that gave us opportunities to discuss matters dealt with in this book with other colleagues.

We are grateful to the Robert Bosch Stiftung for supporting the production of our work. Marion Mehring did a marvellous job of editing and proofreading the chapters, and we thank Joe Greenman for reviewing the linguistic content of the working draft. The responsibility for the final version lies solely with us.

Finally, we are extremely thankful for the support and patience of our respective partners, Heinz Kleemann and Esther Hoffmann, during the writing and editing process (in particular on weekends). The book is dedicated to our daughters, Luisa and Elina. Both were born in the midst of the genesis of the book.

Susanne Stoll-Kleemann Martin Welp

Part I
Setting the Scene

Foreword

Participatory Processes for Natural Resource Management

Ortwin Renn

University of Stuttgart, Stuttgart, Germany

Need for analytic-deliberative processes

Inviting the public to be part of the decision making process in natural resource management has been a major objective in European and American environmental policy arenas. The US-National Academy of Sciences has encouraged environmental protection agencies to foster citizen participation and public involvement for making environmental policy making and natural resource management more effective and democratic (Stern and Fineberg 1996). The report emphasizes the need for a combination of assessment and dialogue which the authors have framed the "analytic-deliberative" approach. Unfortunately, early public involvement of the public in deliberative processes may compromise, however, the objective of efficient and effective policy implementation or violate the principle of fairness (Cross 1998, Okrent 1998). Another problem is that the public consists of many groups with different value structures and preferences. Without a systematic procedure to reach consensus on values and preferences, the public's position often appears as unclear (Coglianese 1997, Rossi 1997). Participatory processes are thus needed that combine technical expertise, rational decision making, and public values and preferences.

How can and should natural resource managers collect public preferences, integrate public input into the management process, and assign the appropriate roles to technical experts, stakeholders (i.e., socially organized groups that are or perceive themselves as being affected by the decision) and members of the public? Who represents the public? The elected politicians, administrators, stakeholders, or all persons who will be affected by the decision? There is a large amount of individual variance when lay persons are asked to set environmental priorities or to evaluate different resource management options (Drottz-Sjöberg 1991, Slovic 1992, Boholm 1998).

This introductory paper discusses the potential and requirements for an analytic-deliberative decision making process in the field of natural resource management. It provides some of the theoretical base for the many case studies most of which have been inspired by the model of analytic-deliberative processes. This model of participation attempts to meet two major objectives: first, to enhance the competence in the decision making process and, second, to assign a fair share of responsibility to manage risks to those who are or will be affected by the potential consequences.

The first element: The integration of science

Natural Resource managers are faced with a difficult dilemma: On the one hand, technical and organizational expertise is a necessary but not sufficient condition to make prudent decisions on resource allocation and distribution of opportunities. On the other hand, public perceptions are at least partially driven by biases, anecdotal evidence, false assumptions about resource interactions with the environment, and sensation (Okrent 1998). We live in a pluralist society with different value systems and worldviews. To choose among equally legitimate courses of action becomes an almost insurmountable task since no meta-arguments are available or convincing enough to distinguish valid from invalid claims. This is particularly true for debates on resource management since economic, ecological and social aspects are being affected that have strong links to particular interests. In this situation of value plurality, uncertainty and competing interests, the resolution of scientific debates is particularly difficult to accomplish.

Based on the analyses from theorists of human knowledge and science (see brief reviews in Dietz et al. 1989, Jasonoff 1993, 1998, 2004; Rosa 1998, Wynne 2002) one can draw the following inferences on the required process characteristics that need to be met when making complex choices in resource management:

- Regardless whether one prefers a constructivist or realist perspective on human knowledge about risks (cf. Bradury 1987, Horlick-Jones 1998, Rosa 1998), scientific rationality as framed by methodological consensus among researchers is insufficient in making unambiguous and uncontested claims about the characteristics and potential uses of a specific natural resource management option under investigation (Margolis 1996, Renn 2004).

- In analyzing the potentials of human intervention into natural environments, one needs to include systematic and experiential sources of knowledge (Wynne 1989). Systematic knowledge is necessary to build upon the collected experiences of the past, experiential knowledge to take account of the idiosyncratic features surrounding the specific decision problem and the accumulated expertise of practioners.
- When contemplating about the acceptability of one management option over another option, one needs to be informed about the likely consequences of each decision option and to be cognizant of the potential violations of interests and values connected with each decision option (Gregory 2004). Although both steps, predicting the likely impacts and evaluating the desirability of each of these consequences, can be separated analytically it is counterproductive to run the two processes in parallel and assign these tasks to different agents, since the answers of the first task co-determines the answers to the second task and vice versa (Jaeger et al. 2001: 243ff.). What is needed is a procedure that integrates both tasks without sacrificing the necessary precision and quality of factual and value judgments that are inherent in both steps.
- Integrating values into resource management decisions requires the input of those people whose interests and values are affected by the decision options (Kunreuther and Slovic 1996). In many instances, these interests and values are so obvious that agencies can act on their behalf without major reassurance that their action is in accordance with the needs and concerns of those whom they serve (Chess et al. 1998). In many environmental decisions, however, it is less obvious what is in the best interest of the people and plural value input is needed to produce a fair and balanced decision (Creighton 1983). If only interests need to be reconciled, involvement of stakeholders may suffice; if broad value judgments or issues of social justice are addressed, representatives of the affected public ought to be involved (IRGC 2005: 53). In both cases such an input requires direct participation efforts beyond the scope of normal decision making procedures based either on agency rules or majority votes by a representational branch of government (Webler 1999).
- Participation is not only a normative goal of democracy, it is also a requirement for rational decision making in situations in which evaluating uncertainty is part of the management effort (Pidgeon 1997). If all society would care about is to reduce the amount of physical harm done to its members, technical expertise and some form of economic balancing would suffice for effective risk management. However, society is not only concerned about risk minimization (Renn 1997).

People are willing to suffer harm if they feel it is justified or if it serves other goals. At the same time, they may reject even the slightest chance of being exposed to a risk if they feel the decision is imposed on them or violates their other attitudes and values (MacLean 1986, Linnerooth-Bayer and Fitzgerald 1996). Context matters. So does procedure of decision making independent of outcome. "Real" consequences are always mediated through social interpretation and linked with group values and interests. Responsive risk management needs to incorporate public values into the decision making process.

The Requirements for Deliberative Processes

Scientific input into resource management decisions are as explained above not sufficient to make prudent choices. First, scientific knowledge in itself is often ambiguous and contested, second it does not include the values and preferences of those who are or will be affected by the decision outcomes. That is why participatory deliberative methods need to be employed in addition to scientific input (Liberatore and Funtowicz 2003). If that is required, how can one select the values or preferences that should guide environmental decision-making? One of the answers to this question can be derived from the theory and practice of discursive deliberation.

The term deliberation refers to the style and procedure of decision making without specifying which participants are invited to deliberate (Stern and Fineberg 1996, Renn 2004). For a discussion to be called deliberative it is essential that it relies on mutual exchange of arguments and reflections rather than decision-making based on the status of the participants, sublime strategies of persuasion, or social-political pressure. Deliberative processes should include a debate about the relative weight of each argument and a transparent procedure for balancing pros and cons (Tuler and Webler 1999). In addition, deliberative processes should be governed by the established rules of a rational discourse. In the theory of communicative action developed by the German philosopher Juergen Habermas, the term discourse denotes a special form of a dialogue, in which all affected parties have equal rights and duties to present claims and test their validity in a context free of social or political domination (Habermas 1970, 1987b). A discourse is called rational if it meets the following specific requirements (cf. McCarthy 1975, Habermas 1987a, 1991; Kemp 1985, Renn and Webler 1998: 48ff., Webler 1995, 1999). All participants are obliged to:

- seek a consensus on the procedure that they want to employ in order to derive the final decision or compromise, such as voting, sorting of positions, consensual decision making or the involvement of a mediator or arbitrator;

- articulate and critique factual claims on the basis of the "state of the art" of scientific knowledge and other forms of problem-adequate knowledge; (in the case of dissent all relevant camps have the right to be represented),
- interpret factual evidence in accordance with the laws of formal logic and analytical reasoning,
- disclose their relevant values and preferences, thus avoiding hidden agendas and strategic game playing,
- process data, arguments and evaluations in a structured format (for example a decision-analytic procedure) so that norms of procedural rationality are met and transparency can be created.

The rules of deliberation do not necessarily include the demand for stakeholder or public involvement. Deliberation can be organized in closed circles (such as conferences of catholic bishops, where the term has indeed been used since the Council of Nicosea) as well as in public forums. It may be wise to use the term ”deliberative democracy” when one refers to the combination of deliberation and public or stakeholder involvement (see also Cohen 1997, Rossi 1997).

What needs to be deliberated? First, deliberative processes are needed to define the role and relevance of systematic and anecdotal knowledge for making far-reaching choices. Second, deliberation is needed to find the most appropriate way to deal with uncertainty and value plurality in natural resource management and to set efficient and fair trade-offs between conflicting goals. Third, deliberation needs to address the wider concerns of the affected groups and the public at large (Renn 2004).

Why can one expect that deliberative processes are better suited to deal with challenges posed by the demand for economically effective, ecologically friendly and socially fair use of natural resources than using expert judgment, political majority votes or relying on public survey data?

- Deliberation can produce common understanding of the issues or the problems based on the joint learning experience of the participants with respect to systematic and anecdotal knowledge (Webler and Renn 1995, Pidgeon 1997);
- Deliberation can produce a common understanding of each party‘s position and argumentation and thus assist in a mental reconstruction of each actor‘s argumentation (Warren 1993, Tuler 1996). The main driver for gaining mutual understanding is empathy. The theory of communicative action provides further insights in how to mobilize

empathy and how to use the mechanisms of empathy and normative reasoning to explore and generate common moral grounds (Webler 1995).

- Deliberation can produce new options and novel solutions to a problem. This creative process can either be mobilized by finding win-win solutions or by discovering identical moral grounds on which new options can grow (Renn and Webler 1998: 64ff., DEMOS 2004).
- Deliberation has the potential to show and document the full scope of ambiguity associated with environmental problems. Deliberation helps to make a society aware of the options, interpretations, and potential actions that are connected with the issue under investigation (Wynne 1992, De Marchi and Ravetz 1999). Each position within a deliberative discourse can only survive the crossfire of arguments and counter-arguments if it demonstrates internal consistency, compatibility with the legitimate range of knowledge claims and correspondence with the widely accepted norms and values of society. Deliberation clarifies the problem, makes people aware of framing effects, and determines the limits of what could be called reasonable within the plurality of interpretations (Skillington 1997).
- Deliberations can also produce agreements. The minimal agreement may be a consensus about dissent (Raiffa 1994, Jaeger et al.: 236ff.). If all arguments are exchanged, participants know why they disagree. They may not be convinced that the arguments of the other side are true or morally strong enough to change their own position; but they understand the reasons why the opponents came to their conclusion. At the end the deliberative process produces several consistent and - in their own domain- optimized positions that can be offered as package options to legal decision-makers or the public. Once these options have been subjected to public discourse and debate, political bodies such as agencies or parliaments can make the final selection in accordance with the legitimate rules and institutional arrangements such a majority vote or executive order. Final selections could also be performed by popular vote or referendum (Wehrli-Schindler 1987).
- Deliberation may result in consensus. Often deliberative processes are used synonymously with consensus seeking activities (Coglianese 1997). This is a major misunderstanding. Consensus is a possible outcome of deliberation but not a mandatory requirement. If all participants find a new option that they all value more than the one option that they preferred when entering the deliberation, a "true" consensus is reached (Renn 2004). It is clear that finding such a consensus is the exception rather than the rule. Consensus is either

based on a win-win solution (examples in Waldo 1987) or a solution that serves the "common good" and each participant's interests and values better than any other solution (Dryzek 1994). Less stringent is the requirement of a tolerated consensus. Such a consensus rests on the recognition that the selected decision option might serve the "common good" best but on the expense of some interest violations or additional costs. In a tolerated consensus some participants voluntarily accept personal or group-specific losses in exchange for providing benefits to all members of society. Case studies have provided sufficient evidence that deliberation has produced a tolerated consensus solution, particularly in siting conflicts (one example in Schneider et al. 1998). Consensus and tolerated consensus should be distinguished from compromise. A compromise is a product of bargaining where each side gradually reduces its claim to the opposing party until they reach an agreement (Raiffa 1994). All parties involved would rather choose the option that they preferred before starting deliberations, but since they cannot find a win-win situation or a morally superior alternative they look for a solution that they can "live with" knowing that it is the second or third best solution for them. Compromising on an issue relies on full representation of all vested interests.

In summary many desirable products and accomplishments are associated with deliberation (Chess et al. 1998). Depending on the structure of the discourse and the underlying rationale deliberative processes can:

- enhance understanding,
- generate new options,
- decrease hostility and aggressive attitudes among the participants,
- explore new problem framings,
- enlighten legal policy makers,
- produce competent, fair and optimized solution packages and
- facilitate consensus, tolerated consensus and compromise.

Commitment matters

The objective of this paper was to address and discuss the need and potential for analytic-deliberative processes in natural resource management. Organizing and structuring discourses to guide resource management decisions goes beyond the good intention to have the public involved in decision making. The mere desire to initiate a two-way-communication process and the willingness to listen to public concerns are not sufficient. Discursive processes need a structure that assures the integration of technical expertise, regulatory requirements, and public values. These different inputs should be combined in such a fashion that they contribute to the deliberation process the type of expertise and knowledge that can claim legitimacy within a rational decision making procedure (von Schomberg 1995). It does not make sense to replace technical expertise with vague public perceptions nor is it justified to have the experts insert their own value judgments into what ought to be a democratic process.

The much cherished solution of the past has been to have expert panels feed in the facts and have democratically elected representatives to reflect these facts on the basis of public values and make informed decisions (Webler and Renn 1995). This so called decisionistic model of communication has several major flaws: The selection of facts relies largely on the choice of concerns, and the value preferences of the elected representatives are at least partially dependent on the knowledge about the likely consequences of each decision option. Separating facts from values by division of labor leads to a vicious cycle. In addition, uncertainty about consequences, ambiguity of the knowledge base, and dissent among experts make it necessary that decision makers interact directly with experts and get an impression of the present state of the art. At the same time, those groups and individuals who are exposed to the consequences of natural resource management decisions demand that their values and preferences are taking into account directly by resource managers without the detour of activating the often only remotely affected political representatives. These arguments have motivated the U.S. Academy of Sciences to advocate the analytic-deliberative approach to decision making in the environmental arena (Stern and Fineberg 1996).

Organizing a common platform for mutual exchange of ideas, arguments, and concerns does not suffice, however, in order to assure fair and competent results. Mixing all these knowledge and value sources into one implies the danger that each group trespasses its legitimate boundary of expertise. If perceptions replace assessments and the rhetoric of powerful agents replace value input by those who have to bear the potential impacts the discourse goes into the wrong direction. An organizational model is needed that assigns specific roles to each contributor but makes sure, at the same time, that each contribution is embedded in a dialogue setting that guarantees mutual exchange of arguments and information, provides all participants with opportunities to insert and challenge claims, and to create active understanding among all participants (Webler 1995: one example for such a model in Renn 1999).

There is no universal recipe for combining expertise, interests and public values into one process model. But the chapters of this volume provide sufficient evidence and material demonstrating both the feasibility of a analytic-deliberative process and the robustness of such a process even if the participatory process runs through major difficulties and experiences several organizational flaws. This impression has also been shared by the empirical analysis of Beierle and Cayford (2002) who were able to demonstrate that professional quality of participation had only a slight influence on overall success rate, it was rather the dedication of the decision maker to involve the public and the intensity of the process that were more or less decisive for the outcome of the whole exercise. The following chapters are in line with this empirical insight. They show that there has been a diversity of approaches and models in using deliberative methods for natural resource management. Regardless of the format or the mix of analytical and deliberative elements, the main driver for success or failure is commitment: Commitment by the agency that has to take decisions, commitment by the organizers of the participatory process, commitment by the stakeholders and the public and commitment by staff members and facilitators. So the main lesson of all these case studies is that we can trust deliberative methods to provide what they promise to perform if all those involved are dedicated to make them successful.

References

Beierle, T.C., Cayford, J. (2002). Democracy in Practice. Public Participation in Environmental Decisions. Resources for the Future, Washington

Boholm, A., (1998). Comparative Studies of Risk Perception: A Review of Twenty Years of Research. Journal of Risk Research 1(2), 135-163

Bradbury, J.A. (1989). The Policy Implications of Differing Concepts of Risk. Science, Technology, and Human Values 14 (4), 380-399

Chess, C., Dietz, Th., Shannon, M. (1998). Who Should Deliberate When? Human Ecology Review 5(1), 60-68

Coglianese, C. (1997). Assessing Consensus; The Promise and Performance of Negotiated Rule Making. Duke Law Journal 46, 1255-1333

Cohen, J. (1997). Procedure and Substance in Deliberative Democracy. In: Bohman, J., Rehg W., (eds.), Deliberative Democracy. Essays on Reason and Politics. MIT Press, Cambridge, 407-437

Creighton, J.L. (1983). The Use of Values: Public Participation in the Planning Process. In: Daneke, G.A., Garcia, M.W., Delli Priscoli, J. (eds.). Public Involvement and Social Impact Assessment. Westview Press, Boulder, 143-160

Cross, F.B. (1998). Facts and Values in Risk Assessment. Reliability Engineering and Systems Safety 59, 27-45

DEMOS (2004). See-Through Science. Why Public Engagement Needs to Move Upstream. Monograph by Wisdon, J., Willis, R., Demos, London

De-Marchi, B., Ravetz, J.R. (1999). Risk Management and Governance: A Post-Normal Science Approach. Futures 31, 743-757

Dietz, T., Stern, P.C., Rycroft, R.W. (1989). Definitions of Conflict and the Legitimization of Resources: The Case of Environmental Risk. Sociological Forum 4, 47-69

Dryzek, J.S. (1994). Discursive Democracy. Politics, Policy, and Political Science. Second Edition. Cambridge University Press, Cambridge

Drottz-Sjöberg, B.-M. (1991). Perception of Risk. Studies of Risk Attitudes, Perceptions, and Definitions. Center for Risk Research, Stockholm

Gregory, R.S. (2004). Valuing Risk Management Choices. In: McDaniels, T., Small, M.J. (eds). Risk Analysis and Society. An Interdisciplinary Characterization of the Field. Cambridge University Press, Cambridge, 213-250

Habermas, J. (1970). 'Towards a Theory of Communicative Competence. Inquiry 13, 363-372

Habermas, J. (1987a). Theory of Communicative Action. Vol. II: Reason and the Rationalization of Society. Beacon Press, Boston

Habermas, J. (1987b). The Philosophical Discourse of Modernity. Polity Press, Cambridge

Habermas, J. (1991). Moral Consciousness and Communicative Action. translated by Lenhardt, C., Weber, S. Nicholson. Second edition. MIT Press, Cambridge

Horlick-Jones, T. (1998). Meaning and Contextualization in Risk Assessment. Reliability Engineering and Systems Safety 59, 79-89

IRGC (2005). White Paper on Risk Governance: Towards an Integrative Approach. International Risk Governance Council, Geneva

Jaeger, C., Renn, O., Rosa, E., Webler, Th. (2001). Risk, Uncertainty and Rational Action. Earthscan, London

Jasanoff, S. (1993). Bridging the Two Cultures of Risk Analysis. Risk Analysis 13 (2), 123-129

Jasanoff, S. (1998). The Political Science of Risk Perception. Reliability Engineering and Systems Safety 59, 91-99

Jasanoff, S. (2004). Ordering Knowledge, Ordering Society. In: Jasanoff, S. (ed.). States of Knowledge: The Co-Production of Science and Social Order. Routledge, London, 31-54

Kemp, R. (1985). Planning, Political Hearings, and the Politics of Discourse. In: Forester, J. (ed). Critical Theory and Public Life. MIT Press: Cambridge, 177-201

Kunreuther, H., Slovic, P. (1996). Science, Values, and Risk. In: Kunreuther, H., Slovic, P. (eds). Annals of the American Academy of Political and Social Science, Special Issue. Challenges in Risk Assessment and Risk Management. Sage, Thousand Oaks, 116-125

Liberatore, A., Funtowicz, S. (2003). Democratizing Expertise, Expertising Democracy: What Does This Mean, and Why Bother? Science and Public Policy 30 (3), 146-150

Linnerooth-Bayer, J., Fitzgerald, K.B. (1996). Conflicting Views on Fair Siting Processes: Evidence from Austria and the U.S. Risk: Health, Safety & Environment 7 (2), 119-134

MacLean, D. (1986). Social Values and the Distribution of Risk. In: MacLean, D. (ed). Values at Risk. Rowman and Allanheld, Totowa, 75-93

Margolis, H. (1996). Dealing with Risk. Why the Public and the Experts Disagree on Environmental Issues. University of Chicago Press, Chicago

McCarthy, T. (1975). Translator's Introduction. In: Habermas, J. Legitimation Crisis. Beacon Press, Boston

Okrent, D. (1998). Risk Perception and Risk Management: On Knowledge, Resource Allocation and Equity. Reliability Engineering and Systems Safety 59, 17-25

Pidgeon, N.F. (1997). The Limits to Safety? Culture, Politics, Learning and Manmade Disasters. Journal of Contingencies and Crisis Management 5 (1), 1-14

Raiffa, H. (1994). The Art and Science of Negotiation. 12th edition. Cambridge University Press, Cambridge

Renn, O. (1997). Three Decades of Risk Research: Accomplishments and New Challenges. Journal of Risk Research 1 (1), 49-71

Renn, O. (1999). A Model for an Analytic-Deliberative Process in Risk Management, Environmental Science and Technology 33 (18), 3049-3055

Renn, O. (2004). The Challenge of Integrating Deliberation and Expertise: Participation and Discourse in Risk Management. In: MacDaniels, T. L., Small, M.J. (eds). Risk Analysis and Society: An Interdisciplinary Characterization of the Field. Cambridge University Press, Cambridge, 289-366

Renn, O., Webler, Th. (1998). Der kooperative Diskurs - Theoretische Grundlagen, Anforderungen, Möglichkeiten. In: Renn, O., Kastenholz, H., Schild, P., Wilhelm, U., (eds). Abfallpolitik im kooperativen Diskurs. Bürgerbeteiligung bei der Standortsuche für eine Deponie im Kanton Aargau. Hochschulverlag AG an der ETH Zürich, 3-103

Rosa, E.A. (1998). Metatheoretical Foundations for Post-Normal Risk. Journal of Risk Research 1, 15-44

Rossi, J. (1997). Participation Run Amok: The Costs of Mass Participation for Deliberative Agency Decisionmaking. Northwestern University Law Review 92, 173-249

Schneider, E., Oppermann, B., Renn, O. (1998). Experiences from Germany: Application of a Structured Model of Public Participation in Waste Management Planning. Interact. Journal of Public Participation 4 (1), 63-72

Skillington, T. (1997). Politics and the Struggle to Define: A Discourse Analysis of the Framing Strategies of Competing Actors in a 'New' Participatory Forum. British Journal of Sociology 48 (3), 493-513

Slovic, P. (1992). Perception of Risk: Reflections on the Psychometric Paradigm. In: Krimsky, S., Golding, D. (eds). Social Theories of Risk. Praeger, Westport and London, 153-178

Stern, P.C., Fineberg,V. (1996). Understanding Risk: Informing Decisions in a Democratic Society. National Research Council. Committee on Risk Characterization. National Academy Press, Washington, D.C.

Tuler, S. (1996). Meanings, Understandings, and Interpersonal Relationships in Environmental Policy Discourse. Doctoral Dissertation. Clark University, Worcester

Tuler, S., Webler, Th. (1999). Designing an Analytic Deliberative Process for Environmental Health Policy Making in the U.S. Nuclear Weapons Complex. Risk: Health, Safety & Environment 10 (65), 65-87

Schomberg, von, R. (1995). The Erosion of the Valuespheres. The Ways in which Society Copes with Scientific, Moral and Ethical Uncertainty. In: Schomberg, von, R., (ed). Contested Technology. Ethics, Risk and Public Debate. International Centre for Human and Public Affairs, Tilburg, 13-28

Waldo, J. (1987). Win/Win Does Work. Timber-Fish-Wildlife. A Report from the Northwest Renewable Resources Center 1 (1), 7

Warren, M.E. (1993). Can Participatory Democracy Produce Better Selves? Psychological Dimensions of Habermas' Discursive Model of Democracy. Political Psychology 14, 209- 234

Webler, Th. (1995). 'Right' Discourse in Citizen Participation. An Evaluative Yardstick. In: Renn, O., Webler, Th., Wiedemann, P. (eds). Fairness and

Competence in Citizen Participation. Evaluating New Models for Environmental Discourse. Kluwer, Dordrecht and Boston, 35-86

Webler, Th. (1999). The Craft and Theory of Public Participation: A Dialectical Process. Risk Research 2 (1), 55-71

Webler, Th., Renn, O. (1995). A Brief Primer on Participation: Philosophy and Practice. In: Renn, O., Webler, Th., Wiedemann, P. (eds). Fairness and Competence in Citizen Participation. Evaluating New Models for Environmental Discourse. Kluwer, Dordrecht and Boston, 17-34

Wehrli-Schindler, B. (1987). Demokratische Mitwirkung in der Raumplanung. SVPW, Bern

Wynne, B. (1989). Sheepfarming after Chernobyl. Environment 31 (11-15), 33-37

Wynne, B. (1992). Risk and Social Learning: Reification to Engagement. In: Krimsky, S., Golding, D. (eds). Social Theories of Risk. Praeger, Westport, 275-297

Wynne, B. (2002). Risk and Environment as Legitimatory Discourses of Technology: Reflexivity Inside Out? Current Sociology 50 (30), 459-477

1 Towards a More Effective and Democratic Natural Resources Management

Susanne Stoll-Kleemann[1], Martin Welp[2]

1 Humboldt University of Berlin, Germany
2 University of Applied Sciences Eberswalde, Germany

Democracies have influenced and been a model for political systems all around the world. Some claim that representative democratic systems are in a crisis, due to disinterest of citizen's in politics. This is mirrored for example in decreasing voting rates. On the other hand many critiques of more direct involvement of citizens are disillusioned by participatory processes and claim that these procedures are too time-consuming and costly. Both claims strengthen the view that democracy and its relationship to participatory procedures need constant updating and learning.

In many countries, participation and stakeholder dialogues are recognised as important elements of management, planning, and policy-making and increasingly of knowledge creation in the field of natural resources management. Approaches such as adaptive management, participatory planning, and participatory integrated assessment have been developed and practiced by many private and public sector organisations.

There are various reasons why organisations in natural resources management want to engage in such dialogues, the three main underlying ones being: First, there is a perceived need for further development of representative decision-making by providing a broader range of actors the opportunity to get involved in processes affecting their lives. This is an important motivation for participatory practises in planning and policy-making. It can be seen as a part of a broader democratisation process that is taking place in many societies throughout the world.

The second motivation is related to effectiveness: decisions and management practices are more likely to be implemented and accepted if key actors support them. Early involvement of actors helps to avoid surprises and usually leads to a more sustained commitment on their part. The opposite is often the case with decisions that are imposed from higher levels of hierarchy without any consultation.

The third reason is related to quality. Problems in today's world are increasingly complex, and proposed solutions demand knowledge from many different knowledge domains; no single agent possesses all relevant knowledge. Rather many different actors have specialised knowledge bases, which need to be brought together (see e.g. Yosie and Herbst 1998, and Renn 2006 in this book).

Facilitating high quality stakeholder dialogues and participation in natural resources management requires many different skills and the use of appropriate methods. Most of the required skills can be acquired through training. Numerous handbooks provide practical guidance for the use of moderation techniques, visualisation techniques, etc. Renn (2006 in this book) also outlines various requirements for stakeholder dialogues und participatory processes in natural resources management in the Foreword of this book. We believe that the practice of stakeholder dialogues would benefit from a practical theoretical framework. In the absence of an integrative theory, the practice of stakeholder dialogues has remained heterogeneous and the objectives sometimes unclear.

1.1 Objectives and structure of the book

This book outlines an integrative theoretical framework and examines examples of stakeholder dialogues and public participation in natural resources management in three areas: science, policy and management. Current practice has generally been to analyse these separately. We, in contrast, feel the three areas should ideally be closely interrelated and therefore have attempted to integrate them in the work by using case studies as examples and by developing an integrative theory of reflexive dialogues that can be applied in all three domains. We exclude stakeholder dialogues that have been conducted by the private sector. The concept of stakeholder dialogues originates from management literature and thus in Part II of this book, the theoretical part, we will discuss this body of literature as well (e.g. Senge 1998).

Public participation and stakeholder dialogues play a crucial role both in environmental policy and management as well as in integrated assessment studies. In the policy and management of natural resources, participatory procedures are implemented at different stages: in defining objectives, in choosing between alternative courses of action, in implementation and finally in evaluation. In integrated assessments, stakeholder dialogues are needed for integrating all relevant knowledge bases. Furthermore, stakeholder dialogues are a reality check that academic studies often lack.

Methods and formal techniques for participation in natural resources management and planning have only been developed during the last few decades. Some of these are applicable for stakeholder dialogues in research and integrated assessments as well (Pahl-Wostl 2002). This is not true in all cases, though, since the process of generating new and relevant knowledge is not as straightforward as, for example, a routine planning procedure.

Natural resources management problems tend to be ill structured, and management options and their impacts are often characterised by uncertainty. The most prominent example of a natural resource problem caused by uncoordinated collective action was described by Hardin (1968) as the "Tragedy of the commons" and has been followed up in several publications by the political scientist Elinor Ostrom (e.g., Ostrom 1990, 1994).

Part II of this book deals with theories and tools. We first outline the Integrative Theory of Reflexive Dialogues, which seeks to integrate the key theoretical approaches that are relevant in dialogues (see Chapter 2). The aim of this chapter is to review social scientific theories and to outline a new integrative theory of stakeholder dialogues. The theory is then used as a framework for the case studies. In particular, we focus on the following theoretical approaches: Social Psychological Theories, Organisational Learning, and Rational Actor Paradigm.

In Chapter 3 Berghöfer and Berghöfer extend the theoretical base by focusing on participation in development thinking. Participation has played an important role in development cooperation for a long time. There is an independent body of theoretical approaches related to development aid and stakeholder dialogues.

Evaluation of stakeholder dialogues and participation are closely related to several theories of participation. Chapter 4, written by Oels, gives an overview of the criteria used in evaluating the success of stakeholder dialogues and public participation. The criteria that are usually used focus either on the processes or the contents. The paper seeks to integrate the two approaches at a conceptual level.

In Chapter 5, Scheffran presents a selection of tools that can be applied in stakeholder dialogues. These tools can be divided in two main categories: communication and analytical tools. The first is important for initiating and fuelling debate as well as for building trust among participants. The analytical tools are relevant when exploring the positions the stakeholders have or when analysing how different measures would affect the dynamics of the system at hand. Analytical tools can, for example, be useful in making areas of agreements and disagreement more explicit. The last chapter in Part II deals with visualisation as a means of facilitating social learning. Maarleveld et al. outline relevant elements of a theory of social learning and how decision support systems, GIS applications, and other visualisation tools can be used in stakeholder dialogues in natural resources management.

Part III has five case studies from three different areas: science, policy, and management. The first case study is about experiences with stakeholder dialogues in the field of climate change studies. Welp et al. present examples of science-based stakeholder dialogues conducted at the Potsdam Institute for Climate Impact Research. These were initiated by scientists and the involved stakeholders included representatives of corporations, companies, NGOs and the public sector. Chapter 8, authored by Hellström, deals with science support in policy-making. The Forest Biodiversity Programme for Southern Finland serves as a case study for analysing how scientists and stakeholders acted in a policy process and what roles they adopted. Chapter 9 deals with public participation in site selection for Natura 2000 in Germany. In her case study from Bavaria, Eben describes what controversies were involved in site selection and what kind of participatory procedures took place there. Experiences with stakeholder dialogues in natural resources management in Ecuador are presented in Chapter 10. Linke, Sturm and Rivera give the reader insights about participation practices in forest and park management as carried out and supported by German development cooperation. Averbeck gives the reader an example of local people's involvement in natural resources management and development cooperation in Uganda.

The above case studies are reflected upon from a theoretical perspective in Chapter 12. The Integrative Theory of Reflexive Dialogues is applied as a conceptual point of reference for the case studies in science, policy and management. The applicability and usefulness of the theoretical framework are discussed. In the Epilogue, O'Riordan emphasises the difficulties in implementing stakeholder dialogues in practice.

In the following part of our introduction, we will set the scene and specify the terms used. Furthermore, we discuss the benefits of stakeholder dialogues and participatory processes as well as factors that endanger these processes as well as problems they can, in some circumstances, engender.

1.2 Context and Definitions

Science's understanding of the dynamics of global environmental change has advanced considerably in recent years. We now understand better what the main drivers of biodiversity loss and water scarcity are. The combination of the improved understanding of the dynamics of societal-environmental relationships and the proliferation of computing power has provided useful models and simulations. Although many global-change problems have been identified and described both in the scientific literature and in the public media, humankind has been less successful in finding solutions to the environmental and development problems it is facing. In our view, a sustainability transition requires more experimentation with participatory methods and a more systematic reflection of success and failure factors in policy and management.

In the following, we clarify key terms that are frequently used in the theory chapter and in the case studies below. These include citizen participation, stakeholder (theory), and different types of stakeholder dialogues, communication, the concept of discourse, and integrated assessments.

Citizen participation is a process that provides individuals with an opportunity to influence public decisions. It has become an accepted component of the democratic decision-making process. The roots of citizen participation can be traced back to ancient Greece. Renn et al. (1995: 2) define public participation[1] as "forums for exchange that are organised for the purpose of facilitating communication between government, citizens, stakeholders and interest groups, and business regarding a specific decision or problem. This definition explicitly excludes non-institutional participation such as citizen protests. Also expert workshops are excluded from this definition. It is inclusive of public hearings, public meetings, focus groups, surveys, citizen advisory committees, referendums and initiatives, and negotiation, among other models" (ibid: 2). Direct involvement of citizens in political decision-making beyond the conventional modes of voting, party involvement, and economic co-determination has been rare in most European countries until fairly recently. Switzerland, with its frequent referendums and citizen initiatives and its still-strong reliance on local governing, is the obvious exception (ibid: 18).

Participation is one of those words that can be interpreted in many different ways. One definition terms it a process of collective learning that

---

1 According to Renn et al. (1995) the word "public" in the singular conveys the untrue impression that "the public" is somehow homogeneous, when in actuality, it is a vast and heterogenous group of individuals. Society consists of individuals with unique sets of interests, some of which are identical to those of others, and collectivities of people who form shared interests. Shared interests are not merely the intersection of individuals` particular interests.

changes the way that people think and act. Pretty (1995) proposes the following distinctions:

- Participation by Consultation: People participate by being consulted or by answering questions. The process does not concede any share in decision-making, and professionals are under no obligation to take on board people's views.
- Functional Participation: Participation is seen by external agencies as a means to achieve their goals, especially reduced costs. People participate by forming groups to meet predetermined objectives.
- Interactive Participation: People participate in joint analysis, development of action plans and formation or strengthening of local groups or institutions. Learning methodologies are used to seek multiple perspectives, and groups determine how available resources are used.
- Self-Mobilisation and Connectedness: People participate by taking initiatives independently to change systems. They develop contacts with external institutions for resources and technical advice they need but retain control over how resources are used.

Pretty further emphasises that the problem with participation as used in the first two types is that any achievements are likely to have no positive lasting effect on people's lives. The latter types, by contrast, involve building of social and human capital (cf. Arnstein's ladder of citizen participation as explained in Arnstein 1971: 177).

There is also a difference between public participation and stakeholder dialogues[2]. The term stakeholder originates from management literature, where a distinction is made between shareholders (owners of the company) and stakeholders. Stakeholder theory, pioneered by Freeman (1984), suggests that an organisation is defined in terms of its relationships with various groups and individuals referred to as stakeholders. A stakeholder is defined as any group or individual who can affect or is affected by the achievement of an organisation's objectives (Freeman 1984: 46). The

---

[2] Delli Priscoli argues that there is a difference in the values behind public participation and conflict management. Although the applied methods can be very similar, conflict management is driven by expectations that are different from public participation. While public participation aims at empowerment, open access to government, and building civic culture, conflict management aims at efficiency, consensus, and speeding up processes (Delli Priscoli 1997). Delli Priscoli views public participation as being more encompassing and as a precondition for conflict management as we know it in Western industrial countries.

stakes of these groups often go beyond property rights, but they do not necessarily have legal obligations to the conflict (Köhn 2002: 344).

Stakeholder theory as outlined by Freeman asserts that it is the responsibility of the management function to select activities to obtain optimal benefits for all identified stakeholder groups regardless of the relative power or interest held by each. In practice, Freeman asserts that the successful implementation of stakeholder management involves "organisational processes to take these groups and their stakes into account routinely as part of the standard operating procedures of the organisation and which implements a set of transactions or bargains to balance the interests of these stakeholders to achieve the organisation's purpose" (1984: 53). The strength of this theory lies in its intrinsic consideration of all stakeholders and the importance of collaborative action.

A stakeholder dialogue can be defined as a process in which a structured exchange of views and reflection on values of stakeholders can take place. The participants may have very divergent assessments regarding the problem at hand or the course of action to be taken. Stakeholder dialogues are increasingly popular in corporate management, in policy-making, in natural resources management, and in integrated assessments.

In contrast to public participation, where, at least in principle, everybody has the possibility to be involved, stakeholder dialogues follow a more "elitist" approach. The selection of stakeholders depends on the issue at hand. Usually the stakeholders are important players or experts in a certain field while in public participation, participants should comprise a broadly representative sample of the affected people (Rowe and Frewer 2000). Public participation aims for representativeness, although not in a statistical sense. Participants should be selected so that the range of different views is collected and reflected.

Connor (1999) proposes a slightly different distinction between stakeholder dialogues and public participation. He assumes that in stakeholder participation some people have a legitimate stake in the outcome and therefore a right to be involved while others, though interested, have no such claims. In his view of stakeholder dialogues, the people involved represent as a rule the interest or the view of one single organisation. They are expected to voice concerns that are relevant for the organisations and should report back to their organisation about the advance of the dialogues.

In public participation, Connor assumes that the issue affects all residents in some way, large or small, and that all have the right to obtain relevant information in an understandable form and to respond to it in a low-risk, low-cost way. Connor's summation is that there is a need to integrate public and stakeholder participation. He applied this concept in a specific project in the way he worked in parallel with stakeholders in regular meetings with a Public Advisory Committee and a public participation programme.

In this book, we do not follow Connor's view, which we believe is not encompassing enough. Although there he makes a distinction between stakeholder dialogues and public participation, the latter should not be reduced to meaning only the informing of the broad public in a non-engaging way.

Welp et al. (2006a) distinguish the following types of stakeholder dialogues: policy dialogues, multi-stakeholder dialogues for governance, corporate dialogues, and science-based stakeholder dialogues. All four types share the basic concept of learning and exchange of knowledge and opinions. The intention is to create a safe space for the exchange of arguments that is based on mutual trust. In such a setting, participants can learn from each other and as a group.

The main objective of policy dialogues is to create support for policies and new pieces of legislation. Although collaborative policy dialogues are far from being the dominant way of policy-making (Innes and Booher 2003), this approach is applied in many different sectors, including water policies, conservation policies, and many others. Therefore in this book, a big emphasis has been put on the reasoning, theory, and practice of policy dialogues; an example of a policy dialogue is outlined by Hellström in Chapter 8.

Multi-stakeholder dialogues for governance are international efforts to create partnerships and voluntary commitments among a broad range of international actors (Hemmati 2002). For example, the Stakeholder Forum for Sustainable Development (www.unedforum.org), which recently became an independent organisation (but is still closely linked with UN organisations), supports the increased involvement of stakeholders in international and national governance processes. Another example is the Forest Stewardship Council, an international network promoting sustainable management of the world's forests (www.fsc.org). The members, including the forest industry and environmental NGOs, have developed an international label for sustainable forest products (Vallejo and Hauselman 2004). Examples of this type of dialogue can also be found in this book, e.g. in Chapter 10.2. by Linke.

The objectives of corporate dialogues are to demonstrate openness and the will for a critical exchange of views. A key objective is to learn about the expectations of different stakeholder groups with regard to the company's business ethics and practices. The insights gained can be important on different levels of corporate decision-making. Today, stakeholder dialogues are a key element in the effort of many corporations to pursue Corporate Social Responsibility (CSR). In such dialogues, private companies reflect on society's views and expectations through consultation with various groups such as consumer associations, suppliers, environmental NGOs, religious organisations, etc.

A science-based stakeholder dialogue is defined as a structured communicative process of linking scientists with selected actors who are relevant for the research problem at hand (Welp et al. 2006a). These actors possess specialised knowledge and have insights relevant to the scientific process. Stakeholders possess knowledge needed by scientists to better comprehend, represent, and analyse global change problems as well as decision-makers', managers' and other stakeholders' mental models. Stakeholder dialogue processes do not always aim at being representative of the full spectrum of interests. The focus is on securing certain competencies (ibid).

An example where not only experts were engaged in the assessment of values, expectations, and risk perception related to climate change was the EU research project ULYSSES (Urban Lifestyles, Sustainability and Integrated Environmental Assessment). Several hundred citizens were engaged in Integrated Assessment Focus Group sessions to learn and debate about the climate change problem (Kasemir et al. 2003, Stoll-Kleemann et al. 2001, Welp et al. in press). Participants were confronted with the latest knowledge on climate change and synthesised their newly gained understanding in citizen assessments of the causes and impacts of climate change, as well as possible solutions. These included suggestions on mitigation and adaptation measures (e.g. within the transport, energy and household sectors) as well on who should act, and where and when.

Our understanding of science accords it an important role in sustainability transition. Thus, in this book we mainly refer to the notion of "sustainability science". Sustainability science seeks to understand the dynamics of global change, i.e., the fundamental character of interactions between nature and society. It also seeks to explore collective ways to create a sustainable world (Kates et al. 2001). When embedded in a transdisciplinary context, sustainability science can play an important role in finding workable solutions for mitigating, and adapting to, global change. However when detached from the "real world" (e.g. from lifestyles, technological innovations, expectations and mental models of actors), it may remain a purely academic endeavour with little social relevance (Welp et al. 2006a). Therefore, science needs to have access to the insights and expertise of various societal actors and incorporate their knowledge bases. On the other hand, scientists need to communicate the results of their inquiries in a comprehensible way (ibid). Examples of science-based stakeholder dialogues are discussed in Chapter 7 (Welp et al 2006b).

In the following, we will clarify definitions referring to different types of communication. Many terms and definitions are used to describe the different ways communication takes place in participation and stakeholder dialogues. According to the Merriam-Webster's Collegiate Dictionary (1995), *deliberation* is "... a discussion and consideration by a group of persons ... of the reasons for and against a measure" (for detailed explanations of the term deliberation, see Renn 2006 in this book).

Bohm (1996) uses somewhat different terms, focussing on a distinction between discussion and dialogue, which in our view is helpful when reflecting the mode of communication in participatory processes. In discussions, individual views are presented and defended. Discussions can be seen as a ping-pong game: the subject of common interest is analysed from many points of view; the purpose of the game is normally to win. Winning means having one's view accepted by the group. Participants in a discussion basically want their view to prevail. In a dialogue, in contrast, the participants are not negotiating positions or trying to reach a consensus. Dialogues are based on mutual respect and on the notion that the others have a valid viewpoint. The word dialogue suggests a free flow of meaning between people. In dialogue, individuals gain insights that could not be achieved individually. Thus dialogues foster interdisciplinarity and a holistic view. Bohm argues that the greatest impact is realised through a synergy between the processes of dialogue and discussion. In the theory chapter, we use the definition used by Bohm (see Chapter 2).

Discourse is defined here as a "specific ensemble of ideas, concepts and categorisations that are produced, reproduced, and transformed in a particular set of practices and through which meaning is given to physical and social realities" (Hajer 1995: 44). Jaeger (2003) makes a distinction between ordinary language, which is indispensable in dialogues and formal domains or discourse, such as computer models. In the following, we often refer to Integrated Assessments as a type of research in which stakeholder dialogues play an important role, e.g. in the CLEAR and ULYSSES research projects, which are described in Jaeger et al. (2000), Stoll-Kleemann et al. (2001), Stoll-Kleemann et al. (2003), and Welp et al. (in press). Rotmans (1998) defines Integrated Assessment (IA) as a structured process of dealing with complex issues using knowledge from various disciplines and/or stakeholders in a way that integrated insights are made available to decision-makers. Integrated assessments have been conducted on Waldsterben, climate change, water availability, as well as in other contexts. Often they are a mixture of stakeholder dialogues in which ordinary language is used as well as more formal ways of conceptualising problems, often in the form of a computer model (see also Pahl-Wostl 2002).

1.3 Benefits of Participation and Stakeholder Dialogues

The benefits of participation and stakeholder dialogues can be discussed on two levels in case studies: the normative level and the practical level of experiences. In this section, we discuss some of the benefits referred to in the participation literature.

Stakeholder dialogues and public participation can be efficient instruments to deal with the following factors (extracted from Messner 1998 and Oels 2003):

- Complexity: The number of collective actors (associations, parties, interest groups, etc.) that have resources, knowledge, and organisational capacity to influence political decision-making processes is rising (Olson 1968, Etzioni 1968). In addition to this, an increasing number of private and public actors participate in the policy-making process, leading to what Jordan and Richardson have called 'overcrowded policy-making' (Jordan and Richardson 1983). At the same time, there is an increasing interdependence of decision-making and a decreasing autonomy of all actors (Scharpf 1993a, b, Mayntz 1987, neo-corporatist theories).
- Colonising of the state by powerful interest groups (Olson 1968, Nozick 1974)
- Increasing individualisation: This leads to a decline in solidarity (i.e. moral decline as raised by neo-conservatives).
- Functional differentiation: This increasingly divides economy and society into sectors and subsystems, and brings with it the danger of fragmentation (Katzenstein 1984, Kenis 1991, Luhmann 1984, 1986, Willke 1998).

The introduction and increased implementation of stakeholder dialogues helps to structure the problems since they are viewed from many different angles. At the same time, the danger of fragmentation can be addressed through participation and dialogues between the actors of the different societal subsystems. The success in finding workable solutions depends on the capacity of these forums to make conflict resolution and integrate diverse viewpoints (Czada 1992, Offe 1987).

A further explanation of why stakeholder dialogues and public participation is useful and important is the increasing global interdependence and globalisation[3]. Oels (2003) notes that "the ephemeral character of economic arrangements in an interdependent world makes people vulnerable, for example to job insecurity. To face this challenge, participation in local networks like alternative economic systems can present a coping strategy of social and economic survival". Participation in the context of development is discussed in Chapter 3 (Berghöfer and Berghöfer 2006) in this book. The basic assumption concerning the benefits of public participation is that the related methodologies "are built on common principles of co-learning and stakeholder involvement that leads to enhanced motivation to act. One thing all these methodologies do is to emphasise people's capacity in their own situation to start and continue change, whilst grounding this in a realistic understanding of what is possible" (Pretty 1995).

Pretty (1995) concludes - on the basis of an evaluation of a number of comparative studies conducted on rural and urban development programmes - that it has become clear in recent years and in a range of sectors is that interactive participation can lead to improvements in performance and outcomes.

Following these studies, Pretty describes the typical impacts of these deliberative and participatory methodologies as follows:

- Enhanced social capital: increased cohesiveness amongst and between groups; greater motivation to act; emergence of new institutional structures for joint action; emergence of new leadership and readiness to revitalise existing structures; increase in self-reliance; greater capacity to negotiate with external bodies and agencies; and increased engagement of youth in civil society;
- Enhanced human capital: greater confidence of individuals; personal empowerment through changes in skills, knowledge, attitudes and action; and the emergence of new social entrepreneurs;

---

[3] Globalisation is defined as "the intensification of world-wide social relations which link distant localities in such a way that local happenings are shaped by events occurring many miles away and vice versa. This is a dialectical process because such local happenings may move in an obverse direction from the very distanciated relations that shape them. Local transformation is as much a part of globalisation as the lateral extension of social connections across time and space." (Giddens 1990: 64, emphasis in original)

– Enhanced natural capital: improvements in natural resource management; increased production of natural resource goods and services[4]; greater added value for local communities; reduced dependence on external resources; and technological improvements based on local resources (Pretty 1995).

Stoll-Kleemann and O`Riordan (2002a, b) summarise the benefits of public participation in the field of biodiversity and protected area management. First, public participation is a democratic necessity. Bringing people into the management process recognises their self-worth, appreciates their vital role, and respects their citizenship credentials. This approach also incorporates the role of local property rights (McNeely 1995, Pretty and Pimbert 1995, Barton et al. 1997).

The second benefit is related to the legitimisation of management. Effective and efficient management requires the understanding and the support of local people. Rigid management structures do not adjust easily to social, economic or ecological changes. Protected areas established authoritatively from above without prior consent may make them "closed territories" with few links to the external worlds of hydrology, ecology and culture on which biodiversity so depends (Batisse 1997: 9). From the point of view of peaceful coexistence, such a desperate state of affairs is not only counterproductive. It may also destroy any chance of long-term co-operation over economic activities such as ecotourism (e.g. McNeely 1995, Borrini-Feyerabend 1996).

The third key argument is that sharing knowledge and understanding is vital for the success of protected areas. All actors have uniquely different perspectives as to what is a problem and what constitutes improvement. Since knowledge and understanding are socially constructed, what each actor knows and believes is a function of unique contexts and experiences. There is, therefore, no single "correct" understanding. What is taken to be "true" depends on the framework of knowledge and assumptions brought in by individuals and their social and occupational settings. It is essential to seek multiple perspectives on any "problem assessment" by ensuring the

[4] There is no convincing evidence (i.e., there is a dearth of scientific studies) that participation really leads to increased natural capital. The GoBi (Governance of Biodiversity) Research Group investigates this question. The project is still ongoing, but first results are summarized in Stoll-Kleemann (2005), Stoll-Kleemann and Bertzky (2006), Stoll-Kleemann et al. (2006). For example a study by Bruner et al. (2001) published in Science Magazine seeks to demonstrate that law enforcement mechanisms (such as the existence of enough guards) produce a more efficient protection of biodiversity in protected areas than participatory strategies.

wide involvement of different actors and groups (Pretty and Pimbert 1995: 10). Cooke and Kothari (2001) remind us that local communities, often targeted for participatory processes, are rarely politically cohesive. They usually do not share a unified view and therefore do not always, or readily, see the need for peacefully linking multiple stakeholders and interests (see also Chapter 3 of this book for a detailed discussion of this issue).

1.4 Difficulties of Participation and Stakeholder Dialogues

However robust the democratic tradition may appear, it is under pressure today from a variety of directions. One is the power of science and technology to transform the world. Another comes from the scale at which democracy must operate. One factor that increasing complexity is that the discourse that underpins the democratic process is attenuated by problems of scale. And while in some sense all politics is local, relating critically to the lived experience of the citizens, politics – even at the local level - is now also to a great degree global. Jasanoff and Martello (2004) argue that institutions are needed that respect the local but transcend localism. Actions taken in one locale often have profound implications across the planet, so the context in which decisions must be assessed is much more complex.

That people should have a say in issues affecting their lives is a well-accepted principle in democratic societies. Although voting is the cornerstone of representative democratic systems, the right to institutional or non-institutional participation is an essential feature. It is considered to complement the work of parliaments, elected decision-makers, and the administrative systems. There is however also fundamental criticism of participation that derives its origin from different worldviews and views of the role of citizen.

Participation is criticised by people who claim that citizens are not able to understand and assess complex issues due to lack of education. Understanding complex issues requires time and specialised knowledge, and average citizens seldom have the opportunity and the means to acquire it. Another point of criticism is that people as a rule do not want to participate. According to this view, people prefer to leave planning and decision-making to bureaucrats and politicians who have been hired or elected to do the job. In case the participatory processes are carried out, the most active and better-educated people tend to participate, and their views do not represent the majority view. In this view, citizens play a political role mainly as voters and not as empowered, active citizens.

There are a number of implementation barriers facing stakeholder dialogues and public participation. The opponents of stakeholders and public involvement in environmental management and/or integrated assessment often criticise these approaches on a very fundamental basis while the problems that arise are more due to implementation barriers that could be improved through, e.g., better management (see also Chapter 3).

But it is true that these implementation barriers are also manifold and that appropriate stakeholder dialogue in natural resource management is not easy to achieve. There is no guarantee that a participatory approach will necessarily be effective, e.g., in delivering the goals one is expecting (see Stoll-Kleemann and O`Riordan 2002a, b on conservation goals). In the book "Participation: The New Tyranny?", the authors suggests that much of the participation literature, particularly in the field of development studies and practice, has been very unreflective and repeats beliefs unverified by empirical studies (Cooke and Kothari 2001).

An important constraint when designing and implementing participatory approaches is the unwillingness to share power. Governments and/or regional or local managers may not support stakeholder involvement or public participation, especially if they regard these as a threat to their own authority or as an encouragement to opposition groups.

Connected to this issue, a failure to adopt a "bottom-up" participatory approach with genuine local involvement and understanding is an important implementation barrier for example to participatory nature conservation projects. Much of this is due to an absence of empathy amongst the implementing agencies. There is also an inability to address basic community needs and to distribute benefits equitably, as well as corruption amongst local officials and community leaders involved in the management of project funds and wildlife benefits (Songorwa 1999). In addition, environmental managers are often uninformed about the citizens' concerns and neglect the experiences and preferences of the public in setting policy or making decisions.

In general, as well as because of bad experiences, most people have limited trust in public institutions and limited confidence in the decision-making process. As possible consequences, they either demand to oversee the process and define objectives or are unwilling to co-operate. The participatory approach may also not be viable because of local political opposition or sheer lack of institutional or peoples' support (Agrawal and Gibson 1999, Cleaver 1999). Other factors include, for example, unclear or contested property rights, which are of particular interest in the use of natural resources. In such a situation, management agreements are usually difficult to establish.

Furthermore, participation by certain disadvantaged groups may clash with local customs (e.g. the participation of women, the landless, ethnic minorities, etc.) and may be quite alien (see also GEF 2000: 47). In this context, Pretty (1995) notes that it is a great challenge to involve those people who have not been involved in the past, or those whose views have not been incorporated into traditional decision-making, such as the less articulate, non-literate, those in poverty, and young people. He further states the dilemma for authorities is that they both need and fear people's participation. They need people's agreement and support, but they fear that such wider involvement is less controllable and less precise. But if this fear permits only stage-managed forms of participation, distrust and greater alienation are the most likely outcomes. This makes it all the more crucial that judgements can be made on the type of participation in use.

Participatory processes require specific investments of time and resources. In particular, the process of participation needs expert facilitation and clear objectives in order to avoid chaotic meetings and a general loss of direction. Furthermore, commitment over time is required, and encouraging results may take a while to appear. This can tax the patience of donors, managers, staff, and local people alike. For example, threats against natural resources may be escalating, and the urgency of taking action may discourage actors from undertaking lengthy participatory processes. Some compromises in the original objectives may need to be made. For instance, a conservation initiative designed by outsiders may propose a total ban on local access to natural resources, which may simply be unacceptable to residents (Barton et al. 1997).

1.5 Lack and need of Theory

As the world becomes more complex and more vulnerable to human action, new risks and uncertainties emerge at the science-society interface. The notion of uncertainty is fundamentally changing our perception of science since scientists have traditionally been keen to present research results as undisputed truth.

Let us take climate change as an example of the new understanding of science. There are many different kinds of uncertainties associated with climate models. These include uncertainty about climate sensitivity, i.e. how much warming is caused by a doubling of the atmospheric (equivalent) CO_2, uncertainty about critical thresholds in the climate system, and non-linear feedback between the climate and vegetation. These ambiguities need to be communicated, and eventually the public perception of science has to change. For stakeholder dialogues, the inability to speak with certainty presents both a communication challenge as well as a challenge for the theoretical underpinning.

The quality and effectiveness of stakeholder dialogues can be improved by better theoretical reflection (see, e.g., Beierle 2000). Currently, this field of societal action and management practice is not well structured, and the objectives and success criteria have remained fuzzy. A pragmatic learning-by-doing approach has advanced the art and practice of stakeholder dialogues considerably, but much remains to be improved in terms of theoretical frameworks. A good theoretical frame can be valuable in guiding practice and in tool development. The "Integrative Theory of Reflexive Dialogue", described in the next and revisited in the last chapter, intends to increase our understanding about processes, the way actors interact and objectives of stakeholder dialogues. It can furthermore fuel tool development, including both communication and analytical tools. The integrative theoretical framework is described in the following section (Chapter 2), followed by related sections focusing on participation in development thinking (Chapter 3), evaluation criteria (Chapter 4), and tools (Chapter 5). In Chapter 12 the Integrative Theory of Reflexive Dialogues is linked to the case studies which are presented and analysed in this book.

With the present book, we seek to demonstrate that stakeholder dialogues play a key role in the process of finding solutions to environmental problems.

References

Agrawal, A., Gibson, C.C. (1999). Enchantment and Disenchantment: The role of community in natural resource conservation. World Development 27 (4), 629-649

Arnstein, S.R. (1971). A ladder of participation in the USA. Journal of the Royal Town Planning Institute April, 176-182

Averbeck, C. (2006). Incorporating Local People through Economic Incentives at Lake Mburo National Park, Uganda – Africa works! In: Stoll-Kleemann, S., Welp, M. (eds). Stakeholder Dialogues in Natural Resources Management. Springer-Verlag, Heidelberg

Barton, T., Borrini-Feyerabend, G., de Sherbinin, A., Warren, P. (1997). Our people, our resources. IUCN (The World Conservation Union), Gland, Switzerland and Cambridge, UK

Batisse, M. (1997). Biosphere reserves: a challenge for biodiversity conservation and regional planning. Environment 39, 7-33

Beierle, T.C. (2000). The Quality of Stakeholder-Based Decisions: Lessons from the Case Study Record. Resources for the Future, Washington

Berghöfer, A., Berghöfer, U. (2006). ‚Participation' in Development Thinking – Coming to Grips with a Truism and its Critiques. In: Stoll-Kleemann, S., Welp, M. (eds). Stakeholder Dialogues in Natural Resources Management. Springer-Verlag, Heidelberg

Bohm, D. (1996). On Dialogue. Routledge, London

Borrini-Feyerabend, G. (1996). Collaborative Management of protected areas: tailoring the approach to the context. IUCN, Gland, Switzerland and Cambridge, UK

Bruner, A.G., Gullison, R.E., Rice R.E., da Fonseca, G.A.P. (2001). Effectiveness of Parks in Protecting Tropical Biodiversity. Science 291 (5501), 125-128

Cleaver, F. (1999). Paradoxes of Participation: Questioning participatory approaches to development. Journal of International Development 11 (4), 597-612

Connor, D.M. (1999). Stakeholder versus Public Participation. Paper prepared for the 19th annual conference of the International Association for Impact Assessment in Glasgow, Scotland, 13. – 19. June, 1999. URL: http://www.islandnet.com/~connor/stakeholder.html

Cooke, B., Kothari, U. (2001). Participation: The New Tyranny? Zed Books, London

Czada, R. (1992). Der Staat als wirtschaftender Akteur. In: Abromeit, H., Jürgens, U. (eds). Die politische Logik wirtschaftlichen Handelns. Edition Sigma, Berlin

Delli Priscoli, J. (1997). Participation and Conflict Management in Natural Resources Decision-Making. In: Solberg, B., Miina, S. (eds). Conflict

Management and Public Participation in Land Management. EFI Proceedings 14, 61-87

Eben, M. (2006). Public Participation during Site-Selections for Natura 2000 in Germany: The Bavarian Case. In: Stoll-Kleemann, S., Welp, M. (eds). Stakeholder Dialogues in Natural Resources Management. Springer-Verlag, Heidelberg

Etzioni, A. (1968). The active society. Macmillan, London

Freeman, R.E. (1984). Strategic management: a stakeholder approach. Boston, Pitman

GEF (Global Environment Facility) (2000). Interim Assessment of Biodiversity Enabling Activities: National Biodiversity Strategies and Action Plans. Global Environment Facility, New York

Giddens, A. (1990). The consequences of modernity. Polity, Cambridge

Hajer, M.A. (1995). The Politics of Environmental Discourse. Ecological Modernization and the Policy Process. Oxford University Press, Oxford

Hardin, G. (1968). The tragedy of the commons, Science 168, 1243-1248

Hellström, E. (2006). Science in Support of the Forest Biodiversity Programme for Southern Finland. In: Stoll-Kleemann, S., Welp, M. (eds). Stakeholder Dialogues in Natural Resources Management. Springer-Verlag, Heidelberg

Hemmati, M. (2002). Multi-Stakeholder Processes for Governance and Sustainability - Beyond Deadlock and Conflict. Earthscan, London

Innes, J.E., Booher, D.E. (2003). Collaborative policymaking: governance through dialogue. In: Hajer, M.A., Wagenaar, H. (eds). Deliberative Policy Analysis: Understanding Governance in the Network Society. Cambridge University Press, Cambridge, 33-59

Jaeger, C.C. (2003). A Note on Domains of Discourse – Logical Know-How for Integrated Modelling. Potsdam Institute for Climate Impact Research (PIK). Working paper No. 86

Jaeger, C. C., Kasemir, B., Stoll-Kleemann, S., Schibli, D., Dahinden, U. (2000). Climate Change and the Voice of the Public. Integrated Assessment 1, 339-349

Jasanoff, S., Martello M.L. (2004). Conclusion: Knowledge and Governance. In Jasanoff, S., Martello, M.L. (eds). Earthly Politics. Local and Global in Environmental Governance. Cambridge, MIT Press, London

Jordan, A.G., Richardson, J. (1983). Overcrowded policy-making. Some British and European reflections. Journal of Theoretical Politics 2, 56-72

Kates, R.W., Clark, W.C., Corell, R., Hall, J.M., Jaeger, C.C., Lowe, I., McCarthy, J.J., Schellnhuber, H.J., Bolin, B., Dickson, N.M., Faucheux, S., Gallopin, G.C., Gruebler, A., Huntley, B., Jager, J., Jodha, N.S., Kasperson, R.E., Mabogunje, A., Matson, P., Mooney, H. III, O'Riordan, T., Svedlin., U. (2001). Sustainability science. Science 292, 641-642

Katzenstein, P.J. (1984). Corporation and change. Cornell University Press, London, Ithaca

Kasemir, B., Jäger, J., Jaeger, C.C., Gardner, M.T. (2003). (eds). Public Participation in Sustainability Science. A Handbook. Cambridge University Press, Cambridge

Kenis, P. (1991). The preconditions for policy networks. Some findings from a three-country study on industrial restructuring. In: Marin, B., Mayntz, R. (eds). Plocy networks. Campus-Verlag: Frankfurt am Main & Westview Press, Boulder, Colorado

Köhn, J. (2002). Conflict Management on Consensus-Building for Integrated Coastal Zone Management – Does Stakeholder Involvement Fit in to the Baltic Sea Region? In: Schernewski, G., Schiewer, U. (eds). Baltic Coastal Ecosystems: Structure, Function and Coastal Zone Management, Springer, Berlin, 341-352

Linke, J. (2006). Community Forest Management in Esmeraldas - Is Constructive Dialogue Possible? In: Stoll-Kleemann, S., Welp, M. (eds). Stakeholder Dialogues in Natural Resources Management. Springer-Verlag, Heidelberg

Luhmann, N. (1984). Soziale Systeme. Grundriß einer allgemeinen Theorie. Suhrkamp, Frankfurt am Main

Luhmann, N. (1986). The autopoiesis of social systems. In: Geyer, F., Zouwen, J.v.D. (eds). Sociocybernetic paradoxes. Sage, London

Maarleveld, M., van de Velde, R., van Uum, J., Pleisier, I. (2006). To See or not to See, that is the Question: Geo-information Visualization Tools as a Means to Facilitate Stakeholder Dialogues in Land and Water Management Planning. In: Stoll-Kleemann, S., Welp, M. (eds). Stakeholder Dialogues in Natural Resources Management. Springer-Verlag, Heidelberg

Mayntz, R. (1987). Politische Steuerung und gesellschaftliche Steuerungsprobleme. Anmerkungen zu einem theoretischen Paradigma. In: Ellwein, T., Hesse, J.J. (eds). Jahrbuch zur Staats- und Verwaltungswissenschaft. Nomos-Verlags-Gesellschaft, Baden-Baden

McNeely, J. A. (1995). Expanding partnerships in conservation. Washington D.C.: Island Press

Merrian-Webster's Colleagiate Dictionary (1995). 10th Edition. Springfield, Massachusetts

Messner, D. (1995, 1998). Die Netzwerkgesellschaft: Wirtschaftliche Entwicklung und internationale Wettbewerbsfähigkeit als Probleme gesellschaftlicher Steuerung. Deutsches Institut für Entwicklungspolitik Band 108. Weltforum Verlag, Köln

Nozick, R. (1974). Anarchy, state and utopia. Basic Books, New York

Oels, A. (2003). Evaluating stakeholder participation in the transition to sustainable development: Methodology, case studies and implications for policymaking. LIT-Verlag, Münster

Oels, A. (2006). Evaluating Stakeholder Dialogues. In: Stoll-Kleemann, S., Welp, M. (eds). Stakeholder Dialogues in Natural Resources Management. Springer-Verlag, Heidelberg

Offe, C. (1987). Die Staatstheorie auf der Suche nach ihrem Gegenstand. Beobachtungen zur aktuellen Diskussion. In: Ellwein, T. et al. (eds). Jahrbuch zur Staats- und Verwaltungswissenschaft. Nomos-Verlags-Gesellschaft, Baden- Baden

Olson, M. (1968). Die Logik kollektiven Handelns. Tübingen

O`Riordan, T., Stoll-Kleemann, S. (eds) (2002). Biodiversity, Sustainability and Human Communities. Protecting Beyond the Protected. Cambridge University Press, Cambridge

O`Riordan, T. (2006). Epilogue: Spreading the Ripples. In: Stoll-Kleemann, S., Welp, M. (eds). Stakeholder Dialogues in Natural Resources Management. Springer-Verlag, Heidelberg

Ostrom, E. (1990). Governing the commons: The evolution of institutions for collective action. Cambridge University Press, Cambridge

Ostrom, E. (1994). Institutional Analysis, Design Principles and Threats to Sustainable Community Governance and Management of Commons. In: Pomeroy, R.S. (ed). Community Management and common property of coastal fisheries in Asia and the Pacific: concepts, methods, experiences. ICLARM Conference Proceedings 45, 34-50

Pahl-Wostl, C. (2002). Participative and Stakeholder-Based Policy Design, Evaluation and Modeling Processes. Integrated Assessment 3 (1), 3-14

Pretty, J.N. (1995). Participatory learning for sustainable agriculture. World Development 23 (8), 1247-1263

Pretty, J.N., Pimbert, M.P. (1995). Beyond conservation ideology and the wilderness myth. Natural Resources Forum 19, 5-14

Renn, O. (2006). Participatory Processes for Natural Resource Management. In: Stoll-Kleemann, S., Welp, M. (eds). Stakeholder Dialogues in Natural Resources Management. Springer-Verlag, Heidelberg

Renn, O., Webler, T. Wiedemann, P. (1995). Fairness and Competence in Citizen Participation. Kluwer: Dordrecht

Rotmans, J. 1998. Methods for IA: The challenges and opportunities ahead. Environmental Modelling and Assessment 3 (3), 155–179

Rowe, G., Frewer, L.J. (2000). Public Participation Methods: A Framework for Evaluation. Science, Technology and Human Values 25 (1), 3-29

Scharpf, F.W. (1983a). Games in hierarchies and networks. In: Scharpf, F.W. (ed). Games in hierarchies and networks. Campus-Verlag, Frankfurt am Main & Westview Press, Boulder, Colorado

Scharpf, F.W. (1983b). Coordination in hierarchies and networks. In: Scharpf, F.W. (ed). Games in hierarchies and networks. Campus-Verlag, Fanrkfurt am Main & Westview Press, Boulder, Colorado

Scheffran, J. (2006). Tools for Stakeholder Assessment and Interaction. In: Stoll-Kleemann, S., Welp, M. (eds). Stakeholder Dialogues in Natural Resources Management. Springer-Verlag, Heidelberg

Senge, P. (1998). The Fifth Discipline. The art and practice of the learning organisations. Century Business, London

Songorwa, A.N. (1999). Community based wildlife management (CWM) in Tanzania: Are the Communities interested? World Development 27 (12), 2061-2079

Stoll-Kleemann, S. (2005). Voices for Biodiversity Management in the 21st Century. Environment 47 (10), 24-36

Stoll-Kleemann, S., Bertzky, M. (2006). Biodiversity Management and Monitoring in Protected Areas – State of the Art and Current Trends. In: Sonak, S. (ed). Multiple Dimensions of Global Environmental Change. Teri Press, New Delhi, India, 143-169

Stoll-Kleemann, S., O`Riordan, T. (2002a). From participation to partnership in biodiversity protection: experience from Germany and South Africa. Society and Natural Resources, 15 (2), 157-173

Stoll-Kleemann, S., O`Riordan, T. (2002b). Enhancing Biodiversity and Humanity. In: O`Riordan, T., Stoll-Kleemann, S. (eds). Biodiversity, Sustainability and Human Communities. Protecting Beyond the Protected. Cambridge University Press: Cambridge, 295-310

Stoll-Kleemann, S., Welp, M. (2006). Linking Case Studies to the Integrative Theory of Reflexive Dialogues. In: Stoll-Kleemann, S., Welp, M. (eds). Stakeholder Dialogues in Natural Resources Management. Springer-Verlag, Heidelberg

Stoll-Kleemann, S., O'Riordan, T., Jaeger, C. C. (2001). The psychology of denial concerning climate mitigation measures: evidence from Swiss focus groups. Global Environmental Change 11(2), 107-117

Stoll-Kleemann, S., O`Riordan, T., Burns, T. (2003). Linking the citizen to governance for sustainable climate futures. In: Kasemir, B., Gardner, M., Jäger, J., Jaeger, C.C. (eds). Public Participation in Sustainability Science. Cambridge University Press, Cambridge, 239-248

Stoll-Kleemann, S., Bender, S., Berghöfer, A., Bertzky, M., Fritz-Vietta, N., Schliep, R., Thierfelder, B. (2006). Linking Governance and Management Perspectives with Conservation Success. Perspectives on Biodiversity Governance and Management 1 (1), 1-40

Sturm, M., Samaniego Rivera, J. (2006). Participation in the Machalilla National Park, Ecuador. In: Stoll-Kleemann, S., Welp, M. (eds). Stakeholder Dialogues in Natural Resources Management. Springer-Verlag, Heidelberg

Vallejo, N., Hauselman, P. (2004). Governance and Multistakeholder Processes. International Institute for Sustainable Development. URL: http://www.iisd.org/pdf/2004/sci_governance.pdf

Welp, M., Stoll-Kleemann, S. (2006). Integrative Theory of Reflexive Dialogues. In: Stoll-Kleemann, S., Welp, M. (eds). Stakeholder Dialogues in Natural Resources Management. Springer-Verlag, Heidelberg

Welp, M., de la Vega-Leinert, A.C., Stoll-Kleemann, S., Jaeger, C.C. (2006a). Science-based stakeholder dialogues: Theories and tools. Global Environmental Change 16 (2), 170-181

Welp, M., de la Vega-Leinert, A.C., Stoll-Kleemann, S., Fürstenau, C. (2006b). Science-based stakeholder dialogues in climate change research. In: Stoll-Kleemann, S., Welp, M. (eds). Stakeholder Dialogues in Natural Resources Management. Springer-Verlag, Heidelberg

Welp, M., Kasemir, B., Jaeger, C.C. (in press). Citizens' Voices in Environmental Policy: The Contribution of Integrated Assessment Focus Groups to Accountable Decision Making. In: Coenen, F.H.J.M., Paterson, R. (Eds). The Promise and Limits of Participatory Processes for the Quality of Environmentally Related Decision-making. Springer-Verlag, Heidelberg

Willke, H. (1998). Systematisches Wissensmanagement. Lucius & Lucius UTB, Stuttgart

Yosie, T.F., Herbst, T.D. (1998). Using Stakeholder Processes in Environmental Decisionmaking. An Evaluation of Lessons Learned, Key Issues, and Future Challenges. URL: http://www.riskworld.com/Nreports/1998/STAKEHOLD/PDF/nr98ab01.pdf

Part II
Theories and Tools

2 Integrative Theory of Reflexive Dialogues

Martin Welp[1], Susanne Stoll-Kleemann[2]

1 University of Applied Sciences Eberswalde, Germany
2 Humboldt University of Berlin, Germany

2.1 The need for an integrative theory

The environmental problems of today's world are increasingly complex and include, among others, the loss of biodiversity, global climate change and water scarcity. Efforts to tackle these problems are exacerbated by uncertainties, facts under dispute, and varying values that lead to ongoing conflicts. We argue that as a consequence of the collapse of the fact/value-dichotomy, dialogues and the exchange of arguments become important at the intersection of science, society, and decision-making (see Putnam 2002). Stakeholder dialogues are one important approach to address these challenges. Experiences have been collected extensively by international organisations, research institutions, corporations, governmental agencies, as well as NGOs (Non-Governmental Organisations).

A key requirement for a practical theory relevant for stakeholder dialogues is that it integrates the different domains and layers of a dialogue (Jaeger 2003). Firstly, a dialogue is about exchanging arguments and creating common meaning. Secondly, dialogues also have a layer of personal relationships where trust building, empathy, antipathy, etc. play a major role.

In science we also face the challenge of attempting to build a bridge between an individual's mental model and conceptual/computer models, which may be used to create and test arguments. However, few efforts have been made to link these models with theoretical approaches. This is why the objectives and success criteria have remained unclear, and the dialogues as such have been exposed to criticism. This chapter outlines a conceptual framework based on the integration of selected scientific traditions on the one hand and on the evaluation of practical experiences on the other. We call this conceptual framework the Integrative Theory of Reflexive Dialogues. The new theory has practical implications for conducting stakeholder dialogues in environmental management, science and policy and for the development of appropriate tools. The concept of learning is the cornerstone of the integrative theory. It draws attention to learning in groups and organisations, but also puts stakeholder dialogues into the broader context of social learning.

2.2 The conceptualisation of the Integrative Theory of Reflexive Dialogues

The lack of a conceptual framework has affected the practice of stakeholder dialogues and the evaluation of the processes and outcomes. Therefore we have made an attempt to select the most relevant theoretical approaches and synthesise them. The relevance of a particular theory depends on the angle the scientist wants to choose and his or her worldview. Our selection criteria have been influenced by practical experiences with various kinds of dialogues and our own intellectual and professional background in the field of sustainability science. A more pragmatic selection criterion is the applicability of a particular theory for tool development.

For our purposes, the three most relevant theoretical backgrounds are social psychological approaches, organisational learning, and formal mathematical approaches for decision support. Social psychological theories were chosen because they cast light on issues such as group processes, social identity, communication, and perception barriers. Organisational theory - and especially organisational learning - is useful for understanding and fostering dialogues in teams and groups. Systems thinking is a key feature for organisational learning because it helps to identify factors that influence the behaviour of the system and its potential change. The third approach establishes links to mathematical representation of stakeholders' assessments that can help to structure debates on complex issues (Welp et al. 2006a, b). The three theoretical approaches are discussed in detail in sections 2.3 - 2.5. In section 2.6, we briefly summarise some other important theoretical traditions. Section 2.7 synthesises the above-mentioned approaches to the new Integrative Theory of Reflexive Dialogues.

Thus, the new theory unifies what has so far appeared to be unrelated and disparate. By linking these theories in a meaningful way, we expect to link various facets of stakeholder dialogues, which cannot be tackled adequately within the confines of a single theory. Such a theory is expected to contribute to tool development in a much more substantial way than others have done so far. A combined use of different tools, for example, by linking formal representation of stakeholders' assessments (analytical tools) with procedures for group work (communication tools) will address the deficits outlined above.

2.3 Social Psychological Theories

2.3.1 Impacts of group diversity and group processes on stakeholder dialogues

To understand better what determines the functioning of stakeholder dialogues in natural resources management, it seems to be useful to consider social-psychological theories, which explain how attitudes, outlooks, and behaviour are shaped in these processes (Stoll-Kleemann 2003). The major drivers that influence stakeholder dialogues from a social-psychological viewpoint are group diversity together with group processes encouraging social identity.

Stakeholder dialogues usually consist of participants who belong to different groups. Therefore, stakeholder dialogues are characterised by high group diversity, which is especially the case in the fields of our book, namely Natural Resources Management and Integrated Assessments. Diversity can be framed in terms of cultural diversity or diversity characterised by different demographic characteristics such as gender, age, education, income, etc.

Enayati (2002) emphasises that it is important to note that NGOs, the academic world, business and industry, indigenous peoples, trade unions, and the like are also "cultures" that can differ with regard to cultural characteristics. In the framework of stakeholder dialogues, she furthermore suggests a view of culture as "the way we do things around here". Members of a culture understand those ways and generally honour them, although without necessarily being conscious of doing so. Enayati (2002: 86) concludes that since stakeholder dialogues "bring people with different cultural orientations into interaction with one another, sensitivity to cultural differences is essential and involves awareness of norms (standards of behaviour) and beliefs (assumptions about the way things are) and values (standards of importance) on which the cultural norm are based".

Stakeholder dialogues are initiated in order to create new knowledge, and due to the diversity of participant perspectives, this outcome is easier to achieve than by individuals working in isolation. Triandis et al. (1965) notes that a diverse group provides a more comprehensive view on possible issues on the agenda. Diverse groups offer immense potential for increased quality of group performance, knowledge creation and innovative decision-making (Jackson 1996, Pavitt 1993, Phillips and Wood 1984, Seibold 1999). The direct involvement in such processes is likely to lead to a change of attitudes and to individual commitment. However, benefits from group diversity in stakeholder dialogues are not automatic.

Group membership itself is an important feature that has to be considered in a theory of reflexive dialogues. Internal bonding processes within social groups (NGOs, scientists, managers, etc.) may account for a decisive rejection of a technically correct compromise. An explanation for expressing more negative attitudes toward a certain issue in stakeholder dialogues may lie in group expectations regarding the roles of the stakeholders involved (Stoll-Kleemann 2001a).

Social psychologists have long looked at the effects and consequences of how people treat members of their own group compared with members of other groups to which they do not belong or identify with; their research takes on two perspectives. The first is where two or more groups are in competition for resources (Sherif 1966) ("Realistic group conflict"), and the second is "how group membership per se affects a person's attitude and behaviour" (Pennington et al. 1999: 326). The latter is the Social Identity Theory[1] (Tajfel 1978, Tajfel and Turner 1979) and provides a better explanation of group-related aspects of stakeholder dialogues.

The basic assumption of Social Identity Theory is that social categorisation results in social discrimination because people make social comparisons between in-groups and out-groups. The four main concepts of Social Identity Theory are social categorisation, social identity, social comparison and psychological group distinctiveness (Tajfel 1978). This distinction between "in-group" and "out-group" suffices to provoke the rejection of the out-group without any competition for resources existing among the groups (Stoll-Kleemann 2001a). Tajfel and Turner (1979: 46) point out that in practice "it is nearly impossible in most natural social situations to distinguish between discriminatory inter-group behaviour based on real or perceived conflict of 'objective' interests between the

---

1 Tajfel and Turner (1979: 34) emphasise that the Social Identity Theory is "intended not to replace the Realistic Group Conflict Theory, but to supplement it in some respects". For an adequate social psychology of inter-group conflict, they regard it as essential to focus on "the processes underlying the development and maintenance of group identity and possibly autonomous effects upon the in-group and inter-group behaviour of these "subjective" aspects of group membership" (ibid: 34).

groups and discrimination based on attempts to establish a positively-valued distinctiveness for one's own group".

Social Identity Theory further states that people make social comparisons because they need to provide themselves with a positive identity (Tajfel 1978, Tajfel and Turner 1979, Turner 1982). Positive social identity is important for a person since it enhances self-esteem and self-worth. Comparisons made between in-groups and out-groups in relation to status, value, and perceived worth lead to social competition. This reflects people's desire to put the groups with which they identify in such a light as to believe their group is "better" than the out-group (Pennington et al. 1999, Turner 1982). Membership in a group relates to external criteria (e.g. being a business leader or representative of an NGO). Furthermore, identification with this group depends on internal criteria, among them cognitive factors (like the awareness of being a group member), evaluative factors (like the social prestige of group membership), and emotional factors (like positive or negative feelings associated with group membership). In this context, Turner (1982: 27) suggests that in order "to understand how social groups are formed", one should also focus on variables such as "common fate" or "shared threat" (ibid: 27).

One lesson that can be learned from social dilemma[2] research in this context is that people in such situations "attend more to the groups' payoffs than to their own, either automatically or to behave appropriately". But whereas social identity elicits co-operative behaviour in dilemmas, it is generally only for the benefit of an "in-group". "Dilemmas between groups (requiring self-sacrificial behaviour within) are often the most extreme. Consequently the framing and manipulation of group identity is critical to co-operation rate" (Dawes and Messick 2000: 111). When people act as individuals who are interacting with other individuals, they are far more co-operative than when they form groups that interact with other groups (ibid: 114).

There is overwhelming evidence that "favouring the in-group over the out-group is extremely common in inter-group relations" (Turner 1982:34, see also Doise 1978, Tajfel 1978). Negative values and exaggerated stereotypes, are attributed to the out-group, while the in-group is perceived to have positive characteristics and values (Pennington et al. 1999).

---

2 In social dilemma situations, each individual always receives a higher payoff for defecting than for co-operating, but all are better off if all co-operate than if all defect (Dawes and Messick 2000).

2.3.2 Stereotyping as a limiting factor for group learning

The fact that stereotypes of out-groups are one important consequence of social identity processes has been alluded to in the foregoing section. A social stereotype is "a set of beliefs about the personal attributes of a group of people" (Ashmore and Del Boca 1981: 30). Such sets of belief are "activated" (that is, start influencing perception in a given situation) through identification of an individual's group membership (Enayati 2002). All members of the out-group are seen as possessing those stereotypical characteristics, and the individual's unique personal characteristics are ignored (Pennington et al. 1999). Stereotypes are usually highly simplified images, and those that refer to out-groups are often of a derogatory nature and based on, or refer to, clearly visible differences between groups, e.g. in terms of physical appearance (Hogg and Vaughan 1998). From this description of stereotypes, it can be seen that they are similar to prejudicial attitudes that people hold about social groups. A person holding a stereotype will show a tendency to note and recall subsequent information about the social group that fits the stereotype (Pennington et al. 1999).

For example conflicts between business leaders and environmental NGO representatives in stakeholder dialogues have to be understood in the context of these stereotyped relationships between environmentalism and other social and economic interests. They are extremely difficult to modify when social tensions and conflicts have arisen among groups (Hogg and Vaughan 1998). Recent research has acknowledged that stereotypes have both cognitive and emotional undercurrents that inflame judgements about social groups. Situations which include strong, negative emotions, such as anger or anxiety, have been found to increase a person's use and reliance on stereotypical thinking (Mackie and Hamilton 1993, Pennington et al. 1999). Therefore, on the one hand, stereotypes can negatively affect communication among opposite groups but on the other hand, according to Enayati (2002), it is important to note that stereotyping is not just a "bad habit"; it is inherent in our cognitive structure. It makes our perception quicker and more economic; we simply cannot meet everybody as a completely "new person", a blank sheet. Nor are stereotypes necessarily completely wrong. Having our perceptions and expectations shaped through stereotyping can indeed have positive effects.

As discussed, stereotyping does not necessarily imply negative evaluation but often it does, and then it implies social prejudice (negative attitudes) and discrimination (negative behaviour): people are judged negatively merely because they belong to a certain social group. Impacts on behaviour can include avoidance, exclusion, fear, and aggression. It is important to note that being discriminated against can elicit "counter-discrimination" and hence further increase distance between social groups (Hemmati 2002).

It is difficult to change attitudes based on stereotypes because information concerning the features on which the stereotypes focus are absorbed and processed in a very one-sided manner. For example, stakeholders such as political decisions-makers, business leaders, and environmentalists select sources of information, e.g. about climate change, from which they can expect (e.g. because the title of a journal article seems to promise it) that their current attitudes, values and knowledge will be confirmed. As mentioned above, the attitudes, values and emotions of business leaders and environmentalists are sometimes biased against each other. Therefore they seek information that reinforces this bias, while challenging the credibility of any information that contradicts their attitudes. Values and emotions act as powerful criteria for the selection and processing of information (Lantermann and Döring-Seipel 1990, Ernst et al. 1992, Lantermann et al. 1992). Especially in conflict situations, values and emotions have a negative influence on learning and mutual understanding and thus have a negative impact on the quality of the outcomes of stakeholder dialogues.

However contact with members of the stereotyped group might be the first step in overcoming stereotyping if it happens repeatedly and with more than one "typical" group member (Pettigrew 1989). In many cases, the best strategy to overcome prejudice has proved to be to engage both groups in a common activity, e.g. working together. Particularly if the activity is successful, it can significantly contribute to reducing prejudice and improving relations between different groups (Sherif and Sherif 1953, Smith and Mackie 2000). Such processes have to be taken into account when searching for adequate strategies to deal with environmental problems in stakeholder dialogues and for creating a reflexive theory of stakeholder dialogues.

A different danger in the information-gathering process in stakeholder dialogues can arise when information is held by only one member of the group and this information is ignored, e.g. because of the relatively low status of that person. Research on social influence and conformity indicates that when a person's private judgement differs from the opinions expressed by others, that judgement is soon abandoned, even when it proves to be verifiably correct. However, in the presence of just one other person who agrees with them, people persevere in the face of opposition (Asch 1956). Also, just as an individual is likely to lack confidence, the group may lack confidence that, in an ambiguous situation, a deviant opinion could be correct. The evidence suggests that for diverse groups to fulfil their potential, group members should have overlapping areas of expertise instead of a sole expert for each relevant knowledge domain (Jackson 1996). This is a challenge for the selection of the right stakeholders.

2.3.3 The Theory of Psychological Reactance

A further well-known social-psychological theory has to be considered in the Reflexive Theory of Stakeholder Dialogues. Brehm's Theory of Psychological Reactance provides a useful explanation of why stakeholder dialogues and public participation are very important instruments to avoid reactance that is counter-productive for the implementation of sustainable development strategies (Stoll-Kleemann 2001b).

The theory states that reactance arises when personal rights to decide and act are threatened, reduced, or eliminated, for example via regulations, prohibitions and controls (Brehm 1966). This is restricted to behaviours where the person – i.e. from his or her subjective perspective – has a perception of being threatened. If people feel restricted in their influence on decision-making, this can provoke reactance and arouse efforts to gain more influence (Stoll-Kleemann 2001b).

This means that a lack of inclusive and meaningful participation is an important factor that hinders the implementation of sustainable development strategies (Scheffran and Stoll-Kleemann 2003, Stoll-Kleemann and O`Riordan 2002a, b). This can be seen in close connection to the Theory of Psychological Reactance because if decisions are taken without the involvement of affected citizens, reactance can occur (Stoll-Kleemann 2001a, b). The lack of a continuous dialogue and "real" communication between the involved citizens in which various interests and points of view may be understood and accommodated, developed and resolved in face-to-face discussions is especially problematic. Findings from psychological social-dilemma research confirm this evidence in the field of water conservation. These findings suggest that people are more willing to support authorities when these authorities use fair decision-making procedures (Tyler and Degoey 1995).

To summarise the social-psychological dimensions of stakeholder dialogues, it is important to note that processes within them, such as communication, are not merely rational processes and should not be approached as such (see also below "bounded rationality" and "mental models"). Instead "people's feelings, attitudes, irrationalities in information processing, and so on, need to be taken into account and respected" (Enayati 2002: 8, Stoll-Kleemann 2003). While the discussions within stakeholder dialogues need to be based on factual knowledge, trust building is also clearly an essential prerequisite for successful stakeholder dialogues. Overcoming prejudice and stereotyping can be framed as a learning process that will lead to people being able to truly "dialogue" (Enayati 2002). It is a process based on interaction between participants that takes time to evolve.

2.4 Theories of Organisational Learning

Learning[3] is indeed one of the key concepts for stakeholder dialogues and participation. An important question in this context is how groups and organisations learn and how inter-organisational learning can be organised. In stakeholder dialogues and participation, representatives from a wide range of organisational backgrounds and professional cultures meet to debate an issue at hand. A new action-oriented, theoretical framework for public participation and stakeholder dialogues thus needs the input of organisational theories. Management science and organisational learning have greatly influenced business practices in the last decade (Senge 1998), but have in practice been largely neglected in global change research. Organisational learning has been influenced by various scientific traditions, most prominently by psychology, cultural studies, sociology, economics, and history. A primary challenge is to find out how people can work together effectively for the period during which they are together.

A paradigm shift[4] has taken place from rigid hierarchies to an emphasis on working in networks. These organisational innovations have relevance for stakeholder dialogues. Underlying this new perspective on organisations (Wheatley 1992) has pulled together the insights of systems thinking from various academic disciplines and developed recommendations for organisational development theory and practice on that basis. The conceptualisation of organisations based on systems thinking takes the human capacity for purposeful behaviour, reflection, and learning as a starting point. People in a system need shared purpose and meaning in order to make sure that their individual actions are in tune with the system as a whole. This requires visioning, commitment, and passion

[3] For a more thorough and partially complementary discussion of theories of learning and their application in the context of stakeholder dialogues, see Chapter 6 of this book (Maarleveld et al. 2006).

[4] For a detailed overview of the paradigm shift within organisational theories see Oels (2003: 43-47).

for the shared purpose (Pratt et al. 1999). Systems thinking is a key concept in Senge's conceptual framework, which is described below.

Management practice within organisations has in recent decades been greatly influenced by systems thinking. Senge (1998) has described it as the essential 'fifth' discipline for organisational learning. The other four disciplines are mental models, shared visions, team learning and personal mastery. In Senge's systems-dynamics approach, attention is paid to dynamic complexity (i.e. how patterns change over time) rather than detail complexity (i.e. full detail at any one point in time). According to Senge (1998), "the real leverage in most management situations lies in understanding dynamic complexity, not detail complexity." Dynamic complexity implies that interrelationships consist of positive and negative feedback loops and not only linear cause-effect chains. Instead of focusing on short-term effects only, individuals and organisations should become aware of how the effects of actions change over time (Senge 1998). A 'quick fix' addresses only the symptoms, while the underlying fundamental cause remains unaltered. In a longer time range, this 'shifting the burden' structure may worsen the situation. According to the systems-dynamics approach, learning means understanding the complex relations of social systems and their dynamics.

Mental models are deeply held internal images of how the world works. These images can be so powerful that they limit us to familiar ways of thinking and acting. They are especially powerful because they shape our perception. Thus, the discipline of management of mental models is about questioning, testing, and updating these images. In stakeholder dialogues, mental models should be made explicit. Understanding the basic assumptions and worldviews the other person holds is key to accepting the other person's position. If mental models are made explicit, others can challenge them. The business world has applied different ways of institutionalising reflection on mental models. Scenarios, mapping mental models, computer simulations (Sterman 2002), and other tools usually deal with non-quantifiable variables and can be used in the context of stakeholder dialogues as well. We will come back to mental models and a specific application of them in Bayesian Networks in the following section.

The development of shared visions (which differ from consensus) is important for organisations and increases their capabilities to focus their activities. As mentioned earlier, a consensus may be the objective of a stakeholder dialogue. In scientific dialogues for example, exploration of different views and dissent may be a relevant result, too. In stakeholder dialogues, consensus on an issue may be achieved even though the participants do not share the same vision.

Besides systems thinking, mental models, and shared vision, Senge (1998) emphasises team learning as one of the key disciplines in organisational learning. In the context of stakeholder dialogues, team learning is highly influenced by the mode of the communication. David Bohm's distinction between discussion and dialogue is useful in this respect. As outlined in Chapter 1, in discussions, individual views are presented and defended. Discussions can be seen as a ping-pong game: the subject of common interest is analysed from many points of view; the purpose of the game is normally to win (Bohm 1996). Winning means having one's view accepted by the group. The basic goal of participants in a discussion is for their view to prevail. In a dialogue, in contrast, the participants are not negotiating positions or trying to reach a consensus. Dialogues are based on mutual respect and on the notion that the others have a valid viewpoint. The word dialogue suggests a free flow of meaning between people. In dialogue, individuals gain insights that cannot be achieved individually. Thus dialogues foster interdisciplinarity and a holistic view. The concept of dialogues resembles Habermas' ideal speech situation (see section 2.6).

Necessary conditions for a dialogue are that (a) participants treat one another as colleagues, (b) that they "suspend" their assumptions, and (c) that the process is structured by a skilled facilitator. Treating one another as colleagues in practice requires that normal in-group/out-group thinking is put aside and that stereotypes do not create barriers between participants (see section 2.3.2). Suspending assumptions means to hold them as if they were 'hanging in front of you', constantly accessible for questioning and observation. It does not mean throwing the assumptions away or suppressing them. By holding assumptions up for examination, the involved indiduals can learn about their own mental models and the mental models of other participants.

Both Senge (1998) and Bohm (1996) argue that a group can achieve and be more than simply a sum of its parts. This requires, however, that the participants go beyond merely trying to convince each other of their personal views and positions. This kind of communication is not easy to achieve, and it requires trust building and usually some time. According to both Bohm (1996) and Senge (1998), there is a place for both discussions and dialogue. The power of the approach lies in the combination of both. It is however useful to be clear about the right timing of the two.

Senge's (1998) five disciplines have been highly influential in business practices and learning within companies and other organisations. For participation and stakeholder involvement, systems thinking, mental models, and team learning (i.e. the difference between discussion and dialogue) are of particular relevance. Unlike learning within organisations, participation in stakeholder dialogues involves learning between organisations and individuals with different backgrounds. Learning between organisations is a challenging effort for a number of reasons: First, the opportunities for regular interaction are rarer and the process of trust building may thus take longer. Second, unlike a company, which may be able to create a common vision of its activities in a short period of time, a multi-stakeholder dialogue faces the challenge of dealing with a multitude of interests. Third, creating a shared language requires flexibility and some time.

Organisational learning is a relatively new and heterogeneous field of inquiry. Among the wide range of perspectives, we find systems approaches, cognitive approaches, communicative approaches, and cultural approaches. For our purposes - developing a practical theory of stakeholder dialogues - the systems approach seems most promising. In Senge's view, systems thinking is the essential discipline, but his other disciplines have great relevance for stakeholder dialogues as well. For a more detailed discussion of this and further theories of learning and related analytical frameworks (e.g. Kolb´s learning cycle or the learning loops of Argyris and Schön) and their practical application in stakeholder dialogues in natural resource management (e.g. GIS-assisted learning in planning), see Chapter 6 (Maarleveld et al. 2006).

2.5 Formal approaches

2.5.1 Are stakeholders rational actors?

A formal representation of stakeholders' assessments seems to be useful in complementing the theoretical framework for stakeholder dialogues and public participation. Although ordinary language is indispensable in exchanging arguments, advancements in mathematics and computer tools open attractive paths to explore. We believe that the diversity of stakeholders' perspectives can be captured by a formal representation of their preferences and mental models, and that a formal approach offers several advantages. First, the language used is close to that of systems dynamics and modelling. Mathematical models can handle complicated descriptions of how variables relate to each other. According to Cain (2001: 11) the "down side of using a mathematical model is that it is hard for people not involved in its construction to understand it". Second, a formal representation forces clarity in making statements in order to reduce ambiguity. Third, models of how humans or social groups behave or how the formation of expectations takes place can be coupled with modules describing natural systems, socio-economic systems, etc. The disadvantage is that some formal representations may be simplistic and reduced, and that many nuanced aspects of the topic that can be described in words are lost.

The theoretical backgrounds of these formal representations are two-fold: The first is Rational Choice Theory, in particular the Rational Actor Paradigm (RAP), which has been influential in neo-classical economics and sociology in particular because it uses a mathematical language to represent preferences and the behaviour of individuals and groups. Rational Choice Theory (including Game Theory) has frequently been applied, e.g. to analyse negotiation situations. At the core of this theory is that rational actors can choose between different possible actions and order different consequences of possible actions according to their preferences. In a decision-making situation with many actors, the possible actions available to each of them depend on parameters of their joint situation. Rational actors choose a possible action that, depending on their preferences, is optimal given the parameters of the situation (Jaeger et al. 1998). Von Neumann's and Morgenstern's (1944) expected utility axioms and their rational choice model were developed to describe how actors should behave if they were about to act rationally. According to the expected utility theory (EUT), by assessing the probability of different outcomes, actors try to maximise the expected utility taking into account that some are risk averse while others are risk seeking. In stakeholder dialogues, such questions can be of interest as well.

The second theoretical background is called "Bounded rationality", which is the key concept of the so-called "behavioural approach". Since real world actors do not have perfect knowledge of the costs and benefits of different alternatives, as postulated in rational choice theory, they are bound to make decisions under uncertainty. Psychologists, the best known in this field are Kahneman, Slovic, and Tversky (Kahneman et al. 1982) have identified several biases that influence human decision-making because of cognitive limitations and also organisational pressures. As a result, many decisions turn out to be incorrect; choosing the best course of action is just too complicated. Therefore, they have to use reduced mental models of the world (Simon 1957) (see also the description of cognitive limits in section 2.3 and the definition of Senge in section 2.4 and Sterman 2002).

According to Sterman (1991: 2) "Mental models have some powerful advantages. A mental model is flexible: it can take into account a wider range of information than just numerical data; it can be adapted to new situations and be modified as new information becomes available. Mental models are the filters through which we interpret our experiences, evaluate plans, and choose among possible courses of action. The great systems of philosophy, politics and literature are, in a sense, mental models."

It is again Sterman (1991) who emphasises that there are also problems associated with mental models. "They are not easily understood by others; interpretations of them differ. The assumptions on which they are based are usually difficult to examine, so ambiguities and contradictions within them can go undetected, unchallenged, and unresolved" (Sterman 1991: 2). Surprisingly, we are also bad at constructing and understanding our own mental models or using them for decision-making. Psychologists have shown that we can take only a few factors into account in making decisions, which leads to usually extremely simple mental models (Kahneman et al. 1982).

Therefore it seems to be useful, considering this theoretical concept of mental models in stakeholder dialogues, to use certain tools as described below to overcome - or at least deal with - the given limitations. These tools offer improvements insofar as they are assigned to have a clear purpose to solve a particular problem. They make assumptions explicitly open to all for review. These approaches make it possible to interrelate many factors simultaneously. The usefulness of the tools/procedures explained below lies in the fact that they simplify reality, putting it into a form that we can comprehend. In the following, we will focus on Bayesian learning and then multi-criteria decision analysis.

2.5.2 Bayesian learning[5]

Bayesian learning seems partly to be a departure from RAP in its original version (i.e. in RAP there is no place for learning since, as mentioned above, actors have complete information and preferences do not change). Models based on Bayesian learning may, however, better represent true human behaviour, primarily because agents have limited information storage capacity. Similar to Game theory, Bayesian learning acknowledges uncertainty and operates with probabilities.

One definition of Bayesian learning reads as follows: "Bayesian learning constitutes a probabilistic view of learning based on Bayes' Theorem. The underlying assumption is that there is a set of hypotheses, each having a certain probability of being correct. Receiving more information changes the probabilities from a learner's point of view. For instance an observation might contradict a hypothesis or strengthen the

[5] This subsection is a modified and shortened version of the analogous subsection in Welp et al. (2006a).

belief in it. The aim in this setting is to be able to find a hypothesis with the highest probability of being correct, given a specific set of data / piece of information" (University of Dortmund 2006).

Box 2.1

Bayes' theorem is a result in probability theory. Bayes' theorem gives the probability of a random event *A* occurring given that we know a related event *B* occurred. This probability is noted $P(A|B)$ and is read "probability of *A* given *B*". This measure is sometimes called the "posterior" since it is computed after it is known whether *B* is the case or not.

Bayesian belief network: a graphical tool to help make decisions under uncertainty. It can be used to build a Decision Support System (e.g. a Bayesian Expert System). Bayesian networks are composed of three elements: a set of nodes representing system variables, a set of links representing causal relationships between the nodes, and a set of probabilities, for each node specifying the belief that a node will be in a particular state given the states of those nodes that can affect it directly.

Bayesian learning: the process by which a Bayesian belief network updates its set of probabilities (so-called conditional probability tables) as a result of receiving case data about variables in the table.

Adapted from: Cain (2001), Wikipedia

Bayesian learning is represented in mathematical terms in the following way. In a simple example, suppose there are two states of the world *s* and *s'*. Agents are unsure which of them is the actual or true state of the world, but at time *t*, the *i*th agent attaches probability $zi(t)$ to *s'* being the true state of the world and thus believes *s* to be true state with the probability $1\text{-}zi(t)$. Beliefs are thus captured in the single parameter $zi(t)$. In the light of their beliefs, the agents choose a particular course of action. Having acted, they observe a result, which is called *X*. Based on this, they update the probabilities of s' being the true state of the world (Breen 1999). The Bayesian mechanism provides a plausible way in which beliefs can change over time, a process called belief updating.

Developing RAP further and applying the concept of Bayesian learning in particular seem to be promising paths for advancing the stakeholder dialogues in natural resource management. Three main areas of relevance can be found: (a) framing problems, (b) finding differences and inconsistencies, and (c) addressing the question of how actors learn.

Framing problems

Environmental policy-making is often faced with factual uncertainty and political controversy. In conflict literature, this is described as issues being at dispute and values being subject to conflict. Because natural resource problems tend to be complex and subject to both factual uncertainty and conflicts over values, they are not easy to frame in a meaningful way. The inability to construct well-formed problems hampers efforts to find mutually acceptable solutions.

Empirical studies have shown that the framing of an issue by using a positive or negative description (e.g. would you invest in a medicine that saves 70% of the patients? vs. would you invest in a medicine when 30% would still die?) has a strong influence on the answers people give (Gardner and Stern 1996). Other studies have attempted to show how citizens perceive certain complex issues (are there wrong, imprecise, or irrelevant beliefs?), and how risk communication can take these insights into account when aiding the public's understanding about complex issues (Bostrom et al. 1992). Wynne (2005) on the other hand turned the problem upside down and argued that public misunderstanding, mistrust, or scepticism regarding scientific discourse on risk may in fact relate to the way risk issues are defined and the risk discourse constructed, which excludes citizens' views and perceptions.

The author further believes that participation processes and framing methods developed to deal with the resistance of the public or to educate citizens solely focus on downstream risk issues (e.g. risk and impacts of a new technology). They furthermore deny citizens the ability and the possibility to address essential social debates (upstream issues – which human purposes drive science and innovation?).

In this context, one application of Bayesian learning is the use of Bayesian belief networks to visualise the structure of our present knowledge and thus come up with an accepted problem definition. The Bayesian formalism allows for subjective probabilities, which is of interest in stakeholder dialogues. Imprecise information on complex systems can be presented by proceeding from a simple influence diagram to a causal network containing system components (nodes) and causal dependencies (links or arcs). The probabilistic concept underlying a Bayesian approach acknowledges the uncertainty of data and of the conceptualisation of problems and is more likely to be accepted by stakeholders than single predicted results.

Finding differences and inconsistencies

Finding an agreement about an issue may be easier if subjective probabilities and assessments are made explicit. Here Bayesian learning can also be useful since it helps to identify inconsistencies in people's thinking. Key experts and decision-makers may have widely different and inconsistent explanations of the problems at hand or opinions on the course to adopt. Bayesian expert systems can, for example, be applied to help structure the debate on various natural resource management problems such as the exploitation of marine resources (see example below).

Thus a structuring process can greatly benefit from the use of Bayesian belief networks. Cain's (2001) illuminating guidelines provide concrete steps to capture and represent the world as described by different stakeholders in simple conceptual models. Stakeholder interviews or group discussions are conducted to elicit expert information and various subjective probabilities. The stakeholder groups can be very small and, e.g., include members of industry, NGO representatives, and lay people. Stakeholder elicitation may take place by conducting semi-structured interviews and group discussions (if appropriate). It is good to start building a Bayesian network by beginning to think of the variables in certain categories. Cain suggests distinguishing between the following six categories of variables as a starting point for a network structure: objectives, interventions, intermediate factors, controlling factors, implementation factors, and additional impacts. Stakeholder Bayesian networks (BNs) are created. A BN is basically a set of nodes representing system variables and a set of links representing causal relationships between these nodes (see Figure 2.1). At a later stage, stakeholder Bayesian networks can be simplified and merged to master BNs.

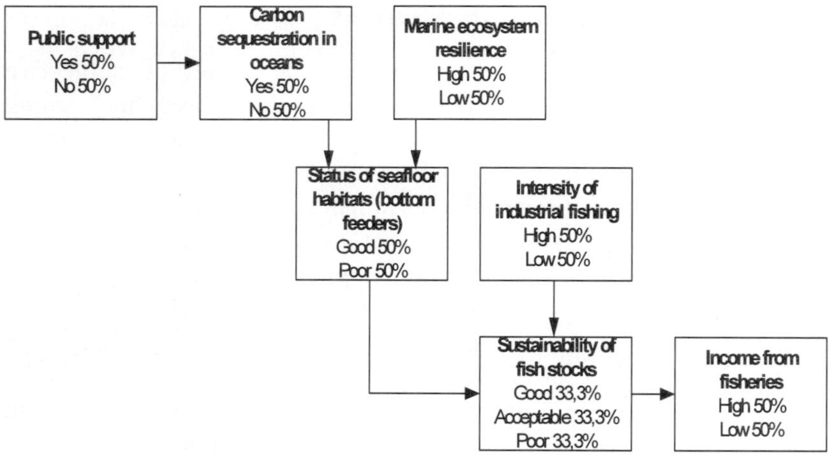

Fig. 2.1. A simple Bayesian belief network

In a next step, Conditional probability tables (CPT) are created: a set of probabilities, one for each node, specifying the belief that a node will be in a particular state given the states of those nodes that affect it directly (its parents). In other words, CPTs express how relationships between nodes operate (see Table 2.1). Each row in a CPT implies a question. Using the belief network in Figure 2.1 as an example, we can ask the following question: "If the status of seafloor habitats (bottom feeders) is poor and the intensity of industrial fishing is high, what is the chance that sustainability of fish stocks is acceptable?" If it appears to be difficult to frame these questions, then it is likely that the master BN is illogical. The structure or the states of the nodes have to be subsequently altered.

Table 2.1. Conditional probability table (CPT) of an imaginary stakeholder.

Intensity of industrial fishing:	Status of seafloor habitats:	Sustainability of fish stocks: Good	Acceptable	Poor
Low	Good	0.60	0.40	0.00
Low	Poor	0.00	0.10	0.90
High	Good	0.40	0.60	0.00
High	Poor	0.00	0.00	1.00

As mentioned above, Stakeholder Bayesian networks can be simplified and merged to master BNs. When the master BN is completed, it can be turned into a fully functioning BN that can be used to help make decisions and to carry out further dialogues with stakeholders. This is done by filling in the CPTs using the best and most appropriate data or expert judgement available and by manipulating the BN (i.e. by changing probabilities).

By building an expert belief system and reviewing it together with stakeholders, a better picture of the problems at hand can be obtained. The whole exercise provides the involved scientists and stakeholders an opportunity to reflect on their basic assumptions, revise their views and learn as individuals and as a team. Such a procedure will reveal gaps in current knowledge and thus point at new research questions.

Expert belief systems can be used to develop empirical explanations (a causing b with a certain probability) but also normative argumentation. Thus both factual uncertainty and conflict about values can be addressed. This helps to identify areas where agreement can be found and where disagreement over issues or values prevails. The possible fields of application encompass a broad range of decision-making situations ranging from natural resources to business management decisions.

How do stakeholders learn? – Constructing a model of learning

As mentioned above, an important aspect of Bayesian learning is that the update of beliefs when new evidence occurs is possible. This takes place formally by experts changing the probabilities of a statement being true (see Figure 2.1). An application of Bayesian learning could be to study "how and on what basis stakeholders update their beliefs when confronted with new, albeit uncertain insights?" It becomes possible to develop formal models of how stakeholders or 'agents' learn. Such models, even though they may remain anecdotal, explicitly aim at simulating more realistic present and future behaviour, such as consumer behaviour, investment decisions, or positions in negotiations. Research in this area, although crucial to improving current research on natural resource problems, is still in its infancy. Agent-based modelling is one approach that is actively being developed and experimented with in sustainability science (see Scheffran 2006, Chapter 5 of this book).

What is the practical relevance of such approaches for stakeholder dialogues and public participation? The method presented above seems promising for exploring stakeholders' mental models and in turning qualitative descriptions into simple quantitative assessments. Each individual's mental models may first appear alien to others, but the visual representation helps them understand the differences. Mental models become clear by interviewing stakeholders and aggregating their views to a Bayesian expert system. The wide application of the approach in tackling management problems and bringing together stakeholder views with scientific models suggests that such an approach helps to identify inconsistencies and differences in stakeholders' assessments. Bayesian learning can be used to develop Internet-based stakeholder tools, such as an Internet-based Bayesian learning model that can be updated online by stakeholders (Ames and Neilson 2001). An encouraging feature of Bayesian networks is that several time steps can be built into the system. Thus, interventions in a management system can be explored in an iterative way.

2.5.3 Multi-criteria decision analysis

Multi-criteria decision analysis (MCDA) is increasingly being used to help resolve emerging goal conflicts in areas such as natural resources management in particular and environmental assessment in general (Jentsch et al. 2003). In stakeholder dialogues, this approach can have a similar structuring effect as the analysis of mental models (cf. Bayesian learning). In a process that uses MCDA, both objectives and measurable criteria are identified to assess the extent to which these objectives are met. Different kinds of objectives can be included, expressing not only economic values but also addressing goals that cannot always be expressed in monetary terms, such as biodiversity, equity, or minimising risk and uncertainty. The factors of a solution are not fixed values but are variable or fuzzy within certain ranges determined by resources availability and socio-economical realities (IIASA 2004). MCDA tools usually provide an explicit relative weighting system for various criteria. In contrast to cost-benefit analysis, where all positive and negative effects are aggregated to a single monetary unit, MCAs are better suited to cope with the fact that not all impacts can be measured using the same unit. Disaggregation thus helps to make explicit what different alternatives mean for different groups.

There are crucial differences between the Bayesian networks and MCDA analysis: MCDA analysis represents only decision criteria, while Bayesian networks help to understand the underlying working of a system. Stakeholders can easily understand the hierarchy of decision criteria, which is a basic concept of MCDA. Cain (2001) argues however that this can sometimes restrict the ways in which stakeholders express themselves. Multi-criteria analysis and Bayesian Networks are thus approaches that are attractive for stakeholder dialogues in natural resources management, especially if there are groups involved whose interests vary greatly.

There is a wide range of MCDA approaches, including commercial software packages. For a detailed description of various tools, see Dogson et al. (1999). In the following, some aspects of MCDAs are highlighted that are relevant for a conceptual framework of stakeholder dialogues.

MCDA can be used for finding areas where stakeholders' interests converge and could potentially lead to building coalitions. One example is NAIADE as an MCDA tool that has been applied in practical management situations (O'Connor 2000). Another software package is the Aspiration-Led Decision Support (ALDS) approach developed at IIASA (International Institute for Applied Systems Analysis). This tool is oriented towards an interactive mode of operation in which a sequence of problems is solved under varying conditions (e.g. different objective functions, reference points, values of constraints and bounds). It also offers many options useful for diagnosis and verification of a problem being solved.

The two formal approaches described above are suited for different kinds of dialogues. Depending on the objectives and the mathematical skills the participants bring in, one can choose between Bayesian networks and MDCAs. A combination of such tools can be a way to move forward.

2.6 Other contributing theories[6]

In the following, selected theoretical traditions such as collaborative planning theories, democratic theories and network theory are summarised. A short review into the history of these three approaches is useful in order to understand the origins of participation and public involvement. Furthermore, selected aspects of these theories will feed into our new Integrative Theory of Reflexive Dialogues (power relations and rules of discourse such as fairness). In the field of planning, different approaches and theories have been competing and moving between the poles of rationality and focus on social processes.

In the heydays of 'rational comprehensive planning' in the 60s' and 70s′ the prescription for planning and policy formation consisted of five stages: identify objectives with weights, identify alternative courses of action, predict consequences, evaluate the consequences on a common scale of value, and finally select the alternative whose net benefit is the highest (Rosenhead 2001). This approach was heavily criticised as being socially undesirable and practically infeasible (Lindblom 1959). The rational comprehensive planning approach neglected the multitude of conflicting

[6] Some of these and other important theories that relate to stakeholder dialogues in natural resources management are discussed (in more detail) in Chapters 3 - 6 (Berghöfer and Berghöfer 2006, Oels 2006, Scheffran 2006, Maarleveld et al. 2006) of this book.

interests and the fact that not all interests are equally represented in the decision-making.

Collaborative planning theories, which encompass critical theory, advocate planning and alternative planning put emphasis on the communicative aspects, power structures and disaggregation of effects (Leskinen 1994). Instead of searching for the optimal solution based on an assessment of net benefits (usually in monetary terms), the alternatives should be made visible by disaggregating the effects on different groups, the environment, the economy, etc. Collaborative planning theory adopted the theory of communicative action of Jürgen Habermas and saw the role of a planner as an active designer of the communicative process in which weak groups are intentionally given the opportunity to voice themselves (Forester 1985, 1993). Present practice of stakeholder dialogues and public participation in many policy fields (e.g. development aid) suggests that a transition has taken place towards more collaborative approaches.

Theories of democracy are also relevant for stakeholder dialogues and public participation. They help to clarify the relationships between representative decision-making and participatory procedures (O`Riordan and Stoll-Kleemann 2002). A difference can also be made, for example, between elitist and populist approaches to stakeholder dialogues. Democratic theories emphasise the importance of power, which different actors such as governments, multinational corporations, NGOs, and others use, which is an important element of our integrative theory. Representative decision-making and stakeholder/public participation do not compete but rather complement each other (Gunderson 1995, Kasemir et al. 2003, Stoll-Kleemann et al. 2001, 2003). In this context, focus groups are a useful communication tool to support democratic decision-making. Focus groups are widely used in public opinion research, marketing (Krüger 1993, Morgan 1988), and studies of mass communication (Merton 1987). In recent years they have also been applied in environmental science, such as in the ULYSSES and CLEAR projects (Jaeger et al. 2000, Kasemir et al. 2000, Stoll-Kleemann et al. 2001, 2003). There are few examples where focus groups have been applied directly to support parliamentary decision-making (Welp et al. in press).

New technologies, such as the Internet also open up new possibilities of citizen involvement and dialogues. The relation of the Internet and democratic decision-making has been discussed, for example, by Beierle (2002). The potential web-based knowledge systems offer for increasing the competence of lay citizen by giving them access to scientific knowledge has been discussed in Kasemir et al. (2003).

The final contributing approach to be considered in the analysis and practice of stakeholder dialogues are theories of networks (including social capital formation). The importance of network theories can briefly be described as follows: Networks are thought to emerge whenever individual actors lack the necessary resources to achieve an output on their own and are required to collaborate with others to mobilise and pool resources (Messner 1998). Networks are understood as co-ordination mechanisms beyond markets and policy hierarchies, i.e. as qualitatively different from these other two mechanisms of co-ordination.

The workings of networks require the building of trust between the actors and are based on the principle of reciprocity. Reciprocity is the outcome of a productive tension between self-interests and solidarity in durable social relations (ibid). Productive networks depend on the actors' capacity for compromise and their respect for the legitimate interests of others. Networks are better equipped to deal with the complexity of 'modern' problems and risks. The communication between the various members of a network increases the system's capacity to take notice of, explore, and describe new problems. Where the resources for addressing the new problem are dispersed amongst diverse actors, an effective network between them is key to making these resources available for a collectively desired outcome (ibid). According to network theories, a crucial factor in the capacity of societies to address pressing problems and achieve collectively desired outcomes by drawing on network structures is the moral resources (ibid) or 'social capital' (Putnam 1993) available for collective action.

We do not claim that the list of theories identified is exhaustive. Some of the more interesting theories, which can only be named here, include post-normal science (see Chapter 7) and theories of power. Some of these, and other important theories that relate to stakeholder dialogues in natural resources management, are discussed in more detail in Chapter 3 of Berghöfer and Berghöfer (2006), Chapter 4 of Oels (2006), Chapter 5 of Scheffran (2006), and Chapter 6 of Maarleveld et al. (2006) of this book. All these Chapters (3 - 6) as well as Chapter 1 mention the "Tragedy of the Commons" (Hardin 1968) as an important starting point for stakeholder dialogues in natural resources management.

2.7 The Integrative Theory of Reflexive Dialogues

As in any research, the choice of theory can make a crucial difference in the kinds of outcomes one can expect to obtain. The practice of stakeholder dialogues is implicitly or explicitly influenced by conceptual frameworks and their underlying theories. In some cases, dialogues have been carried out without reference to a particular theory, but in our view, a good theory is useful and increases the quality of the process and the quality of the output. On the other hand, practice influences theory: much of the theoretical thought is based on our practical work on science-based stakeholder dialogues and public participation in natural resources management. We have been faced with questions related to scientific rigour, relevance of the created knowledge, formal representation of stakeholder assessments, and the direct benefit of the stakeholder exercises for different actors. Our motivation to work in this field is that the present conceptual frameworks do not deal with many factors that are crucial for effective dialogues. We furthermore want to contribute to the further scientific understanding of stakeholder dialogues through theory selection, assessment, and development.

In this section, elements from the above-described scientific traditions (especially social psychological theories, organisational learning and formal approaches) are integrated into a theory of 'Reflexive Dialogues'. We realise that this is an ambitious effort and take note that a profound integration of scientific disciplines is challenging. We argue, however, that developing a "practical" theory of stakeholder dialogues instead of the "grand theory" is sensible and urgently needed in order to link different social scientific and formal ways of representing stakeholders assessment and to foster the development of analytical and communication tools in future research and practice.

The word "reflexive" implies that the rules of the dialogue are not fixed by the initiator or one of the participants, but that these rules are negotiable. This is a key feature and cornerstone of dialogues in the sense of the term used by Bohm. A non-reflexive dialogue would be one in which the initiator or facilitator poses the rules (how is the dialogue carried out) on other participants. In stakeholder dialogues, building mutual trust, knowing each other, and developing a common language requires commitment (time, resources) from all participants (Renn 2006). Commitment is not likely to emerge if the participants do not feel themselves part of the process of creating a dialogue.

We will elaborate the Integrative Theory of Reflexive Dialogues by discussing five key concepts: actors, structures, methods, processes, and outcomes. In terms of actors, our theoretical framework addresses various target groups and acknowledges the different roles individuals may play. So far, theories have variously had a strong focus on individuals, groups, organisations, or the society at large. The Integrative Theory of Reflexive Dialogues however recognises that actors are simultaneously members of very different social groups, different organisations, and part of the society. Some of the main actors that can be identified are scientists, international institutions, governmental bureaucracies, the media, industry, and non-governmental organisations. The role of the stakeholders varies depending on the type of dialogue and the attention cycle of an issue. A stakeholder may act inside or outside of a process.

Furthermore, not only the varying roles of the actors in society but also their different individual preferences, values, and knowledge bases have to be taken into account. Actors can be seen as following the principles of rational decision-making. Rational choice theory provides approaches and tools to study preferences and represents these in formal ways (utility functions). According to an alternative view, decisions are not made by rational considerations of objectives, options, and consequences (Sterman 1991). This is the case because several biases influence human decision-making due to the limited cognitive ability of humans to take more than a few factors into account in making decisions (Kahneman et al. 1982). As a result, many decisions turn out to be incorrect. Therefore, people use mental models, which can be framed both positively and negatively, "as the filters through which we interpret our experiences, evaluate plans, and choose among possible courses of action" (Sterman 1991).

In stakeholder dialogues, people may act as individuals interacting with other individuals or as representatives of a group. In the latter case, they have a mandate to speak according to the group's interests. In the former case, individuals tend to be far more cooperative.

Among involved actors, power relations are usually unequally distributed. Power relations are one of the most important aspects influencing the structures of stakeholder dialogues and thus have to be considered in a theory of reflexive dialogues. In the practice of stakeholder dialogues, the ideal of a powerfree discourse postulated by Habermas will never be met, but it is indispensable in order to be more aware of asymmetric power relations. These asymmetries can be addressed and corrected by applying particular communication tools or other methods. Rules and principles are related to the fairness of the processes and need to be defined and specified by the people involved in the dialogues.

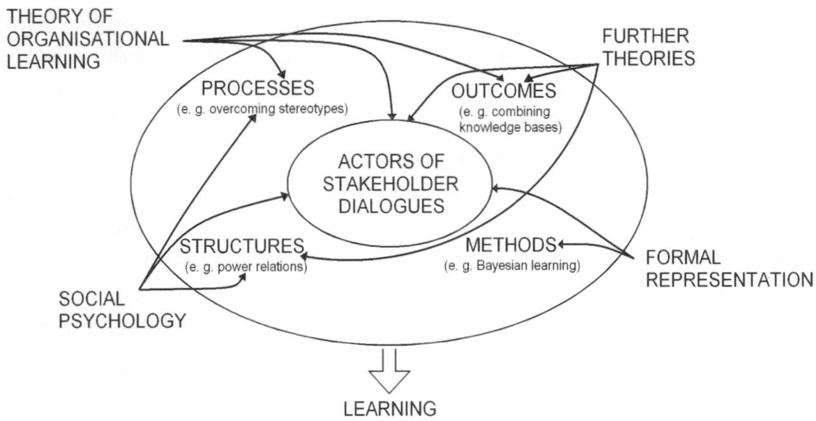

Fig. 2.2. Elements of the Integrative Theory of Reflexive Dialogues

Structures also encompass some of the general conditions in which dialogues take place. An important structural aspect is the (physical) cognitive limits of the human brain. Our ability to deal with complexity as required in stakeholder dialogues is limited and leads to processes such as stereotyping (see below) which in turn negatively affects communication (categorising groups of people into in-groups or out-groups) and learning in stakeholder dialogues. Public understanding of science is also a key component of such structures and has great importance in science-based dialogues but also in policy and management dialogues.

One objective of stakeholder dialogues is to combine different knowledge bases. The amount of attention different ways of knowing (scientific knowledge, expert knowledge and lay knowledge) get depends on the public's understanding of science and the policy process. Each way of knowing has its legitimisation. Lay knowledge is usually defined as being based on casual observations, but it may well be based on long-term experience, for example in natural resources use. In many management situations, scientific data is not available, necessitating knowledge of that kind (such as knowledge of fisheries and forestry). Indigenous people in particular often have detailed knowledge of places and local ecology and therefore can deliver important data relevant for natural resources management.

Processes in our conceptual framework refer to meta-communication, learning, and different modes of communication and stereotyping. Meta-communication is reflection about the process of communication. As mentioned above, a key feature of reflexive dialogue is the reflection on how the process should take place. It is necessary to agree on the rules of the process.

Finally, there is a key difference between consensus-seeking processes (such as policy dialogues, corporate dialogues) and processes that tolerate radically different views (science-based stakeholder dialogues). The exchange of arguments leads the participants to identify areas of agreement and disagreement and thus find new and relevant research questions.

Learning on the individual level, on the group level, and on the organisational level is a key concept in our Integrative Theory of Reflexive Dialogues. A system's capacity to learn is the foundation for self-organisation. Societies have been evolving constantly and have had kingship systems, village systems, empire systems, and national state systems. Monitoring and review practices at all levels are supposed to enable constant improvements in the self-organisation of the individual, the team, and the organisation/ society at large (Weber 1998).

The ability to innovate and create a global learning society may be one of today's greatest challenges. Public participation and stakeholder dialogues, if adopted on a broad basis, can become one way of fostering a global learning society. Social learning is a concept that deals with the question of how societies at large can cope with the changing world and new challenges. Social learning can be described as a cycle of discovering problems or issues, issue framing, drawing public attention to a new issue, debating possible solutions, and creating instruments, policies and management structures to cope with problems. In creating attention and framing issues, public media play a crucial role. The comparative history of three global change challenges - climate change, ozone depletion, and acid rain - as studied and described by the 'social learning group' at MIT provides an interesting overview of the social responses to these challenges (Clark et al. 2001a, b). For this study, the group considered as learning "those processes that deliberatively utilised experience or information to bring about cognitive changes" (Clark et al. 2001a: 14).

Stereotyping has been outlined above as an important process that determines communication and learning in stakeholder dialogues. A social stereotype is "a set of beliefs about the personal attributes of a group of people" (Ashmore and Del Boca 1981). Such sets of belief are "activated" (that is start influencing perception in a given situation) through identifying the group membership of a person (Enayati 2002). The stereotypical characteristics are attributed to all members of the out-group, and the individual's unique personal characteristics are ignored (Pennington et al. 1999).

The methods used in stakeholder dialogues need to be chosen so that they match the objectives of the dialogue. To achieve this, various kinds of tools are needed; we make a distinction between communication tools and analytical tools. Communication tools are needed to inspire and structure interaction between individuals. For example, focus groups or role games provide a setting for people to interact. It is important to create a safe space in which participants feel comfortable to express their views. Each tool applies a certain set of rules that all participants should co-define and follow.

The tool "focus group" combines two social scientific research methods, i.e. the focused interview, in which an interviewer elicits information on a topic, and a group discussion, in which a small number of people from a relatively heterogeneous group discuss a topic raised by a skilled moderator (Dürrenberger 1997). Conventional focus groups are based on a group of people being exposed to some common stimulus, such as a computer model or an expert presentation (Merton 1987). The group then is invited to engage in a free-wheeling conversation about that topic. The point of the exercise lies in the ability to observe social processes of opinion formation in which some new information is taken into account (Jaeger et al. 2000).

Dialogues can greatly benefit from the use of analytical tools as well. Bayesian networks, multi-criteria decision analysis, and computer models can be used for testing arguments, inspiring new ones, and visualising issues and options. Bayesian networks are one way of formalising stakeholders` assessments. They combine a visual presentation of stakeholders` beliefs (mental models) and deal explicitly with uncertainty of information. The Bayesian approach also provides a method to analyse how stakeholders learn, i.e. how they update their beliefs, when confronted with new information and insights. In multi-criteria decision analysis (MCDA), objectives as well as measurable criteria are identified to assess the extent to which these objectives are met. Different options are weighted according to these criteria. MCAs are better suited to cope with the fact that not all impacts can be measured using the same unit. Disaggregation thus helps to make explicit what different alternatives mean for different groups.

The right analytical tools for a specific problem or issue do not always exist or are not always readily available. In such cases, joint model building exercises can be one way forward. The process itself helps clarify the important parameters, the views held by the participating individuals, and the points where consensus exists or disagreements prevail. So far, computer models have been rather inflexible, but new modelling approaches make it possible to link modules programmed in different languages. This contributes to greater flexibility and increases the ability to react to emerging research questions more quickly (Jaeger 2003).

The outcomes of stakeholder dialogues can be framed in different ways. Networking and getting to know interesting people are a type of outcome that usually emerges and that most participants greatly value. Network theories such as those described above explain the superiority of networks in group problem-solving compared to the abilities of individuals. As a result, networks emerge whenever individual actors lack the resources necessary to achieve an output on their own and need to collaborate with others to combine knowledge bases. Thus one important outcome of network formation in stakeholder dialogues is the ability to deal better with complex problems.

Stakeholder dialogues may contribute to attitude and behaviour change (people confronted with new information and experiences) such as more environmentally friendly behaviour[7] or a better acceptance of other groups (e.g. NGOs vs. Corporations or Nature Conservation Agencies vs. Farmers). These attitude and behaviour changes can also change the role of a person in the organisation he or she represents and can become a change agent. While changes in attitude can be assessed by interviewing, changes in behaviour are more difficult to track.

Constructive conflict management is sometimes necessary and requires special skills from the facilitator or moderator of dialogue. This is rarely the case in the area of natural resources management because actors in responsible positions are trained in natural science disciplines (Stoll-Kleemann 2005). Sometimes a consensus view can be the outcome, but especially in scientific dialogues, dissent can be a valuable outcome as well. Conflict management is needed in both cases. Even if consensus is not the ultimate objective of the process, the dialogues have to be managed so that the differences in views can be discussed in a constructive way.

2.8 Conclusions

Our decisions affect the world in a way that has global and lasting results. It is often difficult to determine the consequences of our actions because of

[7] As outlined in Chapter 1, due to a lack of scientific studies, there is no convincing evidence that stakeholder dialogues really lead to more environmentally friendly behaviour. The GoBi (Governance of Biodiversity) Research Group investigates this question in the context of biodiversity management. The project is still ongoing, but first results are summarized in Stoll-Kleemann (2005), Stoll-Kleemann and Bertzky (2006), Stoll-Kleemann et al. (2006).

the increasing interconnectedness of people, organisations, corporations, and states (Keen et al. 2005).

To address these challenges, we have developed an Integrative Theory of Reflexive Dialogue. The innovation of this theory lies in its consideration of social psychological factors - often neglected in this scientific area - and links them to similarly useful concepts of organisational learning and formal approaches. Social psychological approaches aid in fostering a better understanding of what determines the functioning of stakeholder dialogues insofar as they explain how attitudes, outlook, and behaviour are shaped in these processes.

We feel it is important to bring the Theory of Organisational Learning, primarily as outlined by Senge (1998), into the Integrative Theory of Reflexive Dialogues as it points to the conditions necessary for productive stakeholder dialogue. The theory demonstrates how representatives from many different organisational backgrounds and professional cultures can work together effectively for the duration of their joint efforts, and how they can team up in small groups, which provide opportunities for learning and joint problem-solving.

The third part of our theoretical framework, the formal approaches, offer a way to structure complex issues and competing interests. The controversial discussion about the Rational Actor Paradigm plays a key role in the way we see actors in natural resource management. They have varying degrees of risk aversion, have to make decisions under uncertainty, and thus do not have complete knowledge upon which to base their decisions. The Bayesian approach is relevant for framing problems, visualising stakeholders' mental models, and observing how stakeholders learn. Although mathematical and formal applications are useful in participation and dialogues, they need to be embedded in a full cycle of trust building and reflection, i.e. the proper stages of successful dialogues.

The concept of learning is the interface between the theories explained and is thus the cornerstone of the Integrative Theory of Reflexive Dialogue. Stakeholder dialogues benefit from the application of learning as one key concept. It draws attention to learning in groups and organisations but also puts stakeholder dialogues into the broader context of social learning. Stakeholder dialogues have great potential in the assembly, transformation, multiplication, and spread of the knowledge requisite to achieve implementable, successful solutions.

References

Ames, D.P., Neilson, B.T. (2001). A Bayesian Decision Network Engine for Internet-Based Stakeholder Decision-Making. World Water and Environmental Resources Congress. Orlando, Florida

Argyris, C., Schön, D.A. (1996). Organisational Learning II: Theory, method, and practice, Reading. Addison-Wesley

Asch, S. (1956). Status of Independence and Conformity: A Minority of One Against a Unanimous majority. Psychological Monographs 70 (9)

Ashmore, R.D., Del Boca, F.K. (1981). Conceptual approaches to stereotypes and stereotyping. In: Hamilton, D.L. (ed). Cognitive processes in stereotyping and intergroup behavior. Lawrence Erlbaum, Hillsdale, 1-35

Beierle, T.C. (2002). Democracy On-line: An Evaluation of the National Dialogue on Public Involvement in EPA Decisions. RFF Report, Resources for the Future, Washington D.C

Berghöfer, A., Berghöfer, U. (2006). ‚Participation' in Development Thinking – Coming to Grips with a Truism and its Critiques. In: Stoll-Kleemann, S., Welp, M. (eds). Stakeholder Dialogues in Natural Resources Management. Springer-Verlag, Heidelberg

Bohm, D. (1996). On Dialogue. Routledge

Bostrom, A., Fischhoff, B., Morgan, M.G. (1992). Characterizing mental models of hazardous process: A methodology and an application to radon. J. Social Issues 48, 85-100

Breen, R. (1999). Beliefs, rational choice and Bayesian learning. Rationality and Society 11, 463-479

Brehm, J.W. (1966). A Theory of Psychological Reactance. Academic Press, New York

Cain, J. (2001). Planning improvements in natural resources management. Guidelines for using Bayesian networks to support the planning and management of development programmes in the water sectors and beyond. Centre for Ecology and Hydrology CEH, Wallingford, UK

Clark, W.C., Jäger, J., van Eijndhoven, J., Dickson, N.N. (2001a). Learning to Manage Global Environmental Risks: A Comparative History of Social Responses to Climate Change, Ozone Depletion and Acid Rain. The MIT Press, Cambridge, Massachusetts

Clark, W.C., Jäger, J., van Eijndhoven, J., Dickson N.N. (2001b). Learning to Manage Global Environmental Risks: A Functional Analysis of Social Responses to Climate Change, Ozone Depletion, and Acid Rain. The MIT Press, Cambridge, Massachusetts

Dawes, R.M., Messick, D.M. (2000). Social Dilemmas. International Journal of Psychology 35 (2), 111-116

Dogson, J., Spackman, M., Pearman, A., Phillips, L. (1999). DTLR Multi-Criteria Analysis Manual, NERA (National Economic Research Associates)

Doise, W. (1978). Groups and individuals: Explanations in social psychology. Cambridge University Press, Cambridge

Dürrenberger, G., Behringer, J., Dahinden, U., Gerger, Å., Kasemir, B., Querol, C., Schüle, R., Tabara, D., Toth, F., van Asselt, M., Vassilarou, D., Willi, N., Jaeger, C.C. (1997). Focus Groups in Integrated Assessment. A Manual for a participatory tool. ULYSSES WP-97-2. Darmstadt University of Technology, Center for Interdisciplinary Studies in Technology, Darmstadt

Enayati, J. (2002). The research: Effective Communication and Decision-Making in Diverse Groups Beyond Deadlock and Conflict. In: Hemmati, M. (ed). Multi-stakeholder Processes for Governance and Sustainability. Earthscan, London, 73-95

Ernst, A.M., Bayen, U.J., Spada, H. (1992). Informationssuche und -verarbeitung zur Entscheidungsfindung bei einem ökologischen Problem. In: Pawlik, K., Stapf, K.H. (eds). Huber, Bern, 107-127

Forester, J. (1985). Critical Theory and Planning Practice. In: Forester, J. (ed). Critical Theory and Public Life. The MIT Press, Massachusetts, 202-227

Forester, J. (1993). Critical Theory, Public Policy and Planning Practice. Toward a Critical Pragmatism. State University of New York Press, Albany

Gardner, G.T., Stern, P.C. (1996). Environmental Problems and Human Behavior. Allyn and Bacon, Boston

Gunderson, A. (1995). The Environmental Promise of Democratic Deliberation. University of Wisconsin Press, Madison

Hardin, G. (1968). The tragedy of the commons, Science 168, 1243-1248

Hemmati, M. (2002). Multi-stakeholder Processes for Governance and Sustainability. Earthscan, London

Hogg, M.A., Vaughan, G. (1998). Social Psychology. Prentice Hall Europe, Harlow

IIASA (2004). International Institute for Applied Systems Analysis (IIASA). Austria, Vienna. Available from World Wide Web. URL: http://www.iiasa.ac.at/~marek/soft

Jackson, S.E. (1996). The Consequences of Diversity in Multidisciplinary Work Teams. In: West, M. (ed). Handbook of Work Group Psychology. John Wiley & Sons, Chichester, New York

Jaeger, C.C. (2003). A Note on Domains of Discourse – Logical Know-How for Integrated Modelling. Potsdam Institute for Climate Impact Research (PIK). Working paper No. 86

Jaeger, C.C., Renn, O., Rosa, A., Webler, T. (1998). Decision analysis and rational action. In: Rayner, S., Malone, E.L. (eds). Human Choice and Climate Change. Battelle Press, Columbus, Ohio

Jaeger, C.C., Kasemir, B., Stoll-Kleemann, S., Schibli, D., Dahinden, U. (2000). Climate change and the voice of the public. Integrated Assessment 1, 339-349

Jentsch, A., Wittmer, H., Jax, K., Ring, I., Henle, K. (2003). Biodiversity - Emerging Issues for Linking Natural and Social Sciences. GAIA 12 (2), 121-128

Kahneman, D., Slovic, P., Tversky, A. (1982). Judgement under uncertainty: heuristics and biases. Cambridge University Press, Cambridge

Kasemir, B., Dahinden, U., Swartling, Å.G., Schüle, R., Tabara, D., Jaeger, C.C. (2000). Citizens' perspectives on climate change and energy use. Global Environmental Change 10 (3), 169-184

Kasemir, B., Jaeger, C.C., Jäger, J. (2003). Public participation in sustainability assessments. In: Kasemir, B., Jäger, J., Jaeger, C. C., Gardner., M. T. (eds). Public Participation in Sustainability Science. Cambridge University Press, Cambridge

Keen, M., Brown, V.A., Dyball, R. (2005). Social Learning: A New Approach to Environmental Management. In: Social Learning in Environmental Management: building a sustainable future. Earthscan, London

Kolb, D. (1984). Experiential learning: Experiences as the source of learning and development. Prentice-Hall, Englewood Cliffs

Krüger, R.A. (1993). Quality Control in Focus Group Research. In: Morgan, D.L. (ed). Successful Focus Groups: Advancing the state of the art. Sage Publications, Newbury Park, 65-85

Lantermann, E.D., Döring-Seipel, E. (1990). Umwelt und Werte. In: Kruse, L., Graumann, C.F., Lantermann, E.D. (eds). Ökologische Psychologie. Ein Handbuch in Schlüsselbegriffen. Psychologie Verlags-Union, München , 632-639

Lantermann, E.D., Döring-Seipel, E., Schima, P. (1992). Werte, Gefühle und Unbestimmtheit: Kognitiv-emotionale Wechselwirkungen im Umgang mit einem ökologischen System. In: Pawlik, K., Stapf, K.H. (eds). Umwelt und Verhalten: Perspektiven und Ergebnisse ökopsychologischer Forschung. Huber, Bern, 129-144

Leskinen, A. (1994). Environmental Planning as Learning: The Principles of Negotiation, the Disaggregative Decision-making Method and Parallel Organisation in Developing the Road Administration. Publications No. 5. University of Helsinki. Department of Economics and Management, Land Use Economics, Helsinki

Lindblom, C.E. (1959). The science of muddling through. Public Administration Review 19 (2), 79-88

Maarleveld, M., van de Velde, R., van Uum, J., Pleisier, I. (2006). To See or not to See, that is the Question: Geo-information Visualization Tools as a Means to Facilitate Stakeholder Dialogues in Land and Water Management Planning. In: Stoll-Kleemann, S., Welp, M. (eds). Stakeholder Dialogues in Natural Resources Management. Springer-Verlag, Heidelberg

Mackie, D.M., Hamilton, D.L. (1993). Affect, cognition and stereotyping: interactive processes in group perception. Academic Press, San Diego

Merton, R.K. (1987). The focussed interview and focus groups: continuities and discontinuities. Public Opinion Quarterly 51 (4), 550-566

Messner, D. (1998). Die Netzwerkgesellschaft: Wirtschaftliche Entwicklung und internationale Wettbewerbsfähigkeit als Probleme gesellschaftlicher Steuerung. Weltforum Verlag, Köln

Morgan, D.L. (1988). Focus Groups as Qualitative Research. Sage, Newbury Park

O'Connor, M. (2000). The VALSE project - an introduction. Ecological Economics 34 (Special Issue: Social Processes of Environmental Valuation), 165-174

Oels, A. (2003). Evaluating stakeholder participation in the transition to sustainable development: Methodology, case studies and implications for policymaking. LIT-Verlag, Münster

Oels, A. (2006). Evaluating Stakeholder Dialogues. In: Stoll-Kleemann, S., Welp, M. (eds). Stakeholder Dialogues in Natural Resources Management. Springer-Verlag, Heidelberg

O`Riordan, T., Stoll-Kleemann, S. (2002). Deliberative democracy and participatory biodiversity. In: O`Riordan, T., Stoll-Kleemann, S. (eds). Biodiversity, Sustainability and Human Communities. Protecting Beyond the Protected. Cambridge University Press, Cambridge, 87-112

Pavitt, C. (1993). What (little) we know about formal group discussion procedures. A review of relevant research. Small Group Research 24 (2), 217-235

Pennington, D.C., Gillen, K., Hill, P. (1999). Social psychology. Arnold, London

Pettigrew, T. (1989). Intergroup Contact Theory. Annual Reviews of Psychology 49, 65-85

Phillips, G.M., Wood, J.T. (1984). Emergent Issues in Human Decision-making. Southern Illinois University Press, Carbondale

Pratt, J. Gordon, P., Plamping, D. (1999). Working whole systems. Putting theory into practice in organisations. King's Fund, London

Putnam, R. D. (1993). Making Democracy work. Civic Traditions in Modern Italy. Princeton University Press, Princeton

Putnam, R. D. (2002). The Collapse of the Fact/Value Dichotomy and other Essays. Harvard University Press, Cambridge, MA

Renn, O. (2006). Participatory Processes for Natural Resource Management. In: Stoll-Kleemann, S., Welp, M. (eds). Stakeholder Dialogues in Natural Resources Management. Springer-Verlag, Heidelberg

Rosenhead, J. (2001). Introduction: old and new paradigms of analysis. In: Rosenhead, J., Mingers, J. (eds). Rational Analysis for a Problematic World Revisited: Problem Structuring Methods for Complexity, Uncertainty and Conflict. John Wiley and Sons Ltd, Chichester

Scheffran, J. (2006). Tools for Stakeholder Assessment and Interaction. In: Stoll-Kleemann, S., Welp, M. (eds). Stakeholder Dialogues in Natural Resources Management. Springer-Verlag, Heidelberg

Scheffran, J., Stoll-Kleemann, S. (2003). Participatory Governance in Environmental Conflict Resolution: Developing a Framework of Sustainable Action and Interaction. In: Deb, K., Srivastava, L. (eds.). Transitions Towards Sustainable Development in South Asia. The Energy and Resources Institute. India, New Delhi, 307-327

Seibold, D.R. (1999). The Impact of Formal Procedures on Group Processes, Members, and Task Outcomes. In: Frey, L.R. (ed). The Handbook of Group Communication Theory and Research. Sage, Thousand Oaks, CA, 395-431

Senge, P. (1998). The fifth discipline. The art and practice of the learning organisations. Century Business, London
Sherif, M. (1966). Group conflict and cooperation: their social psychology. Routledge and Kegan Paul, London
Sherif, M., Sherif, C.W. (1953). Groups in Harmony and Tension. Harper, New York
Simon, H.A. (1957). Models of Man. Social and Rational. John Wiley & Sons, New York
Smith, E.R., Mackie, D.M. (2000). Social Psychology. Psychology Press, Sussex, Philadelphia
Sterman, J.D. (1991). A Skeptic's Guide to Computer Models. Available from World Wide Web: <http://web.mit.edu/jsterman/www/Skeptic's_Guide.pdf>
Sterman, J.D. (2002). All models are wrong: reflections on becoming a systems scientist. System Dynamics Review 18 (4), 501-531
Stoll-Kleemann, S. (2001a). Reconciling Opposition to Protected Areas Management in Europe: The German experience. Environment 43 (5), 32-44
Stoll-Kleemann, S. (2001b). Barriers to Nature Conservation in Germany: A model explaining opposition to protected areas. Journal of Environmental Psychology, 21 (4), 369-385
Stoll-Kleemann, S. (2003). Integrating social psychological theories for interdisciplinary global change research. GAIA 12 (4), 325-326
Stoll-Kleemann, S. (2005). Voices for Biodiversity Management in the 21st Century. Environment 47 (10), 24-36
Stoll-Kleemann, S., O`Riordan, T. (2002a). From participation to partnership in biodiversity protection: experience from Germany and South Africa. Society and Natural Resources 15 (2), 157-173
Stoll-Kleemann, S., O`Riordan, T. (2002b). Enhancing Biodiversity and Humanity. In: O`Riordan, T., Stoll-Kleemann, S. (eds). Biodiversity, Sustainability and Human Communities. Protecting Beyond the Protected. Cambridge University Press, Cambridge, 295-310
Stoll-Kleemann, S., Bertzky, M. (2006). Biodiversity Management and Monitoring in Protected Areas – State of the Art and Current Trends. In: Sonak, S. (ed). Multiple Dimensions of Global Environmental Change. Teri Press, New Delhi, India, 143-169
Stoll-Kleemann, S., O'Riordan, T., Jaeger, C.C. (2001). The psychology of denial concerning climate mitigation measures: evidence from Swiss focus groups. Global Environmental Change 11 (2), 107-117
Stoll-Kleemann, S., O`Riordan, T., Burns, T. (2003). Linking the citizen to governance for sustainable climate futures. In: Kasemir, B., Gardner, M., Jäger, J., Jaeger, C.C. (eds). Public Participation in Sustainability Science. Cambridge University Press, Cambridge, 239-248
Stoll-Kleemann, S., Bender, S., Berghöfer, A., Bertzky, M., Fritz-Vietta, N., Schliep, R., Thierfelder, B. (2006). Linking Governance and Management Perspectives with Conservation Success. Perspectives on Biodiversity Governance and Management 1 (1), 1-40

Tajfel, H. (1978). Differentiation between Social Groups. Academic Press, London

Tajfel, H., Turner, J.C. (1979). The social identity theory of intergroup conflict. In: Worchel, S., Austin, W.G. (eds). The Social Psychology of Intergroup relations. Brooks/Cole, Monterey, CA, 188-204

Triandis, H.C., Hall, E.R., Ewen, R.B. (1965). Member Heterogeneity and Dyadic Creativity. Human Relations 18 (1), 33-55

Turner, J.C. (1982). Towards a cognitive redefinition of the social group. In: Tajfel, H. (ed). Social identity and intergroup relations. Cambridge University Press, Cambridge, 15-40

Tyler, T.R., Degoey, P. (1995). Collective Restraint in Social Dilemmas: Procedural Justice and Social Identification Effects on Support for Authorities. Journal of Personality and Social Psychology 69, 482-497

University of Dortmund (2006). Available from World Wide Web: URL: http://kiew.cs.uni-dortmund.de:8001/mlnet/instances/81d91eaae441d87555, [07.08.2006]

Von Neumann, J., Morgenstern, O. (1944). Theory of Games and Economic Behavior. Princeton University Press, Princeton

Weber, S. (1998). Organisationsentwicklung und Frauenförderung: Eine exemplarische Untersuchung in 3 Organisationstypen der privaten Wirtschaft. Ulrike Helmer Verlag, Königstein/Taunus

Welp, M., de la Vega-Leinert, A.C., Stoll-Kleemann, S., Jaeger, C.C. (2006a). Science-based stakeholder dialogues: Theories and tools. Global Environmental Change 16 (2), 170-181

Welp, M., de la Vega-Leinert, A.C., Stoll-Kleemann, S., Fürstenau, C. (2006b). Science-based stakeholder dialogues in climate change research. In: Stoll-Kleemann, S., Welp, M. (eds). Stakeholder Dialogues in Natural Resources Management. Springer-Verlag, Heidelberg

Welp, M., Kasemir, B., Jaeger, C. C. (in press). Citizens' Voices in Environmental Policy: The Contribution of Integrated Assessment Focus Groups to Accountable Decision Making. In: Coenen, F.H.J.M., Paterson, R. (eds). The Promise and Limits of Participatory Processes for the Quality of Environmentally Related Decision-making, Springer-Verlag, Heidelberg

Wheatley, M.J. (1992). Leadership and the new science: Learning about organisation from an orderly universe. Berret-Koehler Publishers, San Francisco

Wynne, B. (2005). Risk as globalising 'democratic' discourse? Framing subjects and citizens. In: Leach, M., Scoones, I., Wynne, B. (eds). Science and Citizens. Globalisation and the challenge of engagement. Zed Books, London

3 ‚Participation' in Development Thinking – Coming to Grips with a Truism and its Critiques[1]

Uta Berghöfer[1], Augustin Berghöfer[2]

1 Centre for Environmental Research (UFZ) Leipzig-Halle, Germany
2 Fundación Omora, Puerto Williams, Chile

3.1 Overview

Participation is increasingly popular in environmental discourse as it has been for decades in development thinking: it can take on multiple forms and serve seemingly incompatible interests. If participation is to mean more than a mainstream acclaim, its different levels and objectives need to be made explicit. A vague definition confuses the public and may lead to abuse of the term. Furthermore, a tool-based and solution-oriented understanding of participation risks overlooking and disguising the dimension of political power involved.

Inspired by Cohen and Uphoff (1980), we propose four defining questions to serve as axes of differentiation for the manifold interpretations of 'participation': (1) Who participates?, (2) In what dimension?, (3) How does the process take place?, and (4) For what purpose?. In doing so, we explore definitions and structure recent critiques, laying open the merits and pitfalls of 'participation' in development discourse and practice. The defining questions effectively clarify the haziness around this "warmly persuasive word" (Nelson and Wright 1995: 2). They also put 'stakeholder dialogue' in a wider context.

---

[1] Much of this chapter is inspired by Cohen and Uphoff's early typology of participation (1980). We also benefited greatly from the discussions with Ricardo Rozzi, Heidi Wittmer, Kathrin Blaufuss, Marcus Nüsser, Mark Starmanns and Gero Steup. We thank the editors Susanne Stoll-Kleemann and Martin Welp for their patience, and we thank Jessica Gillingwater for having made these pages readable. We hope L. M. Knopf will consider them worth the inconveniences suffered.

3.2 Introduction: On Doctors and Patients

"The ostensible aim of participatory approaches to development was to make 'people' central to development." (Cooke and Kothari 2001:5)

Why should a book on biodiversity conservation deal with participation in development thinking? We believe there are two reasons: First, it has become clear that in-situ conservation of biodiversity cannot be realised ignoring the development context for practical and ethical reasons (cf Borrini-Feyerabend et al. 2004, Brechin et al. 2003, Berkes and Folke 1998, Pimbert and Pretty 1995, Western and Wright 1994). This context includes all levels from the inhabitants of protected areas to national and international development policies. Secondly, people-oriented policies or interventions have been tested and studied under the title of participation for more than thirty years in the development arena. Hence, conservationists will benefit from looking at that experience and theoretical debate (cf Kapoor 2001). This chapter attempts to provide an introductory overview.

The above cited aim of participation, making 'people' central to development, accounts for the high popularity of the term. It suggests a general consensus about the need to give people a greater say. "Participation" as well as "sustainability", "decentralisation" and "community development" are frequently employed as "umbrella concepts" for complex realities - a fact which makes them easy to use in many contexts. The very fact, however, that "participation" is such a flexible term, means that it is in danger of becoming a "warmly persuasive word" (Nelson and Wright 1995: 2). Already in 1980, Cohen and Uphoff find that "*At present, concern with participation is popular, and one can hardly be against the concept, broadly conceived. When the meaning of development is said to include aspects of popular participation, promoting this becomes good by definition.*" (Cohen and Uphoff 1980: 213).

In the development arena, 'participation' means different things to different people: To some, 'participation' politicises development by linking it with the 'empowerment' of the poor and marginalised; to others, it is not so much political as technical, offering new gains in efficiency and sustainability of development projects. The concepts associated with it can be compared to:

– a doctor, telling his patient that he is sick and asking him to (quietly) cooperate in the healing process by paying part of the bill and following the standard medical instructions;

– a doctor, asking his patient what is wrong with him and then deciding what to prescribe him;
– a doctor, asking "who needs my help?" and then figuring out with those who come to him what they could do about it together;
– Finally a doctor, who is questioned by the people as to how he defines illness and sanity, and on the propriety of his attitude, knowledge and action for their situation.

These four analogies do not represent fixed categories but illustrate the range of interpretations. In this article we want to explore and systemise the different meanings of and contexts for participation. Aware of the large spectrum of interpretations we do not offer one approach to participation as a preferred option we rather seek "clarity through specificity". Inspired by the work of Cohen and Uphoff (1980), who elaborated a typology for participation in development, we present an overview based on the definitions, typologies and critiques of the more recent years.

We begin (Section 3.3) by putting 'participation' in the context of the evolving field of development thinking. In Section 3.4, Participatory Rural Appraisal (PRA) is outlined, a prominent approach to implementing participation in development. Section 3.5 seeks to provide more conceptual clarity around participation by comparing definitions and structuring the most relevant aspects involved. In Section 3.6, we systematically review the different critiques that have been put forward.

3.3 History: Changing Paradigms in Development Thinking

The questions associated with 'participation' have been central to political thinking ever since Aristotle. The organisation of a polity, e.g. a village, a tribe, a state, structures the way in which individuals take part or participate in politics. The effects of political power extend pass infringements of personal freedom to the allocation and utilisation of resources and benefits - issues that are at the centre of development thinking and practice. Hence, development thinking is intertwined with the organisation of the polity, including notions of 'participation'. Different strands of development thinking have come to different conclusions apropos interpretations of 'participation'. What follows is a résumé.

Late colonial thinking is generally thought to be characterised by paternal attitudes towards the inhabitants of overseas territories. Churches and schools served to 'civilise' the new subjects of the British, French and the other Empires. The assimilation of Western thinking and style was for them a pre-condition for politically taking part in any affairs - as second rank citizens. After World War II, the links between political regimes and economic growth were at the centre of national and international politics: The high economic growth rates in the post-war Soviet Union, when compared with western democracies, suggested a trade-off between optimum conditions for economic growth and largely democratic structures (Stiglitz 2002). In this period, participation was primarily interpreted as the adoption of new technologies by formerly 'traditionalist' societies. In his inaugural speech in 1949, President Truman classified most of mankind into the homogenous group of the 'underdeveloped' and prescribed to them Western technical solutions:

"*Greater production is the key to prosperity and peace. And the key to greater production is the wider and more vigorous application of modern scientific and technical knowledge*" (cited in Nustad 1997).

By the 1960's, the focus shifted from technologies to resources and to resource gaps. In order to address the differences between imports and exports or savings and expenditures, the people had to participate as disciplined and rational economic actors, investing, saving and producing goods and capital for their national economies. Rostow's "The Stages of Economic Growth" (1962) epitomised the popular reasoning of that period. He formulated in quasi-biological language the paradigm of a uni-linear course of history, pre-destined and beyond question that leads all mankind from a primitive society to Western industrial modernity. Economic growth is portrayed in terms of an evolutionary process towards civilisation[2].

Hirschman (1958) contested this strand of thought and criticised that development thinking focussed exclusively on aspects like 'capital', 'entrepreneurship' or 'state administration', instead of locating the problems of development "where all difficulties of human action begin and belong: in the mind" (ibid: 11). But the consideration of the roles and opinions of the actual people is diametrically opposed to Truman's or Rostow's concept of ‚underdevelopment'. It is hence peculiar, that one of

[2] However, there is no doubt that imperialist politics had as a primary objective the strengthening in any possible way of the mother-country at the expense of the colonised. For example, Ross (1998) discloses the sophisticated discursive and administrative structures employed to maximise the exploitation of British colonies during the 19th and 20th century. The lack of argument and evidence to support this claim - which also echoes imperialist thinking, e.g. manifest destiny, has been frequently criticised (e.g. Rist 1997, Mohan and Stokke 2000).

the early documents encouraging widespread popular participation in development is the US Foreign Assistance Act of 1966, Title IX:

"(....) there is a close relation between popular participation in the process of development and the effectiveness of that process. (....) it has become increasingly clear that failure to engage all of the available human resources in the task of development not only acts as a brake on economic growth but also does little to cure the basic causes of social and political instability which pose a constant threat to the gains being achieved on economic fronts. (....) Unless the people benefit from development efforts, no meaningful progress can result from foreign aid. It is equally true that unless the people contribute to development efforts, no meaningful progress can result from foreign aid" (cited in Hapgood 1969).

Cohen and Uphoff (1980) suggest that the principal reason for this concern for participation is the disillusionment arising from aid experience coupled with liberal sentiments that were en vogue. The passage displays a shift in social science thinking: Political participation is not assumed anymore to be the fruit of higher stages of development (i.e. a luxury of the well-fed and educated), but it is considered a necessary ingredient, even a pre-condition to development in a more holistic sense.

During the 1970s development thinking shifted further from economic - instrumental participation (based on a strong state with a Western bureaucracy and planning capacity), towards political self-determination and economic self-reliance at grass-root level. The widely experienced disillusionment caused by standardised rural modernisation and industrialisation projects complemented a growing resistance against the Eastern or Western monopoly on the desired end state of development.

Anthropological critiques of development abounded, evolving from dependency theory (Paz, Cardoso), liberation pedagogy (Freire), neo-marxism (Gramsci) and other schools of thought; many critics attacked development aid as the neo-colonial project within the globalisation of capitalism. The idea of trusteeship was increasingly being rejected, as it contravenes a basic assumption of liberal-humanist thinking, already put forward by Kant: that a man has to develop his talents through his own activity (Nohlen and Nuscheler 1993). In that sense, development aid had to be judged by its limiting or enlarging influence on a person's/community's autonomy. Alternative approaches to development argued that mainstream development efforts tended to perpetuate dependency and to reinforce structures of inequality within the Third World – a view that also transcends much of Paulo Freire's work. Especially his 'Pedagogy of the Oppressed' (1970/2000), when translated into the development context, implied a radical critique of mainstream development thinking: "*In the banking concept of education, knowledge is a gift bestowed by those who consider themselves knowledgeable upon those whom they consider to know nothing. Projecting an absolute ignorance onto others, a characteristic of the ideology of oppression, negates education and knowledge as processes of inquiry*" (Freire 2000: 72).

Concerns for project efficiency, for needs orientation, political empowerment and mutual learning were all voiced at some stage during the 1970‘s and often at the same time (Cornwall 2000a). There is a panorama of practical experience and of divergent understandings of participation co-existing in that decade. The Tanzanian statesman Julius Nyerere emphasised the learning aspects associated with participation: "*Rural development is the participation of people in a mutual learning experience involving themselves, their local resources, external change agents and outside resources. People cannot be developed; they can only develop themselves by participation and co-operative activities which affect their well-being. People are not being developed when they are herded like animals into new ventures*" (1968, cited in Oakley 1991).

In 1974, the UN Charter on Economic Rights and Obligations of States instituted economic, political and cultural pluralism at the highest political level[3]. This stressed the interpretation of participation as a mutual learning experience among 'equals'. In the 1980's, non-governmental organisations (NGOs) gained importance for locally organised development efforts. NGOs were considered to be better equipped than (inter-)national organisations for providing needs-oriented and effective development assistance: They were seen as cheaper and 'closer to the people'. This belief coincided with the new trend in downsizing the state apparatus: the neo-liberal stance. New calls for NGO-driven participation provided a convenient background for 'rolling back the state', proposing 'structural adjustment' and privatising the formerly public provision of welfare and services. Participation was then conveniently perceived as a community contributing to an aid project for its own benefit.

During the 1980‘s and 1990's development received far-reaching critiques evolving from post-modern theorising. Some critical theorists disagree with the idea of development itself. Escobar (1984, 1995) applied Foucault‘s dictum - that it is in discourse that knowledge and power are joined together - to development thinking. He contended that the discourse of development sustains processes whereby „the Western developed

[3] Art. 1: „Every State has the sovereign and inalienable right to choose its economic system as well as its political social and cultural systems in accordance with the will of its people, without outside interference, coercion or threat in any form whatsoever“ (GA Res. 3281 (XXIX) 1974).

countries have been able to manage and to control and, in many ways, even create the Third World politically, economically, sociologically and culturally" (Escobar 1984). In a similar perspective, Nustad (1997) distinguished between 'intentional development' (development aid) and 'immanent development' (capitalism), with the former as remedy (or at least a palliative) for the latter, i.e. for the negative consequences of the world's economic system. Such a distinction between intentional and immanent development reaffirmed that development is primarily a political issue - not merely an optimisation problem. This understanding had lost prominence during the early nineties when many development NGOs were becoming professional service deliverers.

The political and economic crises of the 1990's powerfully demonstrated the limits of traditional development approaches. In many countries, issues such as trade relations, credit policies, capital flows or the wars on communism/drugs/terrorism are beyond the local or even national scope of planning, nonetheless they determine much of a country's (development) perspective. This complexity of the task was further enhanced by the search for sustainability in development efforts (Rio 1992) – a quality that encompasses many more criteria than previous development approaches. At the same time, globalisation critics celebrate new forms of participation – here: involvement in political affairs - in international social movements. With the arrival of the internet in developing countries, powerful possibilities have sprung up for the networking and coordination between sheer unlimited numbers of individuals and organisations. It remains to be analysed how this spread of new means of exchange and communication strengthens the political weight of bottom-up involvement and influences the evolution of development practice.

Forty years of development experience have produced a higher awareness of the complexities involved, an ongoing specialisation of tools and methods to deal with them, a sporadic ambivalence about the due roles of external development workers/consultants/experts, and an overall adherence to an a-political management-oriented stance. This is what we could consider the context of participation in development thinking.

3.4 Implementing Participation: The Promise of Participatory Rural Appraisal (PRA)

The FAO's participation website[4] hosts a collection of "187 participatory approaches, methods and field tools" to translate the ideals and principles associated with the term into practice. From the myriad of writings on this matter since the late 1980s a few methods are cited in exemplary fashion. Checkland and Scholes (1990) propose a technique called Soft Systems Analysis, a combination of systems engineering and action research, to understand complex human activity systems from multiple perspectives. This requires the inclusion of stakeholder views as primary sources of information. Adaptive Management (Walters 1986, McLain and Lee 1996) focuses on the value of mutual and continuous learning of stakeholders facilitated by an iterative process of hypothesis testing. Regional workshops serve to collect site-specific information, conduct systems modelling and review results. Composition of workshops and congruence of participants' objectives determine the fruitfulness of the exercise. Rapid Rural Appraisal (RRA) (McCracken et al. 1988) came to be known as a collection of techniques, sensitive towards local knowledge, that allow to gain rapid, informal information from multiple sources. RRA was born out of the realisation that production per hectare was an insufficient criterion for describing the constraints to improving rural livelihood.

Participatory Rural Appraisal (PRA) has been the most prominent approach to implement participation in the last years. PRA evolved from Rapid Rural Appraisal (RRA) and comprises standard tools of qualitative research like Participatory Action Research (PAR)[5]. Unlike the above cited methods, PRA claims its own philosophical foundation in turning the participants into analysts and interpreters of their own situation. PRA offers an array of instruments for analysing a local situation and for building consensus on possible courses of action. A multitude of PRA related manuals have spread throughout the aid community. PRA is conceived as an alternative to the tendency, to impose knowledge and plans on local communities. Chambers, who made this approach prominent in the early 1990's stressed the practice-oriented origins of PRA[6]. Today

4 http://www.fao.org/participation/ (June 2006)

5 See e.g. Fals-Borda and Rahman (1991)

6 "Most of those who have innovated in developing PRA have been practitioners, concerned with what works and what will work better, not academic theorists concerned with why it works. They have been searching not for new theories or principles but for new and better ways of learning and doing" (Chambers 1994b: 1262). See also Chambers (1994 a-c) and (1997).

there are hardly any agencies or NGOs that do not refer to PRA in their strategies and guidelines. Equally, many critical voices about participation focus on PRA and its philosophical underpinnings. For this reason the associated concepts are presented here in more depth. The aims of PRA according to Chambers are: to bring about a shift from top-down to bottom-up in development projects, "from centralized standardization to local diversity and from blueprint to learning process" and by using different participatory tools "to enable rural people to share, enhance, and analyze their knowledge of life and conditions, to plan and to act" (Chambers 1994a: 953). A short overview of the PRA methods is given in Chambers (1994: 959-961). The solution proposed, is to turn the 'outside expert', into a convenor and facilitator and let the 'insiders', the people in the local community, do the analysing and acting themselves. "The thrust of PRA is to reverse dominance, to empower more than extract" (Chambers 1994b: 1265). Chambers calls the principles underlying PRA "reversals": „reversals of frames, reversals of modes, reversals of relations, reversals of power" (ibid: 1262). "Reversals of frames" focus on local knowledge, categories and values instead of the knowledge, categories and values that external professionals present in their investigating of local realities. Chambers sees conventional questionnaire surveys as an example of outsider knowledge with pre-coded responses and reality "forced to fit the professionals' familiar frame" (ibid: 1262). In contrast to this type of investigation Chambers promotes semi-structured interviews and open conversations. "Reversals of modes": the modes of interaction and analysis shall be changed in order to include a larger group of local people and "to empower the weak and disadvantaged" (ibid: 1263). Three directions are considered: (1) emphasis is placed on visual inputs, discussions, story telling, 'transect walks' through a community area, etc., as this makes literacy levels irrelevant, (2) qualitative methods, comparisons, ranking, drawing and diagramming are used instead of absolute measurements, (3) instead of focusing on the individual more attention is paid to groups. "Reversals of relations" shall be reached through confidence and rapport. "This is through outsiders being unhurried, showing respect, explaining who they are, answering questions, being honest and being interested; and asking to be taught, being taught and learning" (ibid: 1264). Local communities are no longer an object of research, but should analyse and investigate themselves. Finally those reversals lead, according to Chambers, to "reversals of power". By "handing over the stick" initiative and control will be passed to local people. The aim is "to reverse dominance and to empower more than extract" (ibid: 1265). At first sight the "empowering" aspect of PRA seems to be seductive and Chambers' rhetoric is charged with calls for a change in personal attitudes[7]. Not surprisingly PRA found its way to the donor agencies: source-books and training courses abound. Nonetheless, the last decade has also witnessed vivid critiques of PRA. Many scientists felt prompted to ask whether the new participatory approaches are actually capable of inducing substantial changes – or whether they are only a surrogate (Rahnema 1992: 124). Doubts seem justified. The OECD's Development Assistance Committee analysed the experiences with implementing participation by reviewing 43 project reports and synthesis documents of eight bilateral aid agencies. Their report concludes: „*Participation in projects is often defined either very generally or interpreted to mean a range of stakeholder roles, few of which actually involve an active and influential hand in shaping development decisions that affect their lives for primary stakeholders*" (DAC/OECD 1997).

More explicitly, "participation" has not yet become operational on a larger scale. The question, why the record of implementation of participatory development is so poor despite its widely recognised benefits, will be addressed in Section "Critiques of Participation".

3.5 Seeking Clarity

3.5.1 Defining Participation

In order to approach the term and its possible interpretations a few of its most recent definitions are given below. They serve as a starting point to subsequently structure its principle aspects. For the Oxford English Dictionary, participation is "the action or fact of partaking, having or forming part of" - leaving open the questions as to whether this is political or material in nature, and whether it is by invitation or by claim, whether justified, deliberate, moral, free - or none of the above. Hence, in some development writing, definitions of participation are not very helpful. For example, Stiglitz writes: "*I will use the term 'participation' in the broadest sense, to encompass transparency, openness, and voice in both public and*

7 "And most of us have ways to empower others, lowers, the weak, poor and vulnerable, to express their realities and make them count. Good change flows from personal decisions and action. There is no need to wait. There is a vanguard to join and new high ground to explore" (Chambers 1997: 237).

corporate settings. There are a variety of institutional arrangements that are consistent with 'participation' in this sense" (2002: 165).

The World Bank Sourcebook on Participation gives a very broad operational definition: "*Participation is a process through which stakeholders influence and share control over development initiatives and the decisions and resources which affect them*" (1996: 3).

From this it is understood that people, representatives of the national government, the private sector, the international aid community and the Bank itself collaborate. This can mean anything from as little as the notification of 'beneficiaries' of a project intended to help them (Cornwall 2002). Notwithstanding, the Sourcebook identifies this 'participatory stance' as a community-driven social learning process. The Bank's objective is, amongst others, to make people better "managers of their own assets" (ibid.). However, it is not clear on what terms the stakeholders are supposed to have an influence and share control. Differences in interests and power are not recognised by this definition. The OECD puts participation closer to empowerment and adds the notion of 'negotiation' to the 'dialogue':

"Participatory development stands for a partnership which is based on dialogue among the various actors (stakeholders), during which the agenda is set jointly, and local views and indigenous knowledge are deliberately sought and respected. This implies negotiation rather than dominance of an externally set project agenda. Thus people become actors instead of being simply beneficiaries" (Schneider and Libercier 1994).

This definition recognises different endowments of power but suggests that the participatory process by itself can overcome this inequality and lead to a harmoniously negotiated consensus. The Swedish Development Agency SIDA employs a definition that links participation explicitly to democratic norms: "*Popular participation (...) can be viewed, with reference to the democracy and equity goals as an objective in itself – that is a basic democratic right that should be promoted in all development projects (....) For political conditions to change in a more fundamental way, a great many social, cultural and even personal relationships must become transformed in a democratic direction*" (cited in Cornwall 2000a).

SIDA is recognising here that relationships have to be transformed – the existence of structural inequalities however, is not made explicit. Nohlen and Nuscheler (1993) strengthen the normative aspect: For them participation is the "opposite of marginality, comprising a set of political and social human rights. It is the co-ownership of the material and cultural goods of a society, and hence the diametric opposite to a "top-down bureaucratic planning, incapacitation and spoon-feeding" of the population. Rahnema concludes in his essay on 'Participation' in Sachs'

Development Dictionary (1992) that participation means "to live and to relate differently. It implies above all the recovery of one's inner freedom, that is to learn to listen and to share free from any fear, or predefined conclusion, belief or judgement" (1992: 127).

This conclusion goes deeper than most other definitions in that it recognises that meaningful participation requires, not only some structure, but first of all a free mind. Arguably, it is a definition difficult to realise within the realm of mainstream development projects.

Many of the above definitions bear positive connotations without specifying the terms or the scope of participation. They can henceforth easily accommodate a variety of interpretations - a condition which accounts for the term's continuing popularity.

3.5.2 Four Axes of Differentiation

Since the 1960's, people have proposed different typologies to allow us to come to grips with levels, aspects and dimensions of participation. While Arnstein (1969) focuses exclusively on levels of power transfer, Cohen and Uphoff (1980) offer a comprehensive typology, differentiating between political and development participation. Pretty (1995) identifies the various degrees of participation during the stages of a project cycle, and White (1996) distinguishes between the respective interests of development workers and intended beneficiaries in participation at these different stages. Borrowing from the insights of these authors – in particular from Cohen and Uphoff – we propose four axes of differentiation that might help to provide a more complete picture (see table 3.1). The use and meaning of participation can be illuminated by examining the basic questions (1) "Who participates?", (2) "In what dimension?", (3) "How?, and (4) "For what purposes?". These questions make different aspects of participation visible and thereby allow us to subsequently systemise the different critiques.

Table 3.1 Basic questions: Axes of differentiation

	Axes of differentiation
1	Who participates?
2	Participation: in what dimension?
3	How does the process of participation take place?
4	What are the purposes of participation?

In the following tables (tables 3.2 – 3.5), we present preliminary answers to these questions in a second column. Aware of the fact, that they are insufficient to adequately describe the complex and sometimes contradictory understandings of participation, we added a third column with related questions that indicate the difficulties involved and can help to direct further exploration. We will address many of them in the following sections. Note that the four axes of differentiation do not describe a space for exact definitions. We *do* want to provide some structure that allows for a more specified and transparent use of 'participation'. But we *do not* intend to "box" it - it is a controversial and dynamic notion. Therefore we emphasise the following: Firstly, it is inappropriate to rank the different aspects as 'axis points' from "light/little participation" to "full/deep participation", because differing objectives make general rankings impossible. Secondly, the axes allow us to view the issue from different perspectives; they are neither complete nor should they be closed, as every situation provides its own insights. Thirdly, as participation is complex, dynamic and specific to context, the axes should adapt to the evolving experience and not vice versa. Otherwise their epistemic utility is lost. We understand Participation as being by itself not a manageable object or process, even though some related aspects may be more so than others.

Who participates?

The first axis "Who participates?" (see table 3.2) seeks to make two things explicit: Firstly, participant selection is a problematic issue. In development jargon, participants are often grouped under terms like "Communities", "Stakeholders", "The poor", "The marginalized" etc. But who are those people? On whose terms are they identified? Secondly, participation can take on very different forms depending on who the participants are. If aid workers select members of village councils to discuss with them and with government officials on some agricultural programme, the kind of interaction, the questions raised and the opinions considered will be very different to those which would arise for e.g. a group of farmers in cooperative to discuss the same issue, purely because of the different compositions of the two groups.

Table 3.2 "Who participates?"

Basic question	First answers	Further questions
Who participates? "Communities" "Local People"	Interested foreign companies and their national partners	How can stakeholders be identified?
	Aid representatives	Who are "the poor" or "the marginalized"?
"Stakeholders" "Beneficiaries"	Representatives of lobby organisations	Who is excluded?
"Women" "The Poor" "The Marginalised"	Scientists Representatives of public services at higher levels	What makes a community? (Guijt and Shah 1998, Agrawal and Gibson 1999, Cornwall 2000a, 2000b, 2003, Cleaver 2001, Mosse 1994, Mayoux 1995)
	Local state or elected authorities Local traditional authorities – recognised as such	What are the limits of a place? (Mohan and Stokke 2000, Mohan 2001, Sommer 2001, Bryant and Bailey 1997)
	Local civil society organisation representatives – not necessarily recognised as authorities but stimulating changes	
	Local population in different sub-groups	

Participation: in what dimension?

People "take part" in different things (see table 3.3): in local politics or in voting for their national government ("political participation"), in digging wells or in evaluation sessions ("project participation"), in sharing the harvest or in developing local tourism ("economic participation"), in playing football or simply in living in the same village ("social participation"). These are extremely different 'activities'. Although the distinctions are obvious, the term participation is applied to all of them, as if they were similar in their connotations.

Table 3.3 "In what dimension?"

Basic question	First answers	Further questions
In what dimension?	Economic participation Political participation Social participation Project participation	In what dimensions of social life can people participate? How to relate the dimensions? What effect does the project as frame have on the process? (Craig and Porter 1997)

How does the process of participation take place?

Table 3.4 "How does the process of participation take place?"

Basic question	First answers	Further questions
How does the process of participation take place?	Type of facilitation/ initiation/ leadership The way to become a participant Activities in which to participate The form/ rules of the activities The consideration of difference and conflict	What legitimisation do the participants have? (Kapoor 2002b, Ribot 2002) What is the role of the facilitator? (Mosse 1994, Rahnema 1992) Why do people exclude themselves?(Cornwall 2000a) What are the costs and benefits for people to participate? (Mayoux 1995, Cleaver 2001) Which institutions are adequate? (Cleaver 2000 and 2001) What are the factors for successful processes? (Webler et. al. 2001)

To structure the aspects characteristic of processes of participation we propose five sub-groups (see table 3.4) here which, again, can be understood as differentiating axes:

- The type of facilitation/ initiation/ leadership: external facilitation, “insider” or local facilitation, traditional leadership, self-governed process, conjoint leadership, invited, elected, nominated, selected, claimed, coerced, voluntary, etc.
- The way in which one becomes a participant: by invitation, election, nomination, selection, claim, coercion, volunteering, own initiative etc.
- The activities in which one participates: information sharing, learning, defining the problem, decision-making, discussion, evaluation, planning, implementation, working, etc.
- The form/ rules of the activities: informal, formal, pre-defined, adaptive, imported, endemic, etc.
- Of central importance within these rules is the consideration of difference and conflict: consensus oriented, negotiation oriented, majority vote, etc.

What is the purpose of participation?

Discussion about the purposes of participation (see table 3.5) is very controversial as it rapidly involves assumptions about development aid which, depending on the perspective taken, seem incompatible. If development aid is pictured as part of today’s imperialist system, ‘participation’ in whatever form – except in struggling for complete autonomy - seems pointless. In this perspective it defuses opposition by creating desirable illusions. On the other hand, at a less fundamental perspective and a more local scale, participation in development may produce concrete improvements for the life conditions and dignity of people. It is at this lower level, or in the greyish areas, that the questions deserve attention: What is stated? What is intended? And what is achieved?

Table 3.5 "What is the purpose of participation?"

Basic question	First answers	Further questions
What are the purposes of participation?	To affirm indispensability of an outside facilitator	What is stated and what remains un- stated? (Cornwall 2000a)
	To get access to relevant (local) information and target groups	What is intended and what is achieved? (Cornwall 2000a)
	To include people in market economies	
	To enhance the acceptance and the long-term effectiveness of a project	Can the intended aim be reached by the adopted strategies?
	To enhance project efficiency and reduce implementation costs	What is "full participation "? (Cornwall 2000a)
	To improve quality of project design by adapting the project to the context and to the people	What is the difference in the perspectives from "above" and from "below"? (White 1996)
	To enhance mutual accountabilities	For what are people empowered? (Taylor 2001, Henkel and Stirrat 2001, Escobar 1984, Cleaver 2001)
	To strengthen capacity for solving collective problems	
	To find innovative solutions through exchange and collaboration with other stakeholders	Who decides what is empowering and what not? (Rahnema 1992)
	To strengthen civil society and create public space	
	To strengthen democracy	
	To strengthen collective self-determination and to challenge the constellation of powers	

3.6 The Pitfalls: Critiques of Participation

In this section we review various critiques of participation that have been put forward in the last decade. We structure them according to the aspect focussed upon, in accordance with the defining questions presented in the tables 3.2 – 3.5 in the previous section.

3.6.1 Who participates?

Three thematic areas related to this question have attracted widespread attention in recent writings: the dangers of localism, the myths about 'community' and the identification of stakeholders (see table 3.2).

The dangers of localism[8]

Discourses about development often revolve around a binary opposition between the state and civil society, where civil society is seen as manifesting itself in the local, i.e. in discrete places. It is often assumed that 'top-down' can be transformed into 'bottom-up' by local participatory projects. Considering the embeddedness of the local dimension into wider political and economic contexts, this assumption must be analysed. Mohan and Stokke (2000) criticise the tendency to 'essentialise and romanticise' the local. They argue that a 'place' is far more than a given locality because it is made up of flows and relations, i.e. flows of people, information, commodities and relations of cultural, economic, social and political dimension. Thus the view of a place as a locality is the reductionist framing of a complex system.

Another critique to the primacy of the local is the isolation from a wider multidimensional context. Localism neglects the formative influences and the constraints imposed on local livelihoods by polical-administrative and economic conditions set at macro-levels. Chambers does note that a political context in favour of participation is necessary, but he seems to be primarily concerned with sympathy for the process, ignoring that the outcomes of community deliberation have the potential to cause conflict with regard to the wider setting. The focus on the local empowerment of a community ignores that wider structural issues such as distribution of property or external trade relations are often important variables in local development. Consequently, there is no recognition that even a fully

[8] c.f. Mohan and Stokke (2000)

capacitated and well-organised community has limited means to go against conditions set at a higher level (Sommer 2001).

As long as any participatory process is not supplemented by attention to the larger structural context, its focus on the local may serve as a disguise for structural injustice set at higher levels. Hence, for example Mohan (2001) argues that for local "empowerment" to become effective, participatory approaches should be linked with efforts for the wider processes of democratisation, anti-imperialism and feminism. Mohan gives no indication of how this should be achieved. But to prevent the risk of disguising structural injustice, a possible starting point would be the politisation of community agendas and the networking of local participatory structures. As a research approach, the Political Ecology's systemic approach[9] (e.g. Bryant and Bailey 1997, Bryant 1998, 1999, Blaikie 1999, Hartmann 1998, Peet and Watts 1996) could provide analyses of inter-connection going beyond simplifying binaries, such as 'village A – village B' or 'village – provincial administration' but corresponding with the multidimensional description of a 'place'.

The Myth of 'Community'

Another assumption concerns the realm of community: Face-to-face interactions are a defining feature of participatory development and it is assumed that just, un-coerced compromises grow within an unbiased arena. Motivations and behaviour in participatory processes are considered authentic. The vision of an integrated community using locally evolved norms and rules to manage their livelihoods and their resources in a sustainable and equitable way is still powerful (Agrawal and Gibson 1999). As people living in the same community are exposed to similar external conditions, and possess common characteristics concerning ethnicity, tradition, language, religion etc., there is a basis of trust and solidarity assumed which would lead to solid change through co-operation if only participatory processes were applied. This understanding of community has provoked varied criticisms. Mayoux (1995) questions the

---

[9] Political Ecology is an analytical approach developed within the field of Geography which deals mainly with the relation between distributions of power and its effect on the natural environment. It is assumed that the understanding of the unequal distribution of power is a key-stone for the understanding of environmental problems (Bryant and Bailey 1997: 38), the main assumption is that of an "politicised environment" (ibid.: 27). One important topic of Political Ecology is the connection between poverty/marginalisation and the degradation of the environment as a counter-argument against the assumption of Neo-Malthusianism, e.g. "*Tragedy of the commons*" (Hardin 1968).

supposed possibility of consensus between participants about needs and aims on two grounds. She argues that different people even in similar circumstances are likely to have different priorities which makes consensus difficult. Secondly she believes that defining needs is problematic without also addressing the underlying intra-community inequalities from which these needs arise.

In most cases, inequality between members of a community can be assumed. This inequality concerns status, means, independence and influence, among other things, which are highly relevant for a participatory process, a dialogue between stakeholders or a negotiation with external actors about, say, a community project. In their analysis of political participation Bachrach and Botwinick (1992) assemble various empirical findings within a US context indicating that participatory approaches to community politics often increase inequality as they favour the active over the apathetic. This judgement needs refinement. Intra-group procedures determine in what way the different interests are reconciled. Allowing more time and an explicit recognition of democratic norms would promote the revision and reapproval of the different internal procedures. However, this would require a learning process longer than most project cycles (Mosse 1994). Furthermore, recent empirical research about political participation of the urban poor in developing countries suggests that collective hardship does not augment mutual solidarity or willingness to co-operate. Instead, the desire to be visibly 'better off' than ones neighbour increases (Berg-Schlosser and Kersting 2000).

As a first step it is important to recognise the differences: the subsequent question of how to deal with differences in a participatory process will be further discussed later. Proponents of participation have to come to grips with the danger of ignoring or even exacerbating inequality among participants. The typical dichotomies between 'insider-outsider', 'local people-outsider', 'the local'- and 'the global' and the generic terms like 'the poor', 'the marginalised' or 'the women' are convenient but risk to obscuring the more complex, less obvious but nonetheless crucial differences.

"If they [the rural poor] are considered in such an aggregated mass, it is very difficult to assess their participation in any respect, since they are a large and heterogeneous group. Their being considered as a group is not, indeed, something they would themselves be likely to suggest. There are significant differences in occupation, location, land tenure status, sex, caste, religion or tribe, which are related in different ways to their poverty. To talk about "the participation of the rural poor" is to compound one complex and ambiguous term with another, even more complicated and amorphous" (Cohen and Uphoff 1980: 222).

A gendered perspective demands even more detailed consideration of social context and categories (e.g. Guijt and Shah 1998, Mayoux 1995, Cornwall 2000b). Cornwall (2000a) admonishes that the category "woman" is often used in such a general way: any woman could come to represent women-in-general. This masks the multiple voices of a heterogeneous group. Guijt and Shah (1998) stress the importance of clarifying what gender means to researchers and professionals - in all its diverse forms and interpretations - so as to avoid a simplistic narrative. The critique is not new, but is no less valid as standard participatory tools and methods seem ill fit to adequately address this complexity.

Who hold the stakes?

If we combine the critiques on assumptions about the local dimension and the community, we can conclude that neither the place nor its human population automatically provide us with a group of adequate participants. Adequacy is determined by the various objectives pursued, but many project designs simply resolve to include "stakeholders", i.e. all those who have - or should have - an interest in the issues at stake. To consider just those actually involved or those potentially interfering is easy and therefore appealing. And to rank them by speaking of primary and secondary stakeholders does, in our view, not duly consider and express any adequate selection

But if the term "stakeholders" is meant to accommodate an ethically informed or more 'rights-oriented' understanding of politics, all those from whom a legitimate voice may be heard have to be included. If the former approach is reifying existing power relations, the latter is prone to entertain endless discussions about stakes and their legitimacies. What stake does a shareholder of a soft drink company have in the villages where these products are sold; and what stake should he have? Even if we could all agree on one set of legitimate stakes and identify their respective holders, on what terms are they supposed to interact? Stakeholders have in common that they claim - or are attributed the right to claim - an interest in one issue: however, this should not create the illusion that their interests are equal, compatible, or equally legitimate, nor that they have equal means to pursue those interests.

3.6.2 Participation: in what dimension?

Participation in projects, in development, in politics, in the local economy or in the community? The way in which these categories are interpreted determines the space for 'participation' within them (see table 3.3). This is very obvious for playing football and for electing the town's mayor, but it is equally determinant in more subtle cases like, for example, poverty reduction strategies of international organisations.

As 'political participation' and 'social participation' are closely related to the critiques of 'empowerment' (cf. Section 3.6.4), we concentrate in this section on economic interpretations and on the project frame which is closely associated with 'development participation'.

Economic Participation

If the World Bank envisions people as "managers of their assets" in the local economy, then participation primarily seeks to improve the collective or individual asset management capacities, and not, say, the capacity to act as critical citizens. Despite a non-political vocabulary, economic interpretations are, in fact, quite political in content. Organisational 'empowerment' to raise productivity, to gain access to credit and to reach new markets are some of the remedies that are frequently recommended to improve participation of the rural poor in their region's economy. The 'participatory stance' of the World Bank suggests coordination of stakeholder interests and sharing of information to realise innovative solutions for the benefit of all (Francis 2001).

A discourse oriented look at the promises and causal links underlying this narrative discloses at least three problematic aspects: Firstly, much use is made of the fear to be excluded. To participate in the market economy is to be 'inside'. 'Insiders' are heading towards growth, wealth and progress whereas 'those excluded from the market' are stagnating in a traditional or archaic environment. Secondly, organisational empowerment for better economic participation concentrates on the feasible and on the aggregate: in order to make a bigger cake, those baking should get the bigger-slice incentive; however, leaving out the details of procedural justice and fair distribution can rapidly increase inequalities and enhance marginalisation. Thirdly, the economic-participation perspective shifts the explanatory focus of poverty away from structural dilemmas, towards agency: Put bluntly, if people were to get active, organised and entrepreneurial, their poverty problems would soon disappear – their apathy/attitude/ignorance is a principle reason for their poverty. The 'poor' are not so much victim of structural injustice - such as imperialism/capitalism/neo-feudalism - but are in fact responsible themselves for their plight. This line of argument is pervasive in various variants, e.g. in association with private property - Tragedy of the Commons (Hardin 1968) - or with neo-Malthusian accounts of overpopulation (Ross 1998). These three aspects illustrate the need to complement notions of economic participation with other, explicitly political interpretations of participation.

The Project Frame

The project approach of development, science and conservation is another focus for critique. Several features inherent in the design of projects pose a challenge to the implementation of 'meaningful participation '. Many projects are limited by a predefined time frame, a fixed amount of funding and a set of predetermined goals. Input and output must be measurable, and the project-cycle has to be maintained. If the intended beneficiaries do not act in accordance with the project approach, their voices will remain unheard, or are disqualified as "project dis-behaviour" (White 1996).[10] In their discussion of structures in development aid Craig and Porter conclude that "participation and effective management are deeply contradictory" (1997: 229).

Many NGOs try to involve themselves in projects to ensure their longer-term commitment in one place: a struggle that is not easy to survive. If the objective goes beyond effective project-management with just a 'scent' of participation, i.e. if NGOs work for changes in the economic, social or political system, than they quickly come to challenge project-based thinking as inhibiting their efforts. However, facilitating processes of 'meaningful participation' is extremely demanding. Cornwall (2000a) argues that the use of PRA and the focus on participatory tools and methods made participation just "another input to be programmed and managed along with other inputs" in the development mainstream. However participatory a development project is designed to be, it cannot escape the limitations on this process that derive from the international system of development aid.

3.6.3 How does the process of participation take place?

Several difficulties have been encountered and concerns voiced that challenge participation as an operational concept. The principal critiques focus on (1) the propriety of imported rules and institutions, (2) the balance between process and output, (3) the role of the facilitator and (4) assumptions about procedural justness (see table 3.4).

---

[10] For a more detailed discussion, see e.g. Craig and Porter (1997) or Reusse (1999)

The propriety of imported rules and institutions

Not all decisions affecting the local living conditions are taken at community level, but every community has functioning, more or less formalised mechanisms for such decisions – the rules of the game.

For a participatory process to be effective, it has to be acceptable within the existing set of rules. Many times they may be in conflict with each other and the facilitator's task is to advance a meaningful and productive process by dealing with these conflicts. However, local institutions are complex and in their complexity not easily understood by the outsider, as many of them are informal. Insiders may be unwilling to help or incapable of explaining to outsiders how the rules function. In addition outside facilitators of a participatory process may not have the time, interest or capacity to understand the local institutions before they start their work. Or project rules make it difficult for them to adapt their concept to local conditions even if they have a good understanding of the rules: Social relations are often more dynamic than designed institutions for participation. (Cleaver 2000, 2001).

Furthermore, participatory processes, such as PRA, emphasise concepts of political articulation which are not always compatible with local custom. Publicly voicing personal opinion, even when invited to do so, may have negative connotations for a member of the community, especially if that opinion is critical of those in power. Thus, the participatory process may become a show, orchestrated by the local elite for the satisfaction of the ill-informed facilitator and his donor agency. This prompts the question of what can actually be expected from the newly established participatory process, if politically relevant interaction is by and large informal.

Balancing process with output: the costs and benefits of participating

As participation is associated with collaboratively seeking better informed and more widely accepted decisions, it bears a highly desirable connotation - to the extent that considering the concrete costs and benefits for those invited or intending to participate would seem to be picking at straws. Nonetheless, from the participant's perspective such issues are important. Furthermore, it seems that the costs are distributed unevenly: participation as in PRA or other processes is time consuming – a high price for those who cannot afford to take time out due to the necessity for working for their daily living (Mosse 1994).

Mayoux (1995) criticises the fact that women's relative absence from participatory processes is often explained as the result of a lack of trust, power or consciousness. Instead, one should ask, to what extent these processes actually bring about tangible improvements for their life situations. Even if projects are designed with an open end to accommodate participatory decision-making, making a project 'time-efficient' remains a challenge. Some initiatives primarily focus on increasing consciousness– such as Paulo Freire's efforts in Latin America – but without tangible improvements they might not be an option for the most marginalised. To address this difficulty, some projects have paid the participants for the time they are spending in the different sessions. This simple remedy may nonetheless worsen the problem of having people participate like actors in a show, drawing and telling according to what they believe is expected from them. This problem becomes all the more acute as participatory processes are outsourced to national NGOs ('being cheaper and closer to the people') who in turn have to deliver presentable results to their northern donors/partners.

Having argued for a closer scrutiny of the actual costs and benefits involved, it remains to be said that such an analysis may well be misleading by itself. Three difficulties arise: Firstly, we can consider (non-) participation as the fruit of either ignorance, habit, apathy, rational strategy or other, without being able to pin down the extent of each. Secondly, calculating the costs and benefits that determine a person's motivation to participate, is an extremely difficult undertaking and is rarely feasible in general terms. Thirdly, such a calculation may seriously misconceive the nature of human motivation. We disagree as much with assumptions about utility maximising human rationality as with assumptions about altruistic spirit frequently suggested to be prevalent in rural communities. Instead, we subscribe to insights from Giddens (1984). He explains human motivation as being embedded within the multiple relations of a man with his environment, and as varying according to them. In that sense human behaviour is part of a process of interaction, and not so much the sum of a persons separate intentions, reasons or motivations.

Thus, we cannot pretend to grasp the motivations of other people so as to adequately consider in the project design. This constitutes an important limit to participation as an operational concept. In this understanding, 'output' refers not to 'quantifiable benefit' but rather to 'compliance with motivation'. As a participatory process largely depends on people's behaviour, balancing process with output seems to remain an exercise of trial and error.

The role of the facilitator: managing expert?

The complexity of the social processes involved in (participatory) development projects challenges any assumption that these processes can be managed, i.e. understood and steered, by an outside facilitator. Bearing in mind his/her only basic understanding of these processes, the task would be extremely demanding for this person. We believe that this requires humility, curiosity, empathy, experience, time, patience and a sharp and open mind – qualities which Chambers also puts forward. But Chambers believes that such a facilitator is then able to identify the weakest members in a community and to apply methods that effectively strengthen the disadvantaged (Chambers 1994c). Moreover, Chambers believes the facilitator can ensure the justness of the participatory process – here PRA (1994a). But social processes within a community are not obvious and power relations even less so – making participation on equal terms challenging.

These difficulties require the facilitator to intervene regularly with the result that his influence in guiding the group processes is strong. Hence, the facilitator's perception of the social context and his understanding of a desirable participatory process determine much of the space within which "bottom-up ideas and action" may gain strength. Here arises the expert status of experienced agencies and professionals. The experts provide the master plan of participation. In positivist management-style documents and source books have been produced that provide tools and strategies for participatory development. According to them, outsiders, backed by scientific insight, shall identify "stakeholders" or "target groups", analyse the local situation and implement institutions for the participatory process. Rahnema in his review of participation in development heavily criticises the new 'participation experts' or "change agents". He writes: "*Acting, in most cases, as a promoter or professional of participation, rather than a sensitive party to a process of mutual learning, he [i.e. the change agent] became sometimes a militant ideologue, sometimes a self-appointed authority on people's needs and strategies to meet them (...) Few were actors genuinely seeking to learn from the people how they defined and perceived change, and how they thought to bring it about. The change, of which they considered themselves agents, was only the projection of a predefined ideal of change (...).*" (Rahnema 1992: 123)

Procedural Justice: Dealing with difference and conflict

One of the central notions in the discourse of participation is that the "marginalised" should be included and given more voice and importance in the process. On top of the problem of adequately defining who precisely is meant by the "marginalised" (c.f. 6.1), there are many examples of well-intended projects where it was impossible to integrate the defined marginalized voices (Mosse 1994: 511).

Kapoor (2002b) argues that especially PRA, as devised by Chambers, does not count on systematic rules and legitimising mechanisms. It is not sufficient to only assemble people with different interests, it is also necessary to have explicit rules for the game in order not to strengthen existing differences in power. There is widespread belief in the possibility to reach consensus, and conflicts are considered an affair of the facilitator. To leave the question to the facilitator and his critical consciousness, respect and patience is questionable in itself. Furthermore, assuming a consensus without taking into account the differences and power relations can lead to further inequalities rather than to a new solution: "(...) PRA, far from providing a neutral vehicle for local knowledge, actually creates a context in which the selective presentation of opinion is likely to be exaggerated, and where minority or deviant views are likely to be suppressed. In practical terms 'community priorities' such as a school, soil and water conservation, social forestry or well deepening conceal private interests" (Mosse 1994: 508).

Mosse argues with Pierre Bourdieus interpretation of authority and dominance. Bourdieu calls it a sign of dominance, if people are able to present their own interests as common ones. Mosse sees the possibility for those ‚officialising strategies' in the context of PRA processes, as defined by Chambers. He argues that PRA can serve as a new means for those in power to legitimise their personal interests (Mosse 1994: 509). As confrontation and negotiation about differing interests are not considered in PRA processes, Kapoor (2002b: 109) argues for the implementation of mechanisms for mediation and moderation. „Without checks against unequal power relationships among participants, without critique, there appears to be little scope for preventing coerced outcomes."

Ribot calls for democratic legitimisation to allow us to deal with pluralistic interests. He argues that without such mechanisms the most organised and influential groups will be the best off (Ribot 2002). Another approach for dealing with conflicts and differences is presented by the negotiation approach or conflict management (Leeuwis 2000). The "ideal speech situation" (Habermas) as a discourse without dominance between equal participants remains an idealistic image. But rules and mechanisms are important to reduce inequalities. The theories of deliberation and deliberative democracy give insights into the question of "How to deal with difference?"[11] especially with regard to the difference in values. The discussions are far from completion as democratic rules and mechanisms have also underlying values which are not automatically shared.[12]

3.6.4 What is the purpose of participation?

Cornwall (2000a) argues that, based on a broad definition, 'participation' can be stretched to include any kind of standard project activity that is intended to benefit somebody. Those who employ the term or engage in acts of participation should therefore lay open their intentions. In table 3.5 we listed several declared or suspected purposes of participation. The list is neither complete nor does it claim to be politically 'balanced'. We do not want to discuss here the objectives of participation on terms of morality or justice – a generalising judgement does not seem feasible to us. As Cohen and Uphoff note, *"Because they [i.e. the purposes of participation] are essentially normative, disagreement on the assessment of purposes is even more likely than with the more descriptive dimensions (...). As with all objectives, they [i.e. the purposes of participation] may be intended or unintended, stated or un-stated, and achieved or unachieved"* (Cohen and Uphoff 1980: 227).

This characterisation throws light on two problems involved, irrespective of the political position: Is the intended also the stated?, and secondly, is the stated actually achievable? We have already described the potential incompatibility of many project designs with the criteria for 'open-ended' or 'meaningful' participatory processes. Furthermore, we have argued that a facilitator's limited understanding of the social processes and power relations involved, make it difficult to 'manage' the process. In addition we have to recognise that he/she is not an uninterested party. But even if that person were fully independent with regard to the outcome of such a participatory process, it has to be questioned whether 'meaningful participation' can be achieved. What then is 'meaningful'? In what follows, two interpretations are reviewed: participation leading to 'aid efficiency' and participation leading to 'empowerment'.

---

[11] See e.g. Habermas (1984), Dryzek (1990), Sanderson (no date). Especially in the field of environmental policy, discussion about deliberation and deliberative democracy have gained a lot of importance (e.g. Owens 2000, Stoll-Kleemann and O'Riordan 2002, Jabobs 1997)

[12] See e.g. Kapoor (2002a)

Towards aid efficiency

If 'meaningful participation' seeks to enhance a project's efficiency, this objective seems relatively feasible to us. Wherever highly capable convenors meet motivated intended beneficiaries within a favourable context, a smoothly conducted process can raise awareness, interest and acceptance, can subsequently improve project design through better understanding of the situation, and can furthermore lead to honest negotiations and effective agreements with those in power to negotiate. In this perspective, 'meaningful participation' is best, if streamlined according to the project's needs. It should not be burdened with special considerations for the 'voiceless', 'disadvantaged' or 'marginalised', but focus on the actually feasible and on sound contractual relations. Apart from the many ifs, we see the principal difficulty with this interpretation in that the term 'participation' suggests much more than effective collaboration within a project, and is therefore miss-leading.

Towards 'empowerment'

When interpreting 'meaningful participation' as 'empowerment', the issue becomes more complicated. Rahnema succinctly describes the conceptual difficulties of externally induced or 'organised' empowerment:

„*When A considers it essential for B to be empowered, A assumes not only that B has no power - or does not have the right kind of power - but also that A has the secret formula of power to which B has to be initiated*" (Rahnema 1992: 123).

Kapoor examines to what extent the processes, which Chambers recommends as empowering, are based on questionable assumptions about power. Informed by Foucault's insight that power manifests itself through rules, institutions and their explanations, Kapoor questions the belief that PRA is a neutral process that can provide the space for empowerment (Kapoor 2002). Associated with power is the understanding of 'knowledge'. In much empowerment thinking, 'knowledge' is perceived as the total of useful or 'empowering' information that needs to be improved and augmented (Kothari 2001). Instead, knowledge can also be understood as being culturally, socially and politically produced and permanently reformulated. In that view, 'knowledge' becomes the essential expression of power, framing our perception and interpretation of the world. Freire's 'pedagogy of the oppressed' (1970) seeks empowerment – or liberation - in making this condition visible. Closely related to this dilemma is the question to what objective or capacity people should actually be empowered. Henkel and Stirrat note: "It seems evident that what people are 'empowered to do' is to take part in the modern sector of 'developing' societies. They are being empowered to be elements in the great project of 'the modern': as citizens of the institutions of the modern state; as consumers in the increasingly global market" (Henkel and Stirrat 2001).

One can conclude from the above that organised participation heading towards mainstream interpretations of empowerment is not quite as liberating. Furthermore, White (1996) believes that many aid organisations cherish 'empowerment', but find it uncomfortable to loose control. A top-down commitment to others' empowerment is not only contradictory in terms (Rahnema 1992), but also conflictive, because it risks to lay bare the power relations involved between the agency and the intended beneficiaries. These inequalities are manifest: First, many communities have to compete with other communities to 'win' a project; second, as project designers employ incentives, the announced material benefits are often subject to a community's 'performance' - he who pays, sets up the rules; third, participatory processes such as PRA provide possibilities for close monitoring of the intended beneficiaries – thereby local opposition to the intervention can be 'kept in control'. In other words, participatory processes can function as a disciplining tool. Cleaver (2001) finds that 'empowering participation' is ever present in development discourse as the ultimate non-political solution. As an 'act of faith' it is presented as an intrinsically good thing, with a focus on 'getting the techniques right' to ensure its success. Such a 'non-political empowerment' describes a self-driven harmonious process that enhances the opportunities of one party without ever threatening any other party. But most understandings of (political) power conceive such power in relative terms, not absolute ones. Hence, 'conflict free', or 'non-political' empowerment is contradictory in itself.

In this context, the distinction between political and development/project participation diminishes when considered in the light of political power. Projects are about resource allocation, and many projects constitute the biggest economic activity at local level. In this perspective, any non-political connotations of participation are in themselves a political issue, celebrating technocracy and proposing 'management' as a substitute for politics. With regard to Chambers' sleeves-up distaste for theorising the political implications of participation, Kapoor (2002) remarks that such a view by itself has political consequences: to privilege the feasible and the most obvious risks to ignore more hidden expressions of inequality. Putting this thought further, Hildyard et al. (2001) justify a radically political interpretation on purely conceptual grounds - inherent in any act of participation – as the only way not to reinforce pre-existing inequalities of power: *"Participation requires wider processes of social relations through which inequalities are reproduced. Behavioural changes, though necessary, are not enough. Addressing the structural causes of inequality demands not only policy changes (...) but, arguably, rethinking the means by which such change is achieved"* (Hildyard et al. 2001: 69).

In our final view on 'empowering participation' the perspective of the radical development critique shall be considered (e.g. Hickey and Mohan 2005, Kapoor 2005, Rahnema 1992, Escobar 1984). In Reusse's systemic analysis of the international aid system 'participation' can be interpreted as a paradigm (like "green revolution" or "needs-based approach") which is capable of indirectly perpetuating long known dilemmas in international development, such as wastefulness, arrogance and ignorance.

Reusse speaks of a paradigm life cycle: A concept is developed, promoted, established and protected until a new one gains strength and gradually replaces the older one. Certain assumptions are moved centre-stage, tools, methods and strategies are developed and implemented until a new sub-paradigm requires different ones. Reusse´s point is that although the new paradigms repeatedly promise to change things altogether, they actually never allow for a questioning of the patterns of interaction between the different actors and actor groups involved. The grand pattern remains as the aid system is defended by the paradigms it purports.

3.7 Conclusion: for a more precise approach to participation

Conclusions from this review will differ – again – according to perspective. The greatest dilemma remains for those who not only think about meaningful participation but try to put it into practice. We have ended up with high barriers which seem paralysing in their sum. Even the local 'natural leader', initiating activities in his/her own community, struggles with local power relations, lack of interest, or with people turning open meetings into one-to-one confrontations. How much more then, does the external facilitator of 'development interventions' struggle with meaningful participation? The pitfalls of participation are very real and should not disappear behind the noble objectives and justifications frequently associated with the term. The critiques should therefore not be underestimated, but the fact that there are no easily available solutions, should not cause us to finish with participation altogether (cf. Hickey and Mohan 2004).

The critiques on aid, its system and the attitude it purports are relevant to participation and have been described. They provide a good starting point for working on changes in the manner participation is employed. Two aspects appear crucial to us here: First, there remains a glaring discrepancy between participation discourse and practice; to acknowledge the existence of this discrepancy would remove the burden of having to maintain a pretentious position (through vague and lofty language). This would also allow focusing more on the context specific difficulties instead of on tools. Secondly, the ideals of humans living together purported by participation, are precious and valid even if they are not fully achieved. These values of empathy, curiosity, honesty, (self-) awareness and humility are central to meaningful participation – not only as a goal but as a baseline. Hence, it is more appropriate to speak about practicing participation than about implementing or achieving it.

Participation refers to the cosmos around the question: how can people be able to peacefully relate and act together in their own best interest? (Rahnema 1992:126). Therefore, participation requires caution in deeds and precision in words. Conceiving it either as a promising solution or as a 'cunning lie' fails, in both ways, to grasp the concrete but limited potential that the concept can offer if it is duly specified. We therefore suggest an approach to practicing participation - be it in biodiversity conservation, natural resource management or development projects - that takes into account the following:

- The ambiguity of the term participation and the potential discrepancy between its discourse and practice should be acknowledged.
- Each approach to participation should specify the underlying definition of participation in terms of Who? In what dimension? How? And for what purpose? The limits arising from this specification should be recognised.
- Every local situation requires a careful analysis of the interactions and power relations within the local context and those between a locality and its wider structural setting, in social, economic, political and ecological terms.
- Each party (group, project etc.) that invites people or wants to mobilise people to participate should consider it's own limits in terms of beliefs, judgements and norms.
- In projects, the actual space for participation - in terms of allocation of funds, timing, and openness of expected results – should be clarified and kept transparent all the way from donors to beneficiaries.

In closing, we would like to emphasise that practicing participation in the context of managing natural resources not only comprises procedural norms for the distribution of resources: Instead, it should also consider the purposes for which natural resources are actually being employed. Visvanathan points out that this concerns nothing less than the diversity of human life itself:

> "*What one needs is not a common future but the future as a commons. A commons is the plurality of life worlds to which all citizens have access. It is not merely the availability of nature as being, but of alternative imaginations, skills that survival in the future might require*" (Visvanathan 1991: 383).

References

Agrawal, A., Gibson, C.C. (1999). Enchantment and Disenchantment: The Role of Community in Natural Resource Conservation. World Development 27 (4), 629-649

Arnstein, S.R. (1969). A ladder of citizen participation. Journal of the American Institute of Planners, 216-224

Berkes, F., Folke, C. (1998). Linking social and ecological systems. Cambridge University Press, Cambridge

Blaikie, P. (1999). A Review of Political Ecology. Issues, Epistemology and Analytical Narratives. Zeitschrift für Wirtschaftsgeographie 43 (3-4), 131-147

Borrini-Feyerabend, G., Pimbert, M., Farvar, M.T., Kothari, A., Renard, Y. (2004). Sharing power. Learning-by-doing in co-management of natural resources throughout the world. IIED and IUCN/ CEESP/ CMWG, Cenesta, Tehran

Brechin, S.R., Wilshusen, P.R., Fortwangler, C.L., West, P.C. (2003). Contested nature. Promoting international biodiversity with social justice in the twenty first century. State University of New York Press, New York

Bryant, R.L. (1998). Power, knowledge and political ecology in the third world: a Review. Progress in Physical Geography 22 (1), 79-94

Bryant, R.L. (1999). A Political Ecology for Developing Countries? Progress and Paradox in the Evolution of a Research Field. Zeitschrift für Wirtschaftsgeographie 43 (3-4), 148-157

Bryant, R.L., Bailey, S. (1997). Third World Political Ecology. Routledge, London

Chambers, R. (1994a). The Origins and Practice of Participatory Rural Appraisal. World Development 22 (7), 953-969

Chambers, R. (1994b). Participatory Rural Appraisal (PRA): Analysis of Experience. World Development 22 (9), 1253-1268

Chambers, R. (1994c). Participatory Rural Appraisal (PRA): Challenges, Potentials and Paradigm. World Development 22 (10), 1437-1454

Chambers, R. (1997). Whose reality counts? Putting the first last. ITDG Publishing, London

Checkland, P., Scholes, J. (1991). Soft Systems Method in Action. John Wiley, New York

Cleaver, F. (2001). Institutions, Agency and the Limitations of Participatory Approaches to Development. In: Cooke, B., Kothari, U. (eds). Participation: The New Tyranny? Zed Books, London, New York, 36-55

Cleaver, F. (2000). Moral Ecological Rationality, Institutions and the Management of Common Property Resources. Development and Change 31, 361-383

Cohen, J.M., Uphoff, N.T. (1980). Participation's place in rural development: seeking clarity through specifity. World Development 8, 213-235

Cooke, B., Kothari, U. (2001). Participation: The New Tyranny? Zed Books, London, New York

Cornwall, A. (2000a). Beneficiary, Consumer, Citizen: Perspectives on Participation for Poverty Reduction. SIDA studies No. 2, Gothenburg, URL: http://www.sida.se/shared/jsp/download.jsp?f=Cornwall+study+Web.pdf&a=2079, [June 2006]

Cornwall, A. (2000b). Making a difference? Gender and participatory development. Discussion Paper 378, IDS (Institute of Development Studies), Brighton, URL: http://www.ids.ac.uk/ids/bookshop/dp/dp378.pdf, [June 2006]

Cornwall, A. (2003). Whose voices? Whose choices? Reflections on Gender and Participatory Development. World Development 31 (8), 1325-1342

Craig, D., Porter, D. (1997). Framing participation: development projects, professionals, and organisations. Development in Practice 7 (3), 229-236

DAC/OECD (1997). Evaluation of Programs Promoting Participatory Development and Good Governance. Synthesis Report of the Development Assistance Committee's Expert Group on Evaluation. OECD, Paris

Dryzek, J.S. (1990). Discursive Democracy. Politics, Policy, and Political Science. (Reprint 1999). Cambridge University Press, Cambrigde

Easterly, W. (2002). The Cartel of Good Intentions - Bureaucracy versus Markets in Foreign Aid. Center for Global Development. Working paper No. 4. URL: http://www.cgdev.org/content/publications/detail/2786, [June 2006]

Escobar, A. (1984). Discourse and power in development: Michel Foucault and the relevance of his work to the Third World. Alternatives 10, 377-400

Escobar, A. (1995). Encountering Development. The Making and Unmaking of the Third World. Princeton University Press, Princeton, New Jersey

Fals-Borda, O., Rahman, M.A. (1991). Action and knowledge: breaking the monopoly with participatory action research. ITDG Publishing, New York, London

Francis, P. (2001). Participatory Development at the World Bank: The Primacy of Process. In: Cooke, B., Kothari, U. (eds). Participation: The New Tyranny? Zed Books, London, New York, 72-87

Freire, P. (2000). The Pedagogy of the Oppressed. 30th Anniversary Edition. Continuum International Publishing Group, New York. (First Edition 1970: Pedagogía del oprimido)

Giddens, A. (1984). Die Konstitution der Gesellschaft: Grundzüge einer Theorie der Strukturierung. Frankfurt am Main

Guijt, I., Shah, M.K. (1998). Waking up to power, conflict and process. In: Guijt, I., Shah, M.K. (eds). The myth of community. Gender issues in participatory development. ITDG Publishing, London, 1-23

Habermas, J. (1984). The theory of communicative action. Vol. 1: Reason and the rationalisation of society. Boston (Original 1981: Theorie des kommunikativen Handelns, Bd.1: Handlungsrationalität und gesellschaftliche Rationalisierung, Bd. 2: Zur Kritik der funktionalistischen Vernunft. Suhrkamp, Frankfurt am Main)

Hapgood, D. (1969). The Role of Popular Participation in Development. Report of a Conference on the Implementation of Title IX of the Foreign Assistance Act. Cambridge

Hardin, G. (1968). The Tragedy of the Commons. Science 162, 1243-1248

Hartmann, F. (1998). Towards a Social Ecological Politics of Sustainability. In: Keil, D., Bell, D.V.J., Penz, P., Fawcett, L. (eds). Political Ecology. Global and Local. Routledge, London, New York, 336-351

Hickey, S., Mohan, G. (2004). Participation: From Tyranny to Transformation? Exploring New Approaches to Participation. Zed Books, London

Hickey, S., Mohan, G. (2005). Relocating Participation within a Radical Politics of Development. Development and Change 36 (2), 237–262

Hildyard, N., Hegde, P., Wolvekamp, P., Reddy, S. (2001). Pluralism, Participation and Power: Joint Forest Management in India. In: Cooke, B., Kothari, U. (eds). Participation: The New Tyranny? Zed Books, London, New York, 56-71

Hirschman, A.O. (1958). The Strategy of Economic Development. New Haven

Jacobs, M. (1997). Environmental valuation, deliberative democracy and public decision-making institutions. In: Foster, J. (ed). Valuing Nature? Economics, Ethics and Environment. Routledge, London, New York, 211-231

Kapoor, I. (2001). Towards participatory environmental management? Journal of Environmental Management 63, 269–279

Kapoor, I. (2002a). Deliberative Democracy or Agonistic Pluralism? The Relevance of the Habermas-Mouffe Debate for Third World Politics. Alternatives 27, 459-487

Kapoor, I. (2002b). The devil´s in the theory: a critical assessment of Robert Chambers' work on participatory development. Third World Quarterly 23 (1), 101-117

Kapoor, I. (2005). Participatory Development, Complicity and Desire. Third World Quarterly 26 (8), 1203-1220

Leeuwis, C. (2000). Reconceptualizing Participation for Sustainable Rural Development: Towards a Negotiation Approach. Development and Change 31, 931-959

Mayoux, L. (1995). Beyond Naivety: Women, Gender Inequality and Participatory Development. Development and Change 26, 235-258

McCracken, J., Pretty, J., Conway, G. (1988). An introduction to rapid rural appraisal for agricultural development. International Institute for Environment and Development, London

McLain, R., Lee, R. (1996). Adaptive management: promises and pitfalls. Environmental Management 20, 437-448

Mohan, G. (2001). Beyond Participation: Strategies for Deeper Empowerment. In: Cooke, B., Kothari, U. (eds). Participation: The New Tyranny? Zed Books, London ,New York, 153-167

Mohan, G., Stokke, K. (2000). Participatory development and empowerment: the danger of localism. Third World Quarterly 21 (2), 247-268

Mosse, D. (1994). Authority, Gender and Knowledge: Theoretical Reflections on the Practice of Participatory Rural Appraisal. Development and Change 25, 497-526

Nelson, N., Wright, S. (1995). Participation and power. In: Nelson, N., Wright, S. (eds). Power and Participatory Development. ITDG Publishing, London, 1-18

Nohlen, D., Nuscheler, F. (1993). Was heißt Entwicklung? In: Nohlen, D., Nuscheler, F. (eds). Handbuch der Dritten Welt, Vol. 1. Dietz, Bonn

Nustad, K.G. (2001). Development: the devil we know? Third World Quarterly 22 (4), 479-489

Oakley, P. (1991). Projects with People: The Practice of Participation in Rural Development. International Labour Organisation (ILO), Geneva

O'Riordan, T., Stoll-Kleemann, S. (2002). Deliberative democracy and participatory biodiversity. In: O'Riordan, T., Stoll-Kleemann, S. (eds). Biodiversity, Sustainability and Human Communities. Protecting beyond the Protected. Cambridge University Press, Cambridge, 87-112

Owens, S. (2000). 'Engaging the public': information and deliberation in environmental policy. Environment and Planning (A) 32, 1141-1148

Peet, R., Watts, M. (1996). Liberation ecologies. Environment, development, social movements. Routledge, London, New York

Pimbert, M.P., Pretty, J.N. (1995). Parks, People and Professionals: Putting 'Participation' into Protected Area Management. UNRISD Discussion Paper No. 57, Geneva

Pretty, J.N. (1995). Participatory learning for sustainable agriculture. World Development 23 (8), 1247-1263

Rahnema, M. (1992). Participation. In: Sachs, W. (ed). The Development Dictionary. A Guide to Knowledge as Power. Zed Books, London, New Jersey, 116-131

Reusse, E. (1999). Interventionist Paradigms and the Ills of Aid. Bochumer Schriften zur Entwicklungsforschung und Entwicklungspolitik, Vol. 44. München

Ribot, J.C. (2002). Democratic Decentralization of Natural Resources. Institutionalizing Popular Participation. World Resource Institute, Washington. URL: http://water.wri.org/decentraldemocratic-pub-3767.html, [June 2006]

Rist, G. (1997). The History of Development: From Western Origins to Global Faith. Zed Book, London

Ross, E.B. (1998). The Malthus Factor: Poverty, Politics and Population in Capitalist Development. Zed Books, London

Rostow, W.W. (1962). The Stages of Economic Growth. Cambridge

Sanderson, I. (no date). Participation and Democratic Renewal: from "instrumental" to "communicative rationality"? Policy and Politics 27 (3), 325-341

Schneider, H., Libercier, M.-H. (1994). Concepts, Issues and Experiences for Building up Participation. In: Participatory Development – from Advocacy to Action. OECD Development Centre, Paris

Sommer, J. (2001). Empowering the Oppressed - Grassroots Advocacy Movements in India. Sage Publications Ltd, New Delhi

Stiglitz, J.E. (2002). Participation and Development: Perspectives from the Comprehensive Development Paradigm. Review of Development Economics 6 (2), 163-182

Taylor, H. (2001). Insights into Participation from Critical Management and Labour Process Perspectives. In: Cooke, B., Kothari, U. (eds). Participation: The New Tyranny? Zed Books, London, New York, 122-138

Visvanathan, S. (1991). Mrs. Brundland´s Disenchanted Cosmos. Alternatives 16 (3), 377-384

Walters, C. (1986). Adaptive management of renewable resources. Macmillan, New York

Webler, T., Tuler, S., Krueger, R. (2001). What Is a Good Public Participation Process? Five Perspectives from the Public. Environmental Management 27 (3), 435–450

Western, D., Wright, R.M. (1994). Natural Connections. Perspectives in Community-based conservation. Island Press, Washington D.C. Covelo, California

White, S.C. (1996). Depoliticising development: the uses and abuses of participation. Development in Practice 6 (1), 6-15

World Bank (1996). The World Bank Participation Sourcebook. Washington. URL: http://www.worldbank.org/wbi/sourcebook/sbhome.htm, [June 2006]

4 Evaluating Stakeholder Dialogues

Angela Oels

University of Hamburg, Germany

Stakeholder dialogue is supposedly a good thing. If successful, it is said to improve the quality of policy decisions, to mobilise urgently needed resources and to increase public acceptance of policy decisions. There is a growing literature that celebrates the expected benefits of involving stakeholders in meaning-making, decision-making and management processes. But do stakeholder dialogues live up to the high expectations that are raised in the literature? What are suitable criteria and indicators of success given that the outcomes are hard to predict? How do the involved stakeholders themselves judge the fruits of their involvement? What can be done to improve the performance of stakeholder dialogues? What are conditions for their success?

The evaluation of participatory processes is a topic that is still in its infancy (Oppermann and Langer 2002: 76, Chess 2000: 769, Rowe and Frewer 2000: 3). Systematic, long-term evaluation studies of stakeholder dialogues are still the exception. The existing evaluation studies vary widely with regards to their purpose, focus, scope and disciplinary perspective. While the methodological and theoretical issues of evaluation have been discussed at length (for example Chess 2000), no set of commonly used indicators for the evaluation has emerged yet. This chapter seeks to make a contribution towards this end by discussing the suitability of criteria sets and procedures for the evaluation of stakeholder dialogues.

The first part of this chapter distinguishes between three types of stakeholder dialogues which are pursued with different purposes in mind. For each of these types, specific indicators of success may be appropriate. The second part of this chapter presents criteria sets from the relevant literature and discusses their strengths and weaknesses. Theory-based and user-based criteria sets are introduced and the possibility of integrating both criteria sets is explored. The third section reviews the most common findings of evaluation studies of stakeholder dialogue. The chapter concludes with a preliminary set of conditions for the success of stakeholder dialogue.

4.1 The case for stakeholder dialogues

4.1.1 Defining stakeholder dialogues

Before engaging with the issue of evaluation, the term stakeholder dialogue needs to be clarified. A wide range of participatory processes is used in environmental policy-making and implementation. Stakeholder dialogues are one of them. They are defined by the fact that they do not involve 'the public' but only those with a stake in the issue at hand. Stakeholders are those with information on the subject at hand, those with the power to influence the decision-making and those affected by the outcome. As explained in Chapter 1, stakeholder dialogues do not pursue the ideal of representative democracy. Instead, the idea is to bring all views into the room, no matter if a view represents 1% or 90% of the population.

Stakeholder dialogues are used for quite different purposes. The selection of participants and working procedures is meant to match the specific purpose of the stakeholder dialogue. In this book, stakeholder dialogues are classified along three purposes:

- clarifying and improving knowledge (stakeholder dialogue for science);
- basing decision-making upon the deliberation of a collective will (stakeholder dialogue for policy-making);
- supporting implementation (stakeholder dialogue for management).

More often than not, stakeholder dialogues for science and stakeholder dialogues for management also aim to resolve a collective action problem by policy-making. It is therefore not always possible to clearly distinguish between the three types of stakeholder dialogue. The next section will introduce the three different types of stakeholder dialogue in more detail.

4.1.2 Stakeholder dialogues for science

Stakeholder dialogues for science aim to improve the knowledge base for decision-making. Their aim is to break the monopoly of "expert science" by providing alternative viewpoints. Most stakeholder dialogues for science involve stakeholders' perspectives in a process of redefining the knowledge base. However, there are many variations in the processes used for stakeholder dialogue. While some aim for consensus and are coupled with the policy process, others are no more than an exchange of arguments with open consequences (and the option of no outcomes at all).

Stakeholder dialogues for science are pitched against the monopoly of science. They take knowledge no longer as something that can be objectively determined by scientists but as socially constructed and inherently value-based (Healey 1997: 29-30). Stakeholder dialogues for science undermine the privileged position of 'experts', whose knowledge is no longer regarded as automatically superior to other ways of knowing. Experts are to be no more than 'specialized citizen[s]' (Fischer 1993: 183). "By demystifying technocratic decision techniques, post-positivist policy inquiry denies the expert's facile claim that there is only one scientific solution to a pressing social or political problem." (Fischer 1993: 167)

Therefore, breaking the hegemony of science requires an end to science's monopoly on knowledge: "Democratize language, ... and other forms of equality will follow" (Barber 1984: 193). Most (but not all) proponents of stakeholder dialogues for science also reject the 'deficit model' according to which the public is considered ignorant and in need of education in scientific ways of knowing (for example Petts 1997: 328, Durant 1995: 75, Street 1997: 142). Instead, lay people's and stakeholders' multiple ways of knowing and communicating knowledge are to be explored, respected and brought together in order to increase the understanding of problematic issues of public concern and to inform action (Innes 1996: 171, Burgess 1995, 1996, Burgess et al. 1988c, Harrison and Burgess 1994).

Deliberation in a stakeholder dialogue for science is unavoidably a political process, in the sense that it involves the careful evaluation of conflicting evidence and decision-making on what should guide the action to be taken (Durant 1995: 77). "In a word, politics is not the application of Truth to the problem of human relations but the application of human relations to the problem of truth" (Barber 1984: 64-65). Depending on the frame of reference applied to a real world problem, the solutions - including the distribution of costs and benefits - will differ. "Issues of risk assessment, for instance, are not simply a matter of discerning scientific risks, but a matter of determining who should bear the risks or costs of a policy choice." (Rossi 1997: 198) This is true even where a stakeholder dialogue is not directly linked to the policy process – the produced knowledge base can never be neutral, even if every effort is made to gather independent, unbiased expertise. French philosopher Michel Foucault has demonstrated how power and knowledge are closely linked, and how all knowledge legitimises certain power relations and ways of making sense of the world at the expense of alternative ones (Hoy 1986). Knowledge is not independent of the world – instead, it actively brings forth the world as it is preconceived (Pretty 2002).

4.1.3 Stakeholder dialogues for policy-making

Stakeholder dialogues for policy-making aim to ground decision-making in a deliberative process that forges the collective will of the stakeholders. Again, the processes used to conduct stakeholder dialogues vary widely. While some require a consensus to be achieved or are directly linked to binding decision-making processes, others have open results and no statutory basis. The later ones should be more correctly called stakeholder dialogues for policy advice.

The major strength of stakeholder dialogues for policy-making is that they foster stakeholder's capacity for genuine public thinking and allow their sense of belonging to a political unit to grow as a result of thinking about the question how they want to live together and what needs to be changed. The underlying assumption is that people's very consciousnesses and preferences are formed in social interactions with others and are subject to constant review in the light of new experiences (Healey 1997). In this process of constant social learning, self-interests can be modified to accommodate public interests.

"The affective power of talk is, then, the power to stretch the human imagination so that the I of private self-interest can be reconceptualized and reconstituted as a we that makes possible civility and common political action." (Barber 1984: 189-190)

Barber argues that this process of reconceptualising one's own interests to embrace the common good requires the active participation of each individual citizen, not just the deliberation amongst elected representatives or amongst chosen few in methods of deliberative opinion polling. However, for practical reasons, most stakeholder dialogues tend to select a tiny sample of all existing stakeholders and involve those in a deliberative process to model what all would think, if they could be involved in the same way. The acceptance of the outcomes often depends upon the transparency and legitimacy of the stakeholder selection process.

The nature of the dialogue that is to facilitate learning amongst the participants has recently been explored by Innes and Booher (1999). They use the metaphor of fantasy role-playing to describe the spirit in which deliberation should take place in order to foster learning. They argue that in role playing and consensus building alike, participants "play with heterogeneous concepts, strategies, and actions with which various individuals in the group have experience, and try combining them until they create a new scenario that they collectively believe will work." (Innes and Booher 1999: 12) Innes and Booher call this process a 'bricolage' which "produces, rather than a solution to a known problem, a new way of framing the situation and of developing unanticipated combinations of actions that are qualitatively different from the options at the table at the outset." (Innes and Booher 1999: 12)

Barber has similarly characterised the strong democratic talk as "an unrehearsed intellectual adventure" drawing on the words of Oakeshott (1962: 198). Of course, the participants' professional and social roles often inhibit the degree to which they can open themselves up to this adventure. Barber argues that participatory processes must ensure the explorative nature of the discourse: "Every expression is both legitimate and provisional, a proximate and temporary position of a consciousness in evolution." (Barber 1984: 183) The learning process is thought to have a real world impact: "Since the players often are the people in a position to have an effect on the resource or the problem, change in their attitudes and knowledge matters and in itself is a major part of the long-term consequences" (Innes and Booher 1999: 11).

4.1.4 Stakeholder dialogues for management

One purpose of stakeholder dialogues for management is to ease the implementation of already decided policy measures by involving those affected by it. Most stakeholder dialogues for management have a rigid frame (for example the management of a nature reserve) but enable the stakeholders to specify the concrete aims and institutions for the implementation. Stakeholder dialogues for management are often employed where conventional approaches to natural resource management have failed. Conventional approaches to natural resource management tend to impose a management scheme top-down, based on the advice of experts but mostly without involving local people. Local people tend to be regarded as a threat to the natural resource – in protected areas they were to be kept out or removed by force. Establishing management schemes for natural resources without the support of local people proved costly (e.g. budget for armed guards) and unsustainable (Pretty 2002).

Stakeholder dialogues for management have established themselves as the favourite alternative to hierarchical approaches to natural resource management. Stakeholder dialogues for management foster processes of social learning and grow social capital. As part of the dialogue, stakeholders are allowed to reflect upon the complexity of their interactions with the natural environment and to talk about this subject on their own terms. They not only contribute their specific local knowledge but may alter it in the interactions of the group. Best practice of natural resources management in other locations is fed into these stakeholder groups and assessed for relevance to local circumstances. Instead of having a management scheme imposed upon them, these stakeholder groups are empowered to experiment with pilot schemes and to establish and be involved in their own management scheme. Changes in attitudes and behaviour which develop as a result of learning processes in the stakeholder group are supposed to be lasting and therefore sustainable. The newly formed networks and shared knowledge between the stakeholders is supposed to increase their capacity to do things for themselves in an effective way. The established management scheme is supposed to be sustainable as it is based upon the newly created social capital and attitude and behaviour changes which were the result of the social learning facilitated by the stakeholder dialogue (Pretty 2002, Averbeck 2006).

Stakeholder dialogues for management are also employed as part of a process of evaluating the effectiveness of the management of protected areas (Hocking et al. 2000, Pomeroy et al. 2003). Management effectiveness refers to design issues (size and shape of individual protected areas or a protected area system), the appropriateness of the established management systems and processes and finally to the delivery of protected area objectives (Hocking et al. 2000: 3-4). The purpose of evaluating management effectiveness according to Hocking et al. (2000: 5) is "promoting adaptive management; improving project planning; and promoting accountability". Hocking et al. (2000: 7) recommend to "involve a broad range of stakeholders, including local and indigenous communities living in or adjacent to protected areas, in the assessment process. (…) It is necessary to take account of the interests and concerns of all such stakeholders if they are to accept changed management priorities that emerge as a result of the evaluation."

4.1.5 Three types, many evaluation strategies

Looking back over all three types of stakeholder dialogue, we can recognise that as each pursues a different purpose, each may lead to a different emphasis in the evaluation criteria. While the success of a stakeholder dialogue for science may depend more on the perceived competence of the process, the success of a stakeholder dialogue for policy-making may depend much more on the perceived legitimacy and effectiveness of the process. When evaluating the success of a stakeholder dialogue for management, the fairness of the process (in the form of inclusiveness and openness for local knowledge) may be key factor(s) of success. The differences between the three types of stakeholder dialogue have been discussed at length to demonstrate that there is not going to be one set of indicators that will fit all types of stakeholder dialogue (unless the purpose is a comparative one). Instead, selecting criteria wisely will be a hallmark of a case-sensitive evaluation. I will now turn to a general introduction to the issue of evaluation, before introducing the most commonly used sets of indicators.

4.2 Evaluating stakeholder dialogue

The following section is an introduction to the basic issues that need to be resolved before engaging in an evaluation. It will provide a brief introduction to the questions why and when to evaluate, what to evaluate, how to evaluate and who should be carrying out the evaluation. (For a more detailled discussion of these issues see Caron Chess 2000). Theory-based and user-based criteria for evaluation are briefly characterised, but the respective criteria sets will be introduced in more detail in the following section (4.3).

4.2.1 Why and when to evaluate

The evaluation of environmental public participation in general and of stakeholder dialogues in particular can be pursued with very different purposes in mind (Chess 2000: 771). The evaluation may be driven by the practitioner's interest to improve practice and process of a stakeholder dialogue. An evaluation may aim to reveal the perceptions of those participating in the stakeholder dialogue in order to measure their satisfaction with the process. An evaluation may be carried out to better understand the intended and unintended effects of a stakeholder dialogue in the short and in the long term. An evaluation during a running stakeholder dialogue can form the basis for mid-course corrections. The justification of expenses for a stakeholder dialogue may be a further motivation for carrying out an evaluation. The results of an evaluation may form the basis for a decision regarding the possible replication of a stakeholder dialogue in the same or another context. Finally, the evaluation may be driven by the academic interest to compare the practice of stakeholder dialogue with ideal type models developed in the theoretical literature.

The timing of an evaluation follows from the purpose pursued (Chess 2000). A formative evaluation that informs the planning of a stakeholder dialogue and forms the basis for mid-course corrections is carried out before and during the stakeholder dialogue. A summative evaluation assesses the worth of a stakeholder dialogue subsequent to completion of it. An evaluation of the long-term impact of a stakeholder dialogue is carried out years after completion of a stakeholder dialogue. Chess (2000: 779) highlights the rewards of formative evaluation as it allows for what she calls 'adaptive participation', namely design changes on the way in order to maximise the benefits of a stakeholder dialogue. Nevertheless, thorough summative and impact evaluations will be needed as a basis for policy recommendations.

4.2.2 Criteria for the evaluation

The criteria for evaluation can be derived from theory (theory-based), from the stakeholders involved in the dialogue (user-based) or the evaluation can be goal-free (Chess 2000: 775-6). A theory-based evaluation uses normative criteria which are universally applied to all stakeholder dialogues, no matter what their contextual differences are. Criteria for theory-based evaluation studies of participatory processes in general and stakeholder dialogue in particular are taken from a wide range of academic disciplines including spatial planning, political theory, psychology, sociology, social geography and organisational management. Criteria for theory-based evaluation have been taken from critical theory (Webler 1995), collaborative planning (Healey 1997), risk communication (Rowe and Frewer 2000, Durant 1995, Rossi 1997), public participation (Fiorino 1990, Webler 1995, Rowe and Frewer 2000) and democratic theory (Fiorino 1990, Barber 1984). Very few evaluation studies combine criteria from different disciplines (for an exception: Innes and Booher 1999). The aim of a theory-based evaluation is to assess to what extent a stakeholder dialogue fulfils the criteria (and related indicators) as spelled out in the theoretical literature. The universal set of criteria eases the comparison of cases. It also helps to highlight the strengths and weaknesses of particular participation methods used to carry out stakeholder dialogue. On the basis of this comparison, certain participation methods can be recommended for certain purposes. One of the problems of theory-based evaluation is that there is no one ideal method of stakeholder dialogue that serves all purposes. Another problem is that the criteria employed may seem highly irrelevant to the practitioners on the ground. The practitioners may reject some or all of the theoretical criteria raised by the academics, thus rendering the acceptance and utility of the evaluation findings problematic. The rejection of evaluation criteria by practitioners may also be caused by cultural differences that lead to different value judgements.

User-based evaluation assesses if the stakeholder dialogue has achieved its broader goals and specific objectives as defined by those with a stake in the process. However, initiators, funders, organisers, participants and those affected by the outcomes may have very different objectives in mind. There may even be conflicting objectives. Therefore, one of the challenges is to either integrate these diverse objectives into a single set of criteria or to evaluate using competing criteria sets from the perspective of a certain stakeholder group, for example the funding agency's. Chess emphasises the need for evaluators to pay sufficient attention to agency interests when deriving criteria sets for the evaluation (Chess 2000: 780). The identified criteria sets and related indicators are time and context specific and therefore vary widely from one case to another. The strength of the stakeholder-based evaluation is its closeness to those actually involved in the stakeholder dialogue. It explores their aspirations, their ways of making sense of their experiences, their subjective interpretations of what constitutes success. The subjectivity of the stakeholder-based evaluation is at the same time its strength and its weakness. The lack of external perspective may lead to a blind eye with regards to power relations. The stakeholder-based evaluation would need to pay attention to hidden agendas that some stakeholders may actually have, in order to compensate for its lack of critical distance. A second shortcoming is the fact that a stakeholder-based evaluation is quite demanding. Defining criteria and indicators of success jointly with the stakeholders is not just time-consuming, but also crucially depends upon the willingness of those stakeholders to invest these extra-hours as well.

The utility of a user-based evaluation can be extended if it is opened up to trace unexpected and unintended outcomes, thereby delivering a much more comprehensive account of the impact of a stakeholder dialogue. This is what is at the heart of the so-called goal-free evaluation (Chess 2000: 776). A goal-free evaluation is liberated from undue bias that might result from a narrow focus on stated objectives or theoretically derived criteria. Instead, a goal-free evaluation is a broad assessment of needs and effects with the aim of providing policy advice. A goal-free evaluation is particularly useful when the objectives pursued by those with a stake in the stakeholder dialogue have stated no clear objectives or when the articulated objectives are in conflict with each other.

In general, a user-based evaluation and a goal-free evaluation are more likely to generate a comprehensive set of criteria, while some theory-driven criteria sets may be quite narrow due to their specific research interest. In fact, a combination of these approaches may deliver the best understanding of success and failure of a stakeholder dialogue (Chess

2000: 780). This may also ease pointing out the differences between theory-based and user-based criteria sets.

Section 4.3 of this chapter will introduce theory-based and user-based sets of criteria and indicators used for the evaluation of stakeholder dialogues.

4.2.3 Process or outcome criteria

The scope of the evaluation studies of stakeholder dialogue varies widely. While some restrict themselves to procedural criteria (Webler 1995), others look at process, outcome and capacity building criteria (Oels 2003). There is general agreement that evaluation criteria can be divided into process and outcome criteria (Chess 2000: 774). Process criteria investigate how a stakeholder dialogue is being carried out. Outcome criteria assess the direct output and long-term outcomes of the stakeholder dialogue, including issues like the influence on policy-making. Capacity building criteria are a subset of outcome criteria which have been highlighted by myself (Oels 2003) as an important category of evaluation criteria. Capacity building criteria include all aspects of social capital building that result from the stakeholder dialogue like the formation of social networks and learning processes which can be used as a resource for future participation processes. The evaluation studies also differ with regards to how much attention they pay to the embeddedness of the stakeholder dialogue in wider society and the formal institutions of government. Again, those interested in evaluating implementation of the outcomes of stakeholder dialogues are more likely to study the institutional context (for example Oels 2003).

4.2.4 Outsider or participatory evaluation

An important decision is who should be carrying out the evaluation. Chess distinguishes between outsider and participatory evaluation (Chess 2000: 776-777). Outsider evaluation brings in an external evaluator who leads and carries out the evaluation process. The argument for an external evaluator is to increase the accuracy and credibility of the evaluation by institutionalising a professional distance between evaluator and those being evaluated. For some, this implies minimizing interaction between evaluator and main stakeholders in order to prevent that friendly bonds bias the evaluation.

By contrast, participatory evaluation considers the evaluator as an educator who facilitates a process of self-reflection and learning among those with a stake in the stakeholder dialogue. This implies the involvement of some or all stakeholders in the design of the evaluation, the criteria, indicators and methods used, sometimes even in the data gathering and analysis itself. By involving the stakeholders, it is assured that the evaluation is useful to them and considered credible (Guba and Lincoln 1989). An extreme variant of this is empowerment evaluation (Fettermann 1996), where stakeholders actively participate in all phases of the evaluation process. Chess (2000: 780) points out that participatory evaluation is particularly prone to undue influence of certain interest groups. She therefore recommends that the issue of possible bias is countered by transparency of how the evaluation was carried out and who was involved in it to what extent.

A third option is a self-evaluation led by the facilitator/organiser of a stakeholder dialogue. Due to a lack of resources and a self-interest of the facilitator, this option is a frequently used type of evaluation (for example Polanyi 2002). However, the credibility of a self-evaluation suffers from the perceived conflict of interests (the need to be successful in order to attract future contracts).

Chess (2000: 780-781) concludes that a combination of outsider evaluation with participatory evaluation is expected to deliver the most insightful results. While the elements of outsider evaluation can minimise the bias, the participatory elements can maximise the utility and resulting mobilisation effects of the evaluation.

4.2.5 Quantitative or qualitative methods

A final consideration to be taken care of is the choice of methodology for gathering and analysing data on the criteria and related indicators. Here once more the choice of methods must match the purposes of the evaluation and the selected criteria and indicator sets in particular. Usually some exploration of the perspectives of those with a stake in the stakeholder dialogue using qualitative methods is recommended in order to grasp the categories in which stakeholders make sense of their experience. It is also important to realise that many substantive outputs and outcomes of stakeholder dialogues can not be quantified at all or only at the expense of losing a lot of insights. Overall, a mix of quantitative and qualitative methods is recommended (Chess 2000: 781). The methodology in general does not have to be limited to positivist criteria of validity. Instead, Chess (2000: 781) recommends that criteria of naturalistic inquiry as introduced by Guba and Lincoln (Guba and Lincoln 1989) could just as well ensure the quality of the evaluation findings.

4.2.6 The use of evaluation findings

The use of evaluation findings very much depends on the purpose and set-up of the evaluation at the outset. If an evaluation is contracted out by a government agency in order to propose mid-course corrections in a stakeholder dialogue programme or in order to inform the decision about replications of a stakeholder dialogue it is more likely to fulfil this function than if no intentions were linked with the evaluation to begin with. The danger has to be avoided though that the contracting agency requires certain evaluation findings, for example a great success of the stakeholder dialogue. An effective way of reaching policy-makers is to involve them in the stakeholder dialogue and in the evaluation process. This empowers policy-makers to pursue their policy-making on the basis of their own learning and reflections (Innes 1995). Overall, evaluation studies are often inconvenient for those with an interest in all things staying the way they have always been. For this reason, evaluation studies have to expect that their findings will not always be welcome and that strong efforts with regards to outreach are required if they are to have any impact. At times an evaluation may fail to find an audience for reasons outside the evaluators' control (Chess 2000: 782).

4.3 Criteria for the evaluation

At the heart of the debate about evaluating stakeholder dialogue is the question of appropriate criteria and indicators of success. The following two sections will review theory-based criteria and user-based criteria in turn. The sections will also make reference to the methodologies employed to collect data on the indicators.

4.3.1 Theory-based criteria

Habermas' ideal speech situation

The most cited set of theory-based criteria for the evaluation of all types of participatory processes used in environmental decision-making has been put forward by German sociologists Ortwin Renn and Thomas Webler (1995). Webler (1995) has taken Habermas' ideal speech situation as a starting point for developing an evaluation framework for deliberative processes. Habermas' definition of fairness is the absence of coercion. Webler operationalises fairness by saying that each person must be able to attend and initiate discourse, to contribute to it by debating and to make decisions about the nature of the process of the discourse itself. Webler defines three key activities of discourse: (i) agenda and rule making, (ii) moderation and rule enforcement and (iii) discussion itself. Competence of speech requires, according to Webler, access to information and its interpretations, and the use of the best available procedures for knowledge selection. Competence of speech also aims to establish communicative reason as the mode of making and challenging validity claims. Webler therefore prescribes investigation of all three dimensions of discourse to establish the competence of a discourse: theoretical discourse (making epistemic or strategic claims about the nature of the objective world), practical discourse (providing a normative value-basis for judgements and positions) and therapeutic discourse (establishing the authenticity and sincerity of the speaker). In addition, he adds explicative discourse (establishing the comprehensibility of communication) to this list, as it can be found in Habermas' theory of pragmatics. A competent discourse employs cooperative reasoning and instrumental reason as opposed to strategic reasoning. The main tenets of the fair and competent 'ideal speech situation' are summarised in Table 4.1. Each of the numbered letters in the table represents a testable criterion that is linked to clearly defined indicators. Table 4.2 provides three examples of the criteria developed by Webler. It is however beyond the scope of this chapter to reproduce the complete set.

Table 4.1 Conditions for the fair and competent ideal speech situation. Source: Webler 1995: 60

EVALUATION FRAMEWORK				
FAIRNESS	NEEDS			
ACTIVITIES	Attend	Inititate	Debate	Decide
Agenda and rule making	A1,A2, A3	A1	A2	A3
Moderation and rule enforcement	B1	B1	B2	B3
Discussion	C1	C2	C2	C3
COMPETENCE	NEEDS			
ACTIVITIES	Access to Knowledge		Best Procedures	
Explicative Discourse	D1		D2, D3, H1, H2	
Theoretical Discourse	E1, E2, E3		E5, E6, E7, H1, H2	
Practical Discourse	F1, F2, F3, F4		F5, F6, F7, F8, H1, H2	
Therapeutic Discourse	G1, G2		G3, G4, G5, H1, H2	

Table 4.2 Three examples of criteria to test for the competence of a discourse. Source: adapted from Webler 1995: 63, 78-86

E6	The model should provide the participants with the option to delegate determinations of factual truth to an outside expert panel.
F1	The model should not contain any implicit barriers that will bias the distribution of interests that participate.
F3	The model should promote both the discovery and the development of mutual understandings of values among all the participants.

However, these criteria are to be applied to an ideal type of a participatory process and are not designed to track the particularities of a time- and context-specific process. The second shortcoming is the focus upon procedural criteria only at the total neglect of context and outcomes. Nevertheless, fairness and competence are the most uncontested criteria put forward for the evaluation of participatory processes and are contained in almost all theory-based criteria sets found in the literature.

The application of theory-based criteria to stakeholder dialogues highlights some of the characteristics of this participation method in comparison to other tools used for participation. Stakeholder dialogues are by definition exclusive and do not grant permission to all those interested and willing to actually attend a stakeholder dialogue. The agenda of a stakeholder dialogue and the choice of a facilitator are often predetermined by those who initiate the dialogue. In my opinion, this failure to match the criteria does not mean that the criteria are not applicable to stakeholder dialogues. The opposite is the case: By applying this set of criteria, we are made aware of the limitations to fairness, which specific types of stakeholder dialogue impose upon the discourse. It is exactly the lack of openness to all willing participants that undermines the legitimacy of stakeholder dialogues in a system of formal governance based on representative democracy. Stakeholder dialogues need to address this issue in order to gain influence in the decision-making process.

Criteria sets which build on the ideal speech situation

Renn et al. (1999) have recently updated the above criteria set (fairness and competence) by adding the political criterion of legitimacy (formal, argumentative and integrative) and the economic criterion of efficiency (time-benefit, cost-benefit, long-term effects) as additional criteria for measuring the success of participatory processes.

A very similar set of criteria has been put forward by Susskind and Cruikshank (1981). Their proposed set of indicators includes most aspects of Renn et al's fairness, competence and efficiency. Instead of the fourth criterion of legitimacy, Susskind and Cruikshank use ‘stability’ to refer to the issue of how likely the implementation of the achieved outcomes is. On the basis of a systematic review of political theories of democracy, Peter H. Feindt (2001) has compiled a set of criteria which also closely resembles Renn et al’s fairness, competence, efficiency and legitimacy. Feindt has broken these criteria down for the planning phase of a participatory process, the facilitation of the event itself and for the follow-up phase. In each phase, fairness, competence, efficiency and legitimacy make specific demands with regards to best practice. For Feindt, the legitimacy of the outcomes also depends on a fair burden sharing under specific consideration of weak interests. Learning is highlighted as an integral part of competence.

Rowe and Frewer (2000) propose a set of nine theory-based criteria in order to evaluate desirable qualities of public participation methods. They distinguish between “acceptance criteria, which concern features of a method that make it acceptable to the wider public, and process criteria, which concern features of the process that are liable to ensure that it takes place in an effective manner” (Rowe and Frewer 2000: 3). Their acceptance criteria include representativeness (of the participants), independence (from control or influence of the sponsoring organisation), early involvement, influence (of the output on policy-making) and transparency (of the participation process). Their process criteria include resource accessibility (information, human, material and time resources), clear task definition, structured decision-making and cost-effectiveness of the procedure. In comparison with the other criteria sets introduced so far, the emphasis on independence from sponsor control and on transparency of the proceedings is striking. The application of Rowe and Frewer’s criteria to stakeholder dialogues highlights their tendency to an exclusive sampling of the stakeholder groups which leads to an elitist bias of the participants. It also highlights the danger of sponsor influence on a stakeholder dialogue and of lack of policy impact of the outcomes of a stakeholder dialogue (Rowe and Frewer 2000: 23).

Collaborative planning criteria

Habermas’ notion of the ideal speech situation has also inspired theory building in the field of planning. A review of the planning literature, known for its interdisciplinary perspective and closeness to local practice, shows the emergence of normative theories of ‘collaborative planning’ (Healey 1997, Fischer and Forester 1993, Innes 1996a, Selle 1996), a normative argument about how local governance in networks should ideally be pursued. Planning theories proved the most developed on the subject of evaluating stakeholder dialogues.

Planning theories are of interest to our issue of stakeholder dialogue as they are most experienced with making decisions in the face of conflicting interests. Bringing the conflicting parties and the wider stakeholders of a locality together in a constructive process of deliberation of the collective will has been the aim of collaborative planning theory. Collaborative planning theory breaks with the supremacy of science and with notions of consumers with fixed preferences. Theories of collaborative planning think highly of the citizens' capacity for learning and genuine public thinking if given a chance to deliberate. Conflicting evidence is to be discussed until a consensus emerges. This is supposed to be the opposite to bargaining between conflicting parties. According to collaborative planning theory, a decision can only be as legitimate as the process that willed it into being. A consensus becomes possible as citizens start to listen to each other and to alter own views in the light of their learning. Deliberative processes which are based on collaborative planning theory should ideally match the following process, outcome and capacity building criteria (Healey 1997, Innes 1996, 1998, Forester 1996a, b) and others:

Collaborative planning theory
(Source: my table on the basis of a literature review)

Process criteria:

- Diversity of stakeholders present
- Constructive dialogue
- Fair process
- Transcending egoistic preferences towards the common good
- Participants are experts on their affairs
- Allowing multiple ways of making validity claims
- Scope for innovation

Outcome criteria:

- A consensus

Capacity building criteria:

- New contacts and partnerships
- Learning amongst the participants
- Systems thinking
- Building trust and reviving local democracy
- Generating community spirit

4.3.2 User-based criteria

A second option for the evaluation of stakeholder dialogue is to base the evaluation on criteria for success as defined by those with a stake in the stakeholder dialogue. Instead of imposing a theoretically derived measurement, a user-based, or more specifially a stakeholder-based evaluation is interested to define criteria and indicators of success together with those carrying out the stakeholder dialogue, with those participating in it and those potentially affected by its outcomes. A stakeholder-based evaluation takes the aspirations of the organisers of a stakeholder dialogue as the starting point for the investigation, but then adds the aspirations of participants and bystanders alike. Even non-participants in a stakeholder dialogue may need to be interviewed in order to understand the political embeddedness of the stakeholder dialogue.

As the outcomes of most stakeholder dialogues are hard to predict, a stakeholder-based evaluation offers the chance to trace unintended and unexpected outcomes if the list of evaluation criteria is kept open until the end. The research question is no longer a narrow 'Did the process match the criteria?'. Instead, the opportunity is taken to ask: 'What was the impact of the stakeholder dialogue in the widest sense?' The evaluation process is no longer the undertaking of a knowledgeable researcher alone, but is instead redefined as a process of joint learning of researcher and those researched. The non-hierarchical approach addresses the stakeholders as experts in their own right as their experience of the stakeholder dialogue is valued. The general approach to evaluation has best been captured by naturalistic inquiry (Guba and Lincoln 1986).

Most instructive in this respect has been the social audit methodology as developed by the New Economics Foundation/London (Zadek and Raynard 1995, Zadek and Evans 1993). A Social Audit "is a means of assessing the social impact and ethical behaviour of an organisation or set of activities in relation to its aims and those of its stakeholders... Stakeholders are individuals and groups who are affected by, or can affect, the activities under review" (Zadek 1994: 632-633). The Social Audit is the most advanced of a number of tools that have been developed in the field of social and ethical accounting. The most frequently used alternatives to the Social Audit are the 'Ethical Accounting Statement' and the 'Social Assessment' (Zadek et al. 1997).

All three approaches involve a broad spectrum of stakeholders in a very participatory evaluation, carry out the accounting on a regular (usually annual) basis and publish the findings for public scrutiny (Zadek and Raynard 1995). The Social Audit as developed by the New Economics Foundation is moreover committed to target setting, systematic bookkeeping, external benchmarking, the establishment of an audit group and external verification of results. The other two approaches only commit to a few of these criteria. Since the Social Audit has been further adapted and successfully used by The Body Shop plc, Happy Computers, Shared Earth and several non-governmental organisations including the New Economics Foundation themselves, it is fair to say that it is the most advanced of the three approaches.

An example of an application of this approach to the area of stakeholder dialogue is my stakeholder-based evaluation of a Future Search Conference that was used to launch a Local Agenda 21 process (Oels 2003). The purpose of my stakeholder-based evaluation was to assess how successful a Future Search Conference was in delivering its stated objectives, perceived both before and after the conference event by those with a stake in it. A first step was therefore to identify the spectrum of stakeholders that should be involved in the evaluation. According to the Social Audit literature, stakeholders are all those core to the mission and values of an organisation/intervention, those who create and affect the organisation/intervention, and those most affected by it. The same stakeholders may appear in more than one category. Who qualifies as a stakeholder group and which individual should be asked to speak on behalf of that stakeholder group is of course contested and the evaluator needs to take precautions to minimize the resulting bias.

In my English case study Rushmoor Borough, the stakeholders to the evaluation were:

Table 4.3 Stakeholders to the evaluation in Rushmoor Borough

Those who affect the intervention	Those core to mission and values of the intervention	Those most affected by the intervention
– Rushmoor Future Search Conference steering group – clerical staff / LA21 officer – conference facilitators – LA21 subcommittee of councillors – Directors Management Board of Rushmoor Borough Council – LA21 officer steering group	– LA21 practitioners – Future Search practitioners – Rushmoor Future Search Conference steering group – LA21 officer – conference facilitators	– conference participants – the conference participants' organisations and sectors – Rushmoor Borough Council /administration – the wider local community in Rushmoor – the local media

The Social Audit approach recommends the merging into a single list of criteria put forward by all those with a stake in the intervention. In the process of doing so, the criteria put forward by those core to mission and values of the intervention are to be given more weight than the criteria put forward by those at the periphery. The major advantage of a single criteria list is that it makes life easier for the researcher and that it makes transparent to all the diverse objectives pursued at the conference event. The downside of merging all criteria into a single list is that it blurs the fact that the stakeholders in the evaluation have different interests and that these interests may be served unequally by the Future Search Conference. A Social Audit methodology directs attention away from a critical assessment of power relations, while emphasising the 'common ground', i.e. those objectives jointly pursued by all.

The resulting list of criteria for the evaluation of my two case studies (Oels 2003) is reproduced in table 4.4.

Table 4.4 Evaluation criteria and data sources generated in a stakeholder-based evaluation of a Future Search Conference in Rushmoor Borough Council, United Kingdom (1997-2000). Source: Oels 2003: 135-136.

PROCESS		
Audit Area	Criteria	Data Sources
inclusive	– broad spectrum of stakeholders present – many people who have not met before, not only the 'usual suspects'	– conference observation – FSC participant list over time – participant focus groups – participant interviews – participant questionnaire
collaborative	– participants able to put forward their heartfelt concerns – all views heard and respected – absence of domination, axe-grinding and polarisation – participants support each other	– conference observation – participant focus groups – participant interviews – participant questionnaire
competent	– participants treated as experts in their own right-required expertise is in the room – discussions go deeper than headline level – all local key issues are put on the table	– conference observation – conference documentation – participant focus groups – participant interviews – participant questionnaire

OUTCOMES		
Audit Area	Criteria	Data Sources
consensus about coherent, innovative vision	– the vision should be capable of guiding action – clear priorities are identified – new solutions to old problems identified	– conference observation – conference documentation – participant focus groups – focus groups with non-participants – stakeholder interviews – participant questionnaire
action groups deliver	– participants take responsibility for seeing their project ideas through – action plans are specific and practical – active Council support for at least some conference outcomes and action plans – visible change on the ground – action groups attract resources – regular progress review	– non-participant observation of action groups – participant questionnaire – stakeholder interviews – participant focus groups – document research – follow-up conference observation and documentation
effective outreach	– each participant gets their organisation and contacts involved in the FSC follow-through – extensive media coverage – some new people join the process – different form of consultation reaches out to the wider community	– participant questionnaire – participant focus groups – document research – stakeholder interviews – follow-up conference observation and documentation
Local Agenda 21 strengthened	– FSC is a demonstration of sustainable development; increased environmental awareness amongst participants – participants carry LA21 into their organisations – media coverage for LA21 – LA21 becomes true umbrella – LA21 gains more influence within the Council	– conference observation – conference documentation – participant focus groups – stakeholder interviews – follow-up conference observation and documentation

CAPACITY BUILDING		
Audit Area	Criteria	Data Sources
networking	– cross-sectoral action groups – new contacts formed across stakeholder group boundaries and valued – new joint projects / alliances set up	– conference documentation – conference observation – participant focus groups – participant interviews – participant questionnaire
learning	– participants genuinely engage with those holding opposite views – participants learn from and with each other – participants let go of prejudices and stereotypes – participants challenge each others' world views – participants recognise the systemic interdependence of their own and others' actions	– conference observation – participant focus groups – participant interviews – participant questionnaire
building trust and community spirit	– more trust between local authority and citizens – participants more optimistic, capable and willing to take on responsibility for local affairs – more things are done 'with' the people, not 'for' or 'to' them – participation methods like Future Search become a common practice locally – community spirit is generated	– stakeholder interviews – document research – non-participant focus groups – participant focus groups – participant questionnaire

The borders between theory-based and user-based criteria sets however are fluid. A set of evaluation criteria like mine (table 4.4) that has been generated in explorative case study work can now be applied as a measuring stick for a theory-based evaluation. Weber (2005) has already used an earlier version of my evaluation criteria for her theory-based evaluation of a Future Search Conference hosted to coordinate and improve youth support services in the Vogelsberg region near Frankfurt a.M., Germany. This may be justified for two reasons. First, anyone conducting another stakeholder-based evaluation is not unlikely to generate a similarly comprehensive list of evaluation criteria as mine, possibly with some minor additions and some minor gaps and some difference in weight of the criteria. Before investing a lot of explorative research work to reproduce a table similar to mine, the short cut may be to start with my table in hand. Of course, this is then theory-based evaluation and not a participatory stakeholder-based evaluation with all the benefits attached. A second reason for starting a theory-based evaluation with a subset or the total of the criteria raised in my table is its comprehensiveness. As the comparison of my stakeholder-based criteria set with the theory-based criteria sets in table 4.5 shows, my list includes all of the criteria raised under the labels of fairness, competence, legitimacy and effectiveness by those following Habermas. In addition, my criteria also match or summarise the criteria raised by collaborative planning theories. For those seeking a comprehensive approach to theory-based evaluation, my stakeholder-based criteria set can therefore be recommended as a starting point for the evaluation. However, as pointed out earlier, the weight given to each respective criterion should differ depending upon the type of stakeholder dialogue under review. While legitimacy is most important for stakeholder dialogues for policy-making, competence may be the hallmark of a stakeholder dialogue for science. Evaluation should be designed sensitive to the respective case, unless it is a comparative endeavor. The final word on this however is that the approach to evaluation must match the purpose of the evaluation. Explorative purposes justify a stakeholder-based evaluation, comparative purposes demand a theory-based evaluation. Now that the possible sets of criteria have been discussed at some length, this chapter will turn towards a review of the findings of those rare evaluation studies of stakeholder dialogues that have been completed and published.

Table 4.5 Comparative view of theory-based and stakeholder-based criteria sets.

Stakeholder-based criteria generated by Oels (2003)	Renn et al. (1999)	Collaborative planning theory
PROCESS		
Inclusive	Fair	– Diversity of stakeholders present – Fair process
Collaborative	Fair	– Constructive dialogue – Transcending egoistic preferences towards the common good
Competent	Competent	– Participants are experts on their affairs – Allowing multiple ways of making validity claims
OUTCOME		
Consensus about coherent, innovative vision	Effectiveness	– A consensus – Scope for innovation
Action groups deliver	Effectiveness (?)	/
Effective outreach	Legitimacy (?)	/
Local Agenda 21 strengthened	/	/
CAPACITY BUILDING		
Networking	/	– New contacts and partnerships
Learning	/	– Learning amongst the participants – Systems thinking
Building trust and community spirit	Legitimacy	– Building trust and reviving local democracy – Generating community spirit

4.4 Common findings of evaluations

There are many stories reporting successes and failures of stakeholder dialogues in the literature. However, the criteria used for evaluation are rarely made explicit and the evidence is often anecdotal. The aim of this section is to provide more background on the relative importance of each of the evaluation criteria on the basis of the findings from case studies. The overall impression from the case study literature is that stakeholder dialogues are very successful at establishing fair and competent processes and thereby at generating capacity building benefits. The main point of failure however is the production of outcomes resulting from stakeholder dialogues, thereby leading to a lack of effectiveness. Let us review the evidence in turn.

The evaluation literature presents evidence that stakeholders are capable of agreeing upon a shared knowledge base, an action plan or a management plan if given suitable conditions for dialogue. Stakeholder dialogues have been reported to make a substantial contribution to capacity building. The evaluation literature carries plenty of evidence that the participants of stakeholder dialogues learn from each other and engage in networking with each other. Judith Innes and her team analysed fourteen stakeholder dialogues in California, each of which sought stakeholder consensus on the future of growth and environmental policy. Out of fourteen cases, eight were classified as suitable examples of the 'new planning paradigm'. For these cases, Innes presents evidence that "The stakeholders in all cases became better informed through the process, and valued and used their new personal and professional networks to coordinate and collaborate. In five of the cases, groups incorporated systematic technical analysis into their deliberations. The breadth of the collective knowledge and interests of group members and the lengthy periods for discussion meant that they explored a wide range of factors and their interrelationships." (Innes 1996a: 465) Similar evidence comes from Amy Helling, who found that when participants in Atlanta's Vision 2020 process were asked to list the accomplishments of the process, "nearly all pertained to the collaborative process itself, most frequently mentioning networking among diverse people concerned about similar issues. Stakeholders also said that the connections made through VISION 2020 had extended beyond the VISION 2020 meetings themselves, and that they had involved people who had not been active before, or brought together people who had not previously met." (Helling 1998: 340)

I conclude that learning and networking are likely results of stakeholder dialogue for policy-makings. The most striking finding of the evaluation literature, however, is the lack of implementation of the outcomes of citizen participation in general and stakeholder dialogues in particular. The following section reviews this failure to deliver for each of the three types of stakeholder dialogues in turn.

4.4.1 Stakeholder dialogues for science

This section reviews evidence in the area of participation processes for the improvement of the knowledge base. Quite a few risk and technology assessments which are conducted with citizen participation fail to influence policy-making. They are mostly ignored by the institutions of representative government that could consider their outcomes when making legally binding decisions.

The first UK 'National Consensus Conference on Plant Biotechnology' which was hosted by the Science Museum and the Biotechnology and Biological Sciences Research Council (BBSRC) in 1994 in London to allow for a citizen assessment of plant biotechnology, has been evaluated by Robin Grove-White and colleagues (1997: 28) as "something of a political cul-de-sac, principally because it was not thought possible to link its findings into other statutory or Parliamentary processes, or to be more systematically diffused. By contrast, Consensus Conferences in Denmark and the Netherlands (on which features of the UK initiative were modelled) have a statutory basis and have already helped shape public policy towards biotechnology and other ethically contentious issues." There is no lack of evidence that reports from similar events gather dust on shelves instead of influencing policy-making.

Welp et al. (in press) have pointed out two ways in which stakeholder dialogues for science could benefit policy-makers. First, they can inform policy-makers how lay people think about complex environmental issues. Second, they can provide feedback on the acceptance of planned policies ('reality check'). I would add as a third point that they can provide policy-makers with new ideas and proposals for policy-making. It remains unresolved though, how policy makers can best be involved in and informed about the outcomes of stakeholder dialogues for science. Welp et al. (in press) suggest as a first step to make policy-makers more aware of the benefits. Secondly, they recommend using the media to distribute the learning and to form opinion. Welp et al. (in press) however resist the demand made by Rowe and Frewer (2000) and many others that the outcomes of stakeholder dialogues should be given legally binding status. They emphasize that political support is much more important for a stakeholder dialogue to gain influence than legal status. They also follow O'Riordan (1998) in arguing that formal decision-making authority should remain with the institutions of representative democracy.

4.4.2 Stakeholder dialogues for policy-making

In the area of stakeholder dialogues for policy-making, the evidence is similar. Amy Helling's evaluation of Atlanta's VISION 2020 process can be summed up in the words of one respondent to her survey: "This process has given the false impression to the public that something is being done, when in fact, all that has resulted … has been the agreement that we need to continue to have more meetings." (Helling 1998: 343) Indeed, Helling found little evidence for progress on the initiatives that originated from the VISION 2020 process "beyond extending desirable networking by continuing to gather people for discussions, meetings, and presentations" (Helling 1998: 342). Out of 41 projects which were initiated by VISION 2020, only eleven were taken forward at all, and out of these, many had achieved no more than to continue to meet. The prospects for the future were not good either, as many interviewees "said they were looking forward to reducing their commitment, but they were nearly unanimous in saying that the most important part of the process was still ahead" (Helling 1998: 342). A major disappointment had also been that "most of the region's political leaders maintained their distance, and many stakeholders complained of their lack of attention." (Helling 1998: 343) Even worse, Atlanta Regional Commission ignored the controversial debates that were led as part of VISION 2020 about Georgia Department of Transportation's plan to build a second, limited-access perimeter highway around Atlanta and simply voted to support this plan (Helling 1998: 343). Helling's evaluation does not forget to mention the tremendous costs of the VISION 2020 exercise, particularly when the volunteer person-hours are added up (total of 25,000) and to contrast them with the failure to deliver.

The key role of the local authority is further illustrated by Penny Street (1997), who reported from a Scenario Workshop used in the UK town of Preston for involving the public in policy formulation on urban sustainable development. As in the case studies presented earlier on, the workshop fell short of enabling participants to make a real input to policy-making. Street identifies the danger that high expectations have been raised while "there was no clear way for participants to take this initiative forward; it was dependent on the Council itself to take action...it is difficult to see how such a range of issues could be dealt with simultaneously and effectively" (Street 1997: 154).

Steelman and Ascher (1997) have argued that while more and more policies require government agencies to provide for public participation, there is a complete lack of clarity about how to obtain public input into decision-making and "how much weight these inputs should be given" (ibid: 72). Left to the discretion of government officials, the scope for manipulation is considerable. Steelman and Ascher (1997) therefore argue for binding forms of direct policy-making by non-governmental representatives, which avoid the polarisation and simplification associated with (legally binding) referenda while keeping the benefits of more explorative proceedings. Hoggett (1995) - with reference to Arnstein's ladder of participation - warns local authorities that "building a ladder of participation" is not "something one can bolt on to or lean against the otherwise unchanged structure of the local authority. Every step up the ladder towards genuine citizen empowerment requires an equivalent change in mainstream practices" (1995: 109). All scholars agree that establishing effective citizen participation requires "the transformation of structures that inhibit collective decision-making" (Kearns 1995: 171).

Contrary to that, Judith Innes has found evidence in some of her case studies that the process of mobilising the stakeholders to an issue was sufficient in itself to then develop the political clout to force through the conclusions of the participatory process "even without support from high elected officials" (Innes 1996a: 468) and without any binding mandate. While this possibility of a conflictive strategy always remains, it looks more like a lucky escape from a situation to be avoided in the first place.

Examples of cases which have been more directly linked with decision-making processes, come from the literature on conflict mediation (Susskind and Cruikshank 1987, Moore 1987, Carpenter and Kennedy 1991, Zilleßen 1998). It is under the weight of high financial (or other) stakes of parties to a multi-party dispute, that the fair and competent exploration of contested issues unfolds its full potential under the strict guidance of a skilled mediator between parties otherwise unable to communicate (Baughman 1995, Nothdurft 1995). As Baughman (1995: 264) has pointed out, parties to a mediation exercise (should and usually do) participate in "full awareness of their best alternative to a negotiated agreement".

4.4.3 Stakeholder dialogues for management

The advantage of stakeholder dialogues for management is, that they usually come in after a policy decision has been taken. They are by definition the policy implementation or at least part of the implementation. Stakeholder dialogues for management have been successfully used in catchment and watershed management, forest management, water management, integrated pest management, wildlife management, farmers research groups and micro-finance delivery (Pretty and Ward 2001). The exception are feasibility and pilot studies which explore the potential of stakeholder dialogues for natural resource management in a specific country, region or nature park. These later ones may once more fall into the trap of raising local people's expectations without then delivering results (see Averbeck 2006 in this book).

The major challenge of stakeholder dialogues for management is to get the right people involved and committed and to secure the resources to sustain such a process. Once this is achieved, there remain much lesser problems. In his review of group-based programmes and initiatives for biodiversity enhancement from industrialised and developing countries, which produced favourable outcomes, Pretty (2002) points towards the following three limitations with regards to the implementation of the evaluated schemes. First of all, the positive effects for biodiversity were often rather small in scope, sometimes limited to the individual farm area. Second, the participation of local people was in several cases not sufficient. This was the case where the programme was voluntary, the incentives not high enough or the infrastructure of spreading information not effective enough. Finally, the implementation was in some cases directly linked to a subsidy scheme, the termination of which threatened the sustainability of the programme.

4.4.4 Criteria for success

The review of literature on case study findings for all three types of stakeholder dialogue has shown that while stakeholder dialogues for science and for policy-making share the problem of producing any changes on the ground, stakeholder dialogues for management may produce changes that are too negligible to matter. These observations from evaluation studies help us to define conditions for the success of stakeholder dialogues. On the basis of her fourteen case studies, Judith Innes (1996) has come up with three conditions for the success of stakeholder dialogues for policy-making:

- a pressing need to come to an agreement / high incentive to participate (i.e. high costs of delay / inaction / imposed solution);
- deliberative process must lead to a clearly defined product (i.e. agreed problem definitions, legislation, clear targets and timetables);
- substantial elements of this product must be formally adopted by the relevant formal political authority.

It is important to note that stakeholder dialogues are mostly carried out without a formal mandate by the elected governmental authorities, they are therefore 'informal' processes outside the sphere of the formal institutions of government. O'Riordan (1998a: 1) not only argues that the formal structures of governance should be "widening their scope for sharing power". He also reminds us that formal and informal institutions are interdependent, and therefore a process of co-evolution of formal and informal governance is required. He argues that the proponents of participatory forms of decision-making are well advised to remain sensitive to the issue of accountability that, according to him, only elected representatives can offer, and the need for transparency of their proceedings. Otherwise, he argues, participatory decision-making processes might in their ignorance reduce the scope for democratic decision-making instead of widening it. I think, O'Riordan rightly criticises the fanatic enthusiasm of many proponents of participatory tools who remain unaware of the consequences of their actions with regards to issues of power and democratic accountability. Nevertheless, without an inroads into formal government structures, stakeholder dialogues are bound to remain ineffective with regards to achieving a policy impact. One practical implication of this is that the organisers of stakeholder dialogues should contact the respective governmental authority as early as possible and foster their support and active involvement. This may enhance the chances for later implementation of the outcome.

4.5 Conclusions

As the popularity of stakeholder dialogues rises, so does the need for evaluation and shared quality standards. The literature on evaluating stakeholder dialogue is growing, but no common standards and criteria for the evaluation have arisen yet. Theory-based evaluations are most often grounded in the Habermasian ideal speech situation and highlight the need for fair and competent proceedings. More recently, efficiency and legitimacy have been added to the list of theory-based evaluation criteria. Stakeholder-based evaluations develop criteria and indicators of success jointly with those with a stake in the participatory process. They are based upon a joint learning process of researcher and stakeholders. Theory-based and stakeholder-based evaluations are not mutually exclusive. In fact, a combination of both may deliver best results as theory-based criteria allow a comparison of many cases and stakeholder-based criteria do justice to the objectives as defined by those involved in the process. This would also help to bridge the gap that divides the theory-based evaluation literature from the stakeholder-based evaluation literature.

The lack of implementation of the outcomes of stakeholder dialogues for science and stakeholder dialogue for policy-making has been identified as the most common failure of stakeholder dialogues. As the user-based evaluation criteria have highlighted, stakeholder dialogues do have a tendency of raising the expectation that something will be done as a result of the process. If nothing happens, stakeholders may be frustrated and unwilling to participate in future stakeholder dialogues. Therefore, the lack of implementation in stakeholder dialogues for science and for policy-making is a serious issue that requires attention. The practice of stakeholder dialogues needs to pay a lot more attention to the power relations between formal structures of government and informal stakeholder dialogues. The interface needs to be improved. Mechanisms need to be explored which provide easier inroads for the outcomes of stakeholder dialogue into the formal decision-making structures of government. A pressing need to come to an agreement, a clearly defined product as outcome and a governmental commitment to formally decide about the adoption of this product are three key conditions for the success of stakeholder dialogues.

References

Agnew, C., Burgess, J., Collins, K., Dando, N., Harrison, C., Hobson, K., Munton, R., Murlis, J. (1999). An analytical and descriptive model of sustainable development. A report prepared for the Environment Agency as part of the national Research and Development Programme, 1998-1999. ESRU/UCL, London

Averbeck, C. (2006). Incorporating Local People through Economic Incentives at Lake Mburo National Park, Uganda – Africa works! In: Stoll-Kleemann, S., Welp, M. (eds). Stakeholder Dialogues in Natural Resources Management. Springer-Verlag, Heidelberg

Barber, B.R. (1984). Strong democracy. University of California Press, London

Baughman, M. (1995). Mediation. In: Renn, O., Webler, T., Wiedemann, P. (eds). Fairness and competence in citizen participation. Evaluating models for environmental discourse. Kluwer Academic Publishers, Dordrecht, The Netherlands, 253-265

Burgess, J. (1995). Growing in confidence: understanding people's perceptions of urban fringe woodlands. Countryside Commission Technical Report CCP 457. Countryside Commission, Cheltenham

Burgess, J. (1996). Focusing on fear: the use of focus groups in a project for the Community Forest Unit. Countryside Commission Area 28 (2), 130-135

Burgess, J., Harrison, C.M., Limb, M. (1988c). People, parks and the urban green: a study of popular meanings and values for open spaces in the city. Urban Studies 25, 455-473

Carpenter, S.L., Kennedy, W.J.D. (1991). Managing public disputes: a practical guide to handling conflict and reaching agreements. Jossey-Bass Management Series. Jossey-Bass, San Francisco, CA

Chess, C. (2000). Evaluating Environmental Public Participation: Methodological Questions. Journal of Environmental Planning and Management 43 (6), 769-784

Durant, J. (1995). An experiment in democracy. In: Joss, S., Durant, J. (eds). Public participation in science. The role of consensus conferences in Europe. Science Museum with the support of the European Commission Directorate General XII, London, 75-80

Eijndhoven, J.v., Wynne, B. et al. (2001). Evaluation in the Mangement of Global Environmental Risks. In: The Social Learning Group: Learning to Manage Global Environmental Risks Vol. 2. MIT Press Cambridge, Massachusetts, USA

Feindt, P. (2001). Regierung durch Diskussion? Diskurs- und Verhandlungsverfahren im Kontext von Demokratietheorie und Steuerungsdiskussion. Peter Lang Verlag, Frankfurt

Fetterman, D.M. (1996). Empowerment evaluation: an introduction to theory and practice. In: Fetterman, D.M., Kafterian, S.J., Wandersman, A. (eds).

Empowerment Evaluation: Knowledge and Tools for Self-assessment and Responsibility. Sage, Thousand Oaks, CA

Fiorino, D.J. (1990). Citizen participation and enviornmental risk: a survey of institutional mechanisms. Science, Technology, and Human Values (15), 226-243

Fischer, F. (1993). Citizen participation and the democratization of policy expertise: From theoretical inquiry to practical cases. Policy Sciences 26, 165-187

Fischer, F., Forester, J. (1993). The argumentative turn in policy analysis and planning. Duke University Press, Durham, NC

Flyvbjerg, B. (1998). Rationality and power: democracy in practice. University of Chicago Press, London

Forester, J. (1989). Planning in the face of power. University of California Press, Berkeley, CA

Forester, J. (1993). Critical theory, public policy and planning practice. State University of New York Press, Albany, NY

Forester, J. (1996a). The rationality of listening, emotional sensitivity, and moral vision. In: Mandelbaum, S.J., Mazza, L., Burchell, R.W. (eds). Explorations in planning theory. Centre for Urban Policy Research, Rutgers. The State University of New Jersey, New Brunswick, 204-224

Forester, J. (1996b). Argument, power, and passion in planning practice. In: Mandelbaum, S.J., Mazza, L., Burchell, R.W. (eds). Explorations in planning theory. Centre for Urban Policy Research, Rutgers. The State University of New Jersey, New Brunswick, 241-262

Grove-White, R., Macnaghten, P., Mayer, S., Wynne, B. (1997). Uncertain world. Genetically modified organisms, food and public attitudes in Britain. Centre for the Study of Environmental Change, Lancaster

Guba, E., Lincoln, Y. (1989). Fourth Generation Evaluation. Sage, Newbury Park, CA

Harrison, C.M., Burgess, J. (1994). Social constructions of nature: a case study of the conflicts over Rainham Marshes SSSI. Transactions, Institute of British Geographers (19), 291-310

Healey, P. (1997). Collaborative planning. Shaping places in fragmented societies. Macmillan Press, Basingstoke

Helling, A. (1998). Collaborative visioning: Proceed with caution! Results from Atlanta's Vision 2020 Project. Journal of the American Planning Association 64 (3), 335-349

Hockings, M., Stolton, S., Dudley, N. (2000). Evaluating effectiveness: A Framework for Assessing the Management of Protected Areas. International Union for the Conservation of Nature and Natural Resources (IUCN). Gland, Switzerland and Cambridge, UK

Hoggett, P. (1995). Does local government want local democracy? Town and Country Planning April 1995, 107-109

Hoy, D.C. (1986). Power, repression, progress: Foucault, Lukes, and the Frankfurt School. In: Hoy, D.C. (ed). Foucault: a critical reader. Basil Blackwell, Oxford, 123-148

Innes, J.E. (1996a). Planning through consensus building. A new view of the comprehensive planning ideal. Journal of the American Planning Association 62 (4), 460-472

Innes, J.E. (1996b). Group processes and the social construction of growth management: Florida, Vermont, and New Jersey. In: Mandelbaum, S.J., Mazza, L., Burchell, R.W. (eds). Explorations in planning theory. Centre for Urban Policy Research, Rutgers. The State University of New Jersey, New Brunswick, 164-187

Innes, J.E. (1998). Information in communicative planning. Journal of the American Planning Association 64 (1), 52-63

Innes, J.E., Booher, D.E. (1999). Consensus building as role playing and bricolage. Toward a theory of collaborative planning. Journal of the American Planning Association 65 (1), 9-26

Innes, J.E., Gruber, J., Neuman, M., Thompson, R. (1994). Coordination growth and environmental management through consensus building. CPS Brief 6 (4). California Policy Seminar, Berkeley, CA

Joss, S. (1995). Evaluating consensus conferences: Necessity or luxury? In: Joss, S., Durant, J. (eds). Public participation in science. The role of consensus conferences in Europe. Science Museum with the support of the European Commission Directorate General XII, London, 89-108

Kearns, A. (1995). Active citizenship and local governance: Political and geographical dimensions. Political Geography 14 (2), 155-175

Kuper, R. (1997). Deliberating waste: The Hertfordshire citizens' jury. Local Environment 2 (2), 139-153

Lauber, T.B., Knuth, B.A. (1997). Fairness in moose mangement decision-making: the citizens' perspective. Wildlife Society Bulletin 25 (4), 776-787

Lincoln, Y.S., Guba, E.G. (1986). But is it rigorous? Trustworthiness and authenticity in naturalistic evaluation. In: Williams, D.D. (ed). Naturalistic evaluation. New directions for program evaluation No. 30. Jossey-Bass, San Francisco, 73-84

Macnaghten, P., Grove-White, R., Jacobs, M., Wynne, B. (1995). Public perceptions and sustainability: indicators, institutions, participation. Lancashire County Planning Department, Preston

Moore, C. (1987). The mediation process: Practical strategies for resolving conflict. 1st edition. Jossey-Bass, San Francisco

Mumpower, J.L. (2001). Selecting and evaluating tools and methods for public participation. International Journal of Technology, Policy and Management 1 (1), 66-77

Nothdurft, W. (1995). Environmental mediation: insights into the microcosm and outlooks for political implications. In: Renn, O., Webler, T., Wiedemann, P. (eds). Fairness and competence in citizen participation. Evaluating models for environmental discourse. Kluwer Academic Publishers, Dordrecht, The Netherlands, 267-282

Oakeshott, M. (1962). Rationalism in politics. Basic Books, New York

Oels, A. (2000). 'Let's get together and feel alright!' Eine kritische Untersuchung von 'Agenda 21'-Prozessen in England und Deutschland. In: Heinelt, H.,

Mühlich, E. (eds). Lokale ‚Agenda 21'-Prozesse. Leske und Budrich, Opladen, 182-200

Oels, A. (2002). Investigating the emotional roller-coaster ride: A case-study-based assessment of the Future Search Conference Design. Systems Research and Behavioral Science. Special Issue: Participatory Planning and Designing. John Wiley & Sons Ltd

Oels, A. (2003). Evaluating stakeholder participation in the transition to sustainable development. Methodology, case studies, policy implications. LIT-Verlag, Münster

Oppermann, B., Langer, K. (2002). Die Qualität partizipativer und kooperativer Projekte in der Technikfolgenabschätzung. Arbeitsbericht der Akademie für Technikfolgenabschätzung Nr. 226, Dezember 2002. Akademie für Technikfolgenabschätzung in Baden-Württemberg, Stuttgart

O'Riordan, T., Voisey, H.D. (1998a). The political economy of the sustainability transition. In: O'Riordan, T., Voisey, H.D. (eds). The transition to sustainability. The politics of Agenda 21 in Europe. Earthscan, London, 3-30

Petts, J. (1997). The public-expert interface in local waste management decisions: Expertise, credibility and process. Public Understanding of Science 6, 359-381

Polanyi, M.F.D. (2002). Communicative action in practice: Future Search and the pursuit of an open, critical and non-coercive large group process. Systems Research and Behavioral Science 19 (4), 357-366

Pomeroy, R.S., Parks, J.E., Watson, L.M. (2003). How is your MPA doing? A Guidebook of Natural and Social Indicators for Evaluating Marine Protected Area Management Effectiveness. International Union for Conservation of Nature and Natural Resources. Gland, Switzerland and Cambridge, UK

Pretty, J. (2002). People, livelihoods and collective action in biodiversity management. In: O'Riordan, T., Stoll-Kleemann, S. (eds). Biodiversity, Sustainability and Human Communities. Protecting beyond the Protected. Cambridge University Press, Cambridge

Pretty, J., Ward, H. (2001). Social capital and the environment. World Development 29 (2), 209-227

Renn, O., Webler, T., Wiedemann, P. (1995). Fairness and competence in citizen participation. Evaluating models for environmental discourse. Kluwer Academic Publishers, Dordrecht, The Netherlands

Rocha, E.M. (1997). A ladder of empowerment. Journal of Planning Education and Research 17, 31-44

Rossi, J. (1997). Participation run amok: The costs of mass particiation for deliberative agency decisionmaking. Northwestern University Law Review 92 (1), 173-249

Rowe, G., Frewer, J.L. (2000). Public Participation Methods: A Framework for Evaluation. Science, Technology and Human Values 25 (1), 3-29

Schattschneider, E.E. (1960). The semisovereign people. Holt, Rinehart and Winston, New York

Selle, K. (1996). Planung und Kommunikation. Gestaltung von Planungsprozessen in Quartier, Stadt und Landschaft. Grundlagen, Methoden, Praxiserfahrungen. Bauverlag, Wiesbaden, Berlin

Steelman, T.A., Ascher, W. (1997). Public involvement methods in natural resource policy making: advantages, disadvantages and trade-offs. Policy Sciences 30, 71-90

Street, P. (1997). Scenario workshops: A participatory approach to sustainable urban living? Futures 29 (2), 139-158

Susskind, L., Cruikshank, J. (1987). Breaking the impasse: Consensual approaches to resolving public disputes. Basic Books, New York

Susskind, L., McKearnen, S., Thomas-Larmer, J. (1999). The consensus building handbook: A comprehensive guide to reaching agreement. Sage, Thousand Oaks, CA

Weber, S.M. (2005). Rituale der Transformation. Großgruppenverfahren als pädagogisches Wissen am Markt. VS Verlag für Sozialwissenschaften, Wiesbaden

Webler, T. (1995). 'Right' discourse in citizen participation: An evaluative yardstick. In: Renn, O., Webler, T., Wiedemann, P. (eds). Fairness and competence in citizen participation. Evaluating models for environmental discourse. Kluwer Academic Publishers, Dordrecht, The Netherlands, 35-86

Welp, M., Kasemir, B., Jaeger, C. C. (in press). Citizens' voices in environmental policy: The contribution of integrated assessment focus groups to accountable decision-making. In: Coenen, F.H.J.M., Paterson, R. (eds). The Promise and Limits of Participatory Processes for the Quality of Environmentally Related Decision Making. Springer-Verlag, Heidelberg

Wilcox, D. (1994). The guide to effective participation. Delta Press, Brighton

Zadek, S. (1994). Trading ethics: Auditing the market. Journal of Economic Issues 28 (2), 631-645

Zadek, S., Evans, R. (1993). Auditing the market: A practical approach to social auditing. Traidcraft/New Economics Foundation, Gateshead

Zadek, S., Raynard, P. (1995). Accounting works: A comparative review of contemporary approaches to social and ethical accounting. Accounting Forum 19 (2/3), 164-175

Zadek, S., Pruznan, P., Evans, R. (1997). Building corporate accountability: Emerging practices in social and ethical accounting, auditing and reporting. Earthscan, London

Zilleßen, H. (ed.) (1998). Mediation, Kooperatives Konfliktmanagement in der Umweltpolitik. Westdeutscher Verlag, Opladen

5 Tools for Stakeholder Assessment and Interaction

Jürgen Scheffran

University of Illinois at Urbana-Champaign, USA

5.1 Introduction

Stakeholders play an increasing role in environmental policy and are key actors in the supposed sustainability transition (O`Riordan and Stoll-Kleemann 2002). Characterising their role as well as the transition itself requires a deeper understanding of the interconnectedness of social networks and the relationship between individual and collective action. In many models, used in integrated assessment, a top-down rational actor drives the world towards a desired direction, optimising a global welfare function, notwithstanding the fact that the development of the real world is affected by numerous actors who act according to their own interests and capabilities. Sustainable development strategies often fail due to their inability to adequately take social interaction into consideration, such as dialogues, negotiations, conflicts, coalition formation and institution building. Usually collective action is seen as a problem in common pool resources (such as fishery), where unregulated resource access by a group of individuals leads to resource depletion and the violation of common interests, as can be observed in the so-called "Tragedy-of-the-Commons" (Hardin 1968, Ostrom 1990).

While such conflicts of individual vs. collective interests can be observed in many social dilemma situations, there was less attention to the possibility that joint action could rather preserve natural resources and support a sustainability transition if individuals cooperate and coordinate their actions. A key question is under which conditions one or the other form of interaction occurs and how it can be realised. The answer depends on the rules of communication and the institutional settings, as well as on the tools available during, before and after the interaction. Such tools, designed as simple or comprehensive models, and established as rules, procedures or programs, could deepen the understanding of the phenomena and manage their complexities and uncertainties. They are an issue in interactive decision-making and communication processes, such as stakeholder dialogues and negotiations, in support of sustainable development.

This article describes some of the modelling and computational tools that can play a role in complementing and supporting dialogues, including mathematical models, negotiation analysis, mediation, experimental games, agent-based models and participatory integrated assessment. Properly designed, these tools can be used as instruments during real-world dialogue or decision-making, with stakeholders being active users or objects of study. By making critical information accessible and facilitating communication in dialogue and negotiations, by structuring complex problems and providing practical models in support of decision-making and risk analysis, such tools would support stakeholders in assessing problems and expressing their views more explicitly. As educational and research tools they can improve the understanding of the issues at stake. To illustrate the use of some of the tools in environmental policy, applications in climate, fishery and water management are mentioned.

5.2 Stakeholder involvement in interactive decision-making

In a broad sense, stakeholders can be considered as those individuals or groups that have an interest or concern in a particular issue. There are a variety of potential stakeholders who can be governmental or non-governmental, pursue their individual or group interests, act on local, national or global scales. Dialogues or interactive decision-making are an opportunity to bring the diversity of stakeholders together for the discussion or resolution of burning societal problems. Such processes can be quite complex, involving multiple stakeholders, issues, interests, disciplines and levels of decision-making. Stakeholder dialogues empower the parties involved and seek to reconcile and integrate divergent interests to reach agreement or consensus.

In environmental assessment and management, stakeholders participate in developing sustainable investment strategies, define work plans for implementation and finally push for and monitor the planning process. Ideally, stakeholder assessment would take into consideration the interdependence between stakeholders and the environment in which they interact, including the institutional frameworks, as characterised by rules and strategies that are embodied in regularised patterns of behaviour or procedures for conflict resolution (Priscoli 1989). With the increasing involvement of stakeholders in social and political processes, decision-making would become more interactive and complex, demanding new types of procedures and tools to manage it. While agreement would be a preferred solution in many negotiations, not all dialogues aim at a consensus. Science-based stakeholder dialogues also seek to explore different arguments and viewpoints. Knowing where the stakeholders disagree is a valuable result too and guides future research.

While a main characteristic of stakeholder dialogues is communication and exchange, not necessarily with a pre-assigned goal, interactive decision-making and negotiation (IDN) implies activities in groups of stakeholders that seek an agreement, either as a whole group, as sub-groups or individually. Generally, IDN is relevant in science, policy and management, dealing with critical issues of technical change and its impact on society, environmental management, and position finding in democratic decision-making processes that require a "critical mass" of supporters. In a specific sense, IDN refers to activities "in which a government involves citizens, societal organisations, private parties and/or other governments into the decision-making-process as soon as possible, in order to interact and/or to co-operate with them to establish the preparation, the determination, the implementation and/or the evaluation of policy" (van der Veen 1999). In this context, IDN closes the gap between the government and its citizens and enlarges the support for its policy and decisions, increasing legitimacy and responsibility and improving the chances for problem solving.

On an organisational level, the concept of "participatory governance" is complementary to hierarchical governance, which is mainly organised at the governmental macro level. At the meso and micro level citizens, together with industries and governments, debate and negotiate public issues and policies. Institutional structures facilitate communication, co-management and the sharing of responsibility. The participant composition and decision rules are crucial for participatory governance as they influence the internal power structure.[1]

IDN involves societal resources, such as time, money and knowledge, to improve the quality of decision-making and create a surplus for its participants (van der Veen 1999). Stakeholder participation is an opportunity for individuals to influence public decisions and shape the policy process into a direction that meets their interests. Citizens can develop their capabilities and creativity, and bring about new innovative ideas for a more direct democracy. Governments, on the other hand, can demonstrate openness and transparency, enlarge their legitimacy and improve their public image. All actors can reduce prejudices about each

---

[1] On participatory governance in a multi-level context see Heinelt et al. 2002.

other, realise a synergy and surplus value (win-win). IDN can be used to locate potential conflicts in an early stage and make the interaction less complex.

Other factors may impede the implementation of interactive decision-making. If a government is involved in a decision-making process (e.g. on national legislation or regionally in the construction of a new airport), the sharing of information and knowledge with citizens may be seen as a loss of power and control. Because of self-interest, inadequate representation, and selective participation, the democratic legitimacy of IDN can be questioned. Interactive processes can cost considerable time and efforts, without producing a reasonable or justifiable outcome. The need to reduce time and effort may diminish the efficiency and effectiveness of the whole process. For difficult and complex issues, or too many stakeholders involved, participants may feel incompetent or overwhelmed by the process. In case of severe disagreement, conflict or inadequate management, there is the risk of complete failure, i.e. the interactive process will not yield any result. In case of success, however, participation can strengthen cooperation, conflict resolution and democracy.

Whether stakeholder dialogues are perceived as successful, depends on the selected evaluation criteria (cf. Chapter 4 Evaluating stakeholder dialogues). Even if no agreement can be achieved, stakeholder dialogues may produce net benefits, simply by the information shared. Successful stakeholder dialogue and interaction is a way of increasing the social capital and thus the productivity of society, in addition to natural, physical and human capital as used in economic theory. Social capital refers to the institutions, relationships, and norms that shape the quality and quantity of a society's interactions. These are embodied in the values, habits, and relations among persons, strengthening social cohesion and the formation of social networks.[2] Integrated sustainability strategies aim at strengthening the links between social and natural capital.

[2] See Putnam (1995). Coleman (1990) describes it as 'the structure of relations between actors and among actors' that encourages productive activities.

5.3 Tools in Stakeholder Interaction and Modelling

5.3.1 The stakeholder concept in management and systems science

Whether stakeholder interaction is a success or a failure may depend on the tools used during the process, which can influence the amount and efficiency of the resources used. From this viewpoint, stakeholder analysis is an issue for the behavioral, organisation and management sciences. Since its first appearance in an international memorandum at the Stanford Research Institute in the early 1960s, the stakeholder concept has found increasing attention in the management literature, first of all with regard to business responsibility (Elias et al. 2000). In a first definition, stakeholders were defined as essential players for an organisation as 'those groups without whose support the organisation would cease to exist' (cited in Freeman 1984).

In the 1970s, the concept diversified into a wide range of fields, including organisation theory, systems theory, corporate planning and social responsibility. The systems model of stakeholders emphasised participation and argued that problems should not just be defined by focussing or analysing, but by enlarging or synthesising, addressing social issues from an open systems point of view. Ackoff (1974) argued that participation of stakeholders is essential for systems design, as they would help in solving societal problems. Many researchers were concerned with the social responsibility of business firms, including nontraditional stakeholders who were having adversarial relationships with the firm.

Since the landmark book by Freeman (1984), the stakeholder concept has become increasingly embedded in the thinking of managers and decision-makers. The literature focused on descriptive and empirical aspects, as well on instrumental and normative aspects which were integrated into the stakeholder theory of corporations. More recent issues are the dynamics of stakeholders as well as stakeholder theories and their validation. These trends indicate the growing complexity of the field, as the mix of stakeholders and their attributes may change, such as responsibility, power, legitimacy and urgency (Mitchell et al. 1997).

Increasingly the stakeholder process and its organisation is subject to analysis in management science. This concerns the different phases in the interaction cycle, including communication, decision-making, negotiation and joint action, as well as the framing conditions, under which it will operate, such as timing, budgeting and staffing (see last section of this article). The procedural conditions determine the rules of the game with regard to the style of exchange and its facilitation, as well as conditions for entrance and withdrawal. The roles, functions and responsibilities of the different participants are essential to maintain support for the process.

5.3.2 Stakeholder modelling and simulation

The analysis of stakeholder interaction is linked to social systems modelling and simulation, and much of the methodology in this area can be applied. Analysing the emergence of collective action from individual action is a dynamic field of current interdisciplinary research, combining natural and social sciences. Modelling and computer simulation can contribute to a deeper understanding and the development of new instruments for decision-making in complex environments. There is a wide range of formal methods and models that are relevant in this context:

1. *Computer simulation and dynamic systems theory* study the time evolution of trajectories in state space, in many cases driven by a set of differential or difference equations that describe local change for given initial conditions and constraints. A major focus are equilibria and stability of system dynamics, as well as phenomena such as order and disorder, chaos, self-organisation and phase transitions which have a symbolic and practical relevance in the natural and social sciences.[3] Stability theory deals with methods to sustain essential system properties and equilibria against disturbances, seeking resistance against change over a given period. Dynamic models can be directly relevant for stakeholder dialogues and interactions dealing with ecological or economic system dynamics. Examples are models of weather, climate and water cycles, the growth of forests, fishery, cities or the economy as a whole. Dynamic competition models (such as the Lotka-Voltera model) describe the interaction and potential conflicts between actors or populations, often with regard to scarce resources. For a fixed parameter set in deterministic dynamic systems, implemented as computer programs, stakeholders can explore future options and scenarios in virtual experiments, simply by changing parameters and initial conditions. By adapting their control variables, they can steer the dynamics towards particular targets or keep the dynamics within given limits satisfying their interests.
2. *Decision analysis* deals with the ranking and selection of actions from a set of options, following certain rules, preferences and criteria. In the

[3] With regards to applications in socio-economics see Gandolfo (1997), Grebogi/York (1997), Epstein (1997).

standard case, a single rational decision-maker chooses the most preferred option, by maximising a utility or value function in a set of constraints. A wide range of methods have been developed to search for the optimum, such as decision trees, the steepest ascent/descent in a utility landscape, linear and non-linear programming (such as the Simplex method), combinatorial optimisation, and others. Benefit-cost analysis seeks the most efficient way of achieving goals with limited resources. Optimal control theory determines optimal paths in dynamic systems with state-dependent feedbacks in the control set, usually by optimising constrained value functions (called Hamiltonians) via the Maximum Principle (Feichtinger and Hartl 1986). In case of multiple independent value functions, generally not all can be optimised at the same time which requires a balancing mechanism between objectives. Multi-criteria decision-making derives solution concepts such as Pareto optimality, seeking the set of combined actions that do not allow further joint improvements for all players. Despite the achievements of economics with the rational actor paradigm (RAP), rational actors have been characterised as "lonely social atoms with infinite computational capacities", optimising their utility with every step (Pahl-Wostl 2001).[4] While they can be adequate in environments with a few number of state and control variables, the limits are more obvious in a complex environment and with human beings of bounded rationality.

3. *Game theory* extends rational decision-making to two and more players, each pursuing their own preferences and values in response to the supposed or observed decisions of other players (von Neumann and Morgenstern 1948, Owen 1982). For a small number of players and discrete options, these can be depicted as matrix games. A situation in which no player has an incentive to change its action is called a Nash equilibrium. The most well-know type is the Prisoners' Dilemma (PD) game, describing situations in which two players individually prefer not to cooperate, even though mutual cooperation would be of advantage for both. While cooperation is potentially possible in such a game, it is excluded in zero-sum games of conflicts in which a player can only gain if the opponent looses. A key issue in conflict resolution is to create incentives for cooperation by seeking joint gains. Cooperative game theory studies the formation and stability of coalitions of players, representing larger social units, as well as the power of individuals in group decision-making, measured by indices such as the Shapley Value in voting processes. If all players' decision criteria are taken into

[4] See also Welp et al. (2006). A discussion of the rational actor paradigm can be found in Jaeger et al. (2001).

consideration, methods of multi-criteria decision-making are applicable in game theory. Game theory is appropriate when the goals and options of players can be clearly laid out, but becomes more difficult to handle when a large number of players interacts in a dynamic environment.

4. *Differential game models* with a few numbers of optimisers in controlled dynamic systems are widely used in economic theory as well as in natural resource management (see Carraro and Filar 1995, Dockner et al. 2000). More general dynamic game models describe the interaction between multiple players according to situation-dependent decision rules and reaction functions (Intriligator 1971, Tuinstra 2000). In repeated games players can learn and adapt their behavior to the strategies of other players, leading to the evolution of cooperation. In a computer competition between strategies in the repeated PD game, the Canadian psychologist Anatol Rapoport succeeded with the simple tit-for-tat strategy, which suggests cooperating in the first place and only switch to non-cooperation if the counterpart does (see Axelrod 1984). Evolutionary games, framed by John Maynard Smith, analyse the selection among competing populations of game strategies according to their fitness in replication (Hofbauer and Sigmund 1998). Selection of strategies and decision rules in computer-based simulation models can be based on observation and include real-world actors, offering a wide field of experimental games for educational and research purposes as well as for decision support and policy advice.
5. *Agent-based modelling* uses computer simulation to analyse complex interaction between multiple agents who follow given action rules and stimulus-response mechanisms to form complex social patterns. Using the approach of cellular automata, agents move like insects in virtual landscapes, acting as laboratories of *artificial societies* (see Epstein and Axtell 1997, Gaylord and D'Andria 1998). For a large number of homogenous agents, methods from statistical physics, non-linear dynamics and complexity science are applicable, building on the terminology of "Synergetics" (Haken 1977), such as self-organisation or micro-macro phase transitions. Such approaches to collective phenomena have been transfered to interdisciplinary fields such as socio-physics and econo-physics (Helbing 1995, Weidlich 2000, Schweitzer 1997). Different from top-down models in decision-making, agent-based models are a bottom-up approach to stakeholder analysis. Observed macroscopic properties emerge from the behaviour and interactions of the component agents. Applications range from moving crowds and traffic systems to urban, demographic and environmental planning (Billari et al. 2006) (see more in Section 5.6 of this article).

6. *Qualitative reasoning* can represent heterogeneous data and dynamic behaviour under uncertainty, integrating knowledge from different disciplines on an aggregated level (Kuipers 1994). Methods of qualitative differential equations (QDEs) analyse the qualitative system properties in dynamic systems under uncertainty. Instead of an exact functional and numerical specification, it is only necessary to formulate qualitative if-then relationships. This approach is suited to classify dynamical systems and solutions with similar properties and formulate rules about the interrelationship of nature and human action which are robust to uncertainties and parameter changes. By clustering system patterns and policy options, qualitative approaches can be useful in stakeholder decision-making dialogues. Qualitative approaches found applications in environmental research through the "syndrome concept". It was developed to describe *patterns of global environmental change*, based on qualities and dynamic interactions, perceived as relevant by stakeholders and decision-makers. An example is the overexploitation of marine resources, a pattern associated with the loss of marine biodiversity, overcapitalisation, and declining coastal economies (Eisenack and Kropp 2001, Kropp et al. 2002, Eisenack et al. 2006). Here statements are of the form '*if harvest increases and stock regeneration decreases, then the fish stock decreases*', or '*the stronger the pressure of the fishing lobby, the higher the total allowable catch*'. This allows clustering harvest regions according to qualitative properties and analyse transitions between them. It provides a basis for negotiations and institutional mechanisms to control allowable catch.
7. Another innovative – even though mathematically challenging - method is *viability theory*, which provides mathematical methods and tools to maintain a controlled system dynamics within given boundaries. To stay within the viability constraints of a system, given by objective limits or value-based judgements, reverse methods are applied for selection of admissible control variables that correspond to a feedback mechanism at the boundary conditions (Aubin 1991). The theory of viable control is a useful instrument to design and control the complex interaction between the economic, environmental and political spheres in natural resource management (see for instance Aubin and Saint-Pierre 2004). The impact of changing crucial couplings is studied to improve viability in resource networks, resolving conflict between environmental damage and expected gains from resource use. Viability theory provides a powerful tool to predict the confinement of the resource dynamics to a pre-defined regime in phase space, e.g. given by tolerable windows for fish catch or guardrails for greenhouse gas concentration (Petschel-Held et al. 1999, Bruckner et al. 1999). This would allow stakeholders to

identify controls necessary to stay within sustainable limits, e.g. for fish catch, and to avoid non-viable regions in which fish resources decline or fishery becomes unprofitable, taking into account uncertainties about fish stocks and catch efficiency. Computational tools for practical implementation of viability theory are under development.

A variety of further methods and tools have been developed in systems science, operations research, artificial intelligence and computer simulation which cannot be mentioned here in detail. These include for instance neural networks, expert systems, complexity theory, social network analysis, spatial modelling, Bayesian learning and statistical methods in all variations. They are applied or potentially applicable to stakeholder assessment in environmental management (see Kropp and Scheffran 2006). Some of the mentioned tools and methods are explained in more detail in the following, in the context of environmental stakeholder assessment and interaction.

5.4 Tools in environmental conflict resolution and mediation

In the environmental sciences, systemic approaches (from eco-systems) and agent-based approaches (in economic-social relations) are directly linked, which facilitates the use of modelling and computational tools in stakeholder assessment. In particular, this applies to environmental conflicts over the use of natural resources, or intensified by their use. Environmental degradation and resource scarcities are relevant sources of conflict in various regions of the world (see for instance Homer-Dixon 1991, Baechler and Spillmann 1996, Carius and Lietzmann 1999, Diehl and Gleditsch 2001). Besides conflicts over exhaustible resources (minerals, fossil fuels, territory), there are also conflicts over the degradation of renewable resources, such as agricultural products, fish stocks, favorable climatic conditions, water, soil and air.

Conflict is a particular form of human interaction, resulting from incompatible objectives or actions of agents. Conflicts can emerge as a result of collective interaction among rational actors which seek their own advantage but fail to achieve potential joint gains. While conflicts are a natural part of social life, they may indicate or contribute to the inefficiency and instability of societies. Thus mechanisms for mediation and conflict resolution could stabilise social interaction and prevent environmental destruction as well as the most destructive forms of conflict. The problems causing a conflict could be denied, taking the risk of conflict escalation, or a conflict could be decided by a court. More attention should be paid to the reduction of potential conflicts and the enlargement of consensus in interactive decision-making and negotiations.

Negotiation is an interactive process in which negotiating parties try to reach an agreement on issues under dispute, usually starting with different interests and information sets. Different types of negotiations can be specified, depending on the degree of conflict, the number of parties, and their willingness to share information or find compromise (Kettunen 1999). An example is the dispute about fishing rights between countries harvesting the same fish stock. Negotiation analysis aims at understanding and supporting negotiation processes, studying negotiation procedures and properties of negotiated settlements. Besides the descriptive aspects, one objective of negotiation analysis is to develop a prescriptive theory of negotiations and provide useful advice for involved parties and negotiators (Raiffa 1982, Pruitt and Carnevale 1993).

Stakeholders can negotiate to find a solution to the conflict themselves, or with the help of a mediator. Mediation is a voluntary part of the negotiation process which aims for a common ground, with a neutral, independent person who monitors and manages the process and assists the disputing parties. During the mediated process, the decision objectives and alternatives as well as the conflict areas have to be specified, based on the preferences of the parties, to find the areas for consensus, compromise and cooperation. The solutions of mediation should find support and be acceptable and binding to all parties. If no mediated solution can be found, conflict parties can seek arbitration by an independent group of experts who analyse the conflict and propose a solution. A crucial issue is whether a solution is widely supported and whether the arbitrators have the power and authority to implement a solution to the conflict, respectively the underlying problems. Communication is an ingredient part of mediation, in dialogues, disputes or formal negotiations between stakeholders throughout the conflict. Particular tools can model the interactions in conflict and facilitate the mediation process, seeking options and strengthening the capabilities of the conflict parties towards cooperation

Environmental communication, mediation and alternative dispute resolution can contribute to finding solutions to environmental conflicts. They can complement governmental or business decision-making and legal procedures in environmental policy. Environmental mediation has received great interest during the past decade in Germany (e.g. Weidner 1998), although the number of implemented procedures is still small. Since 1990 the Social Science Research Centre Berlin (WZB) has been conducting an interdisciplinary research project on mediation procedures in the field of environmental protection, largely on conflicts arising out of waste management (Holzinger 1997).

Similar cases were also studied by Peterson (2002) who analyses and demonstrates the potentials and limits of decision-analytic tools, including value tree analysis and multicriteria methods, which have sporadically been used in environmental mediation. They can be practical instruments in support of environmental mediation, to structure the decision-making process and make it transparent and comprehensible to all parties. To actually apply them, the mediation process needs to be formalised to allow for the specification of steps. In a process-orientated game-theoretic model of negotiations, parties compare their outside options with possible agreements in the mediation process. Taking into account the iterative character of negotiations in computer simulation, the alternatives and claims of the parties change perpetually. By including emotions, such as anger and envy, the approach explains the difference between rational, utility maximising behaviour and supposedly irrational behavior, leading to an escalation of conflicts.

Each simulation consists of several negotiation rounds, during which the mediator proposes a compromise solution calculated in solution space that the parties can either accept or reject, depending on their outside options, the so-called Best Alternative To Negotiated Agreement (BATNA). The claims depend on the parties' negotiation strategies and the attitudes of the opposing parties which change in each negotiation round. The mediation ends when the parties agree to an option, it fails if an agreement cannot be reached, either because parties drop out or the conflict escalates or is deadlocked.

The potential of the approach has been investigated in detail in the case of the German city Bremen where an environmental mediation was held upon the conflict between the environmental and social acceptability of a waste disposal site (Peterson 2002). Using the simulation model, implemented in Mathematica as a programming tool, a potential compromise solution was found, even though in the real mediation no agreement was achieved. The analysis was able to explain the conflict and find possible win-win solutions, taking into account the escalation dynamics and the role of the mediator in deescalation. The simulations underline the importance of outside options as parties with higher BATNAs tend to drop out of the mediation if they have not much to gain from the negotiations.

5.5 Interactive methods for group decision and negotiation support

5.5.1 Basic approaches

Negotiation support research develops tools for finding agreement and producing solutions which would satisfy all involved parties, even in situations where the negotiating parties fail to find a satisfactory agreement by themselves. By structuring and reducing the complexity of the negotiation problem, tools can help mediators to find jointly beneficial proposals. The methodology of interactive optimisation methods has its roots in decision analysis and game theory, with input from various disciplines including mathematics, social psychology, political science, management and computer science (Sebenius 1992, Jelassi et al. 1990).

Raiffa (1982) observes that agreements made in negotiations are frequently inefficient in the sense that alternative agreements preferred by all parties were not reached.[5] An important research issue is to analyse the procedures and conditions that lead to 'good' agreements. The success of negotiations can be measured with several criteria, including fairness and equality as well as rationality and efficiency (Mumpower 1991). Efficiency is usually represented by Pareto optimal solutions, which means that no feasible alternative agreement exists improving two parties at the same time. In the joint utility framework, a good agreement would maximise the (weighted) sum of the negotiating parties' utilities. Giving equal weight to the utilities of all parties would satisfy the criterion of equality.

Negotiation support research develops practical and constructive methods for efficient agreements in multi-party negotiations, reaching joint gains for all parties. Some of the approaches are described in the following, based on the survey in Kettunen (1999):

- *Utility Function Assessment:* Various methods have been developed for constructing a decision-maker's utility (or value) function by eliciting his/her preferences. If preferences can be expressed by utility, efficient agreements can be calculated by increasing and maximising utility. For multiple actors, maximising a weighted sum of the joint utility functions (also referred to as social welfare function) is a commonly used method to produce the `best' agreement. With different weights, different Pareto-optimal agreements are found. Selecting the weights for each

[5] On reasons for inefficiency see Kersten and Noronha (1998).

individual utility function is a political issue, reflecting the power and preferences of actors, and runs into problems of interpersonal utility comparisons. One option is to avoid aggregation of utility functions by lexicographic preferences, an approach that requires a hierarchical ordering of preferences, like words in a dictionary.

- *Constraint Proposal Methods* are more appropriate in cases of partial information of the decision-makers' preferences, in particular when no explicit utility functions can be identified or constructed. Negotiators are asked to select their most preferred alternatives on a plane of constraints similar to a budget constraint. Based on a suitable updating scheme, new constraints are generated until the negotiators' selections coincide. Under certain conditions the process should converge to a Pareto-optimal point, which varies with changing the initial reference point. The method has been generalised to multi-party negotiations with multiple issues.
- *Single Negotiating Text Based Methods* generate a single negotiating text which serves as a tentative agreement, suggested by a mediator or the negotiators themselves. Based on the information the parties give about their preferences, the mediator examines the old text and searches a new one which all parties prefer to the previous one. The step-by-step process, which matches many real-world problems, gradually approaches a Pareto efficient agreement which depends on the initial points chosen.
- *Multi-criteria decision-making* in consensus seeking groups: Group decision-making problems are mathematically similar to problems in multiple criteria decision-making (MCDM). This implies that methods developed for supporting a single decision-maker with multiple objectives could be applied in negotiation settings with multiple decision-makers, each having their own objective. The balance of interests among the different actors is subject to an interactive negotiation process in which not just the actors' preferences come into play but their power as well and the possibility of conflict and coalition formation, shifting the weights of the different criteria. The art of mediation is to find aggregate consensus functions balancing the weights of all criteria.
- *Group Decision Support Systems (GDSS) and Negotiation Support Systems (NSS)* take advantage of computer systems to improve communications among participants, provide group decision modeling and techniques, and include expert systems components. Different decision support tools are used in the decision-making cycle, from problem structuring to consensus seeking. Model-based computational

tools are integrated into real life experiments with stakeholders, which requires the development of a user-friendly negotiation support software.

A method of improving directions has been presented by Ehtamo et al. (1999) which builds on a two-party negotiation procedure.[6] The underlying idea is to search for jointly improving directions in the issue space. Starting from their most preferred positions, the negotiators make concessions in their subsequent offers and counter offers, until all parties have offered the same alternative and agreement is reached. The method is tested in practice by role-playing experiments in which the negotiating parties are required to answer relatively simple questions concerning their preferences.

5.5.2 Internet tools for negotiation analysis

The Internet is a promising platform for exchange and negotiations among distributed individuals located at various places around the world. Because of their interactive nature, models of negotiation analysis are suitable for e-learning. Various e-learning material on negotiation analysis and on related fields exists, such as decision analysis, game theory and experimental games, potentially usable in stakeholder dialogues and negotiations (see the survey in Ehtamo et al. 2003). For instance, Al Roth at Harvard University has a game theory and experimental economics website, which contains electronic books on game theory and interactive applets to play different games against the computer, such as variants of the prisoners' dilemma game.[7] The Decisionarium site provides interactive multicriteria decision support with tools for individual decision-making as well as for group collaboration and negotiation. Access to the Web-HIPRE value tree software allows evaluating negotiations.[8] The Harvard Business School developed a commercial e-learning on-line negotiator, based on the book by Fisher and Ury (1981).[9]

Not many web-tools are yet used to support real decision-making, negotiations and stakeholder dialogues. Specific software systems are required for remote negotiation support, providing possibilities for exchanging messages between distributed negotiating parties and

---

6 A similar idea, using heuristic rules, has been suggested by Teich et al. (1996) for two-party resource allocation negotiations.

7 See www.economics.harvard.edu/~aroth

8 See www.decisionarium.hut.fi.

9 See www.dieu.com/e-learning/Yes_The_Online_Negotiator.asp.

implementing mathematical methods in negotiation analysis. Some recent studies have focused on using computers for teaching negotiation and e-negotiation skills for university students, providing an effective way for testing, evaluating and reinforcing acquired knowledge on negotiations. Köszegi and Kersten (2003) presented a course and experiences in multi-criteria based e-negotiations, including an electronic textbook which focuses on basic concepts of economics, game theory and social psychology, as well as case studies. Related role-playing exercises involve students to negotiate both with a NSS and face to face directly. The software tool INSPIRE has been used for teaching e-negotiations in continuously organised negotiation sessions. At the tournament International Competition for Online Dispute Resolution 2003, the participating students formed local teams to solve a negotiation case by negotiating with another team, selecting from six different NSSs.[10] These systems provide tools for communication through exchange of text messages over the Internet or allow videoconferencing. The systems INSPIRE and SmartSettle help the parties to describe their preferences by constructing additive value functions which are used in negotiation for evaluating offers and counteroffers.[11]

To promote e-negotiations, the Systems Analysis Laboratory at Helsinki University has developed e-learning material to teach e-negotiation methods for university students (Ehtamo et al. 2003). Students practiced the use of NSS in role-playing exercises via "learning by doing". A website on e-learning decision-making contains material and tools for negotiation support and discusses students' experiences of its use.[12] The material consists of sections on negotiation theory, including case studies, software assignments, quizzes for self-evaluation, and multimedia presentations such as video clips, animations and color graphics.

The software tool "Joint Gains" is based on an interactive multiple criteria negotiation method for solving multi-player/multi-issue negotiation problems. It is publicly available online and allows users to create their own customised negotiation cases, requiring local preference information from the parties, without using the additive value function model, and provides aid for reaching Pareto solutions. In its web-based implementation, Joint Gains can serve as an online-interactive negotiation support system for real-life negotiations as well as active learning through role-playing experiments, which allow exploring and learning about the objectives and power of different parties and the potential for coalition

---

[10] See www.enegotiation.org.

[11] See www.smartsettle.com, http://interneg.org.

[12] See www.dm.hut.fi.

building (Kettunen et al. 1999). The teacher not only acts as a lecturer but also as an instructor, who guides the students to work with the material more independently than usual, and takes care for technical arrangements of the software accessibility and maintenance.

Sotware-based NSSs can be useful for education purposes, making negotiation analysis a promising field for e-learning in application areas such as political or environmental decision-making, international affairs, e-business, etc. It is particulary relevant for environmental negotiations and has been used for lake-river regulation policy problems and in a workshop on environmental negotiations at the Caspian Sea area (Ehtamo et al. 2003).

5.6 Agent-Based Modelling

5.6.1 Structure and behavior of agents

An agent-based model (ABM) is based on a set of autonomous agents capable to interact with each other as well as with the environment according to rules of behavior. In this context, an agent has been defined as "*an object in a computer program that encapsulates a particular behaviour when interacting with other agents within an environment. The behaviour may be simple or complex; deterministic, stochastic or adaptive; and the system as a whole may be homogeneous (all agents are of the same type) or heterogeneous (more than one type of agent present)*" (Hood 2003). In this understanding, agents may be endowed with cognitive capabilities "to perceive signals, react, act, making decisions, etc according to a set of rules". Cognitive agents are characterised by (Conte and Castelfranchi 1995):

- beliefs: what agents think to know about the world (based on experience and perception);
- goals: what agents would like to achieve (desired states of the world);
- intents: which specific actions will agents undertake to achieve the desires.

To achieve desired goals, agents may try new intents, alter their desires or change their beliefs. With regard to their action abilities, agents can be

- autonomous: they act independently of any controlling agency;
- social: they interact with other agents;

- communicative: they can communicate with other agents explicitly via some language;
- pro-active: they are driven by goals and objectives;
- reactive and adaptive: they observe and respond to changes in the environment;
- rational: they can follow a well-defined and logical set of decision rules.

Agents can both change their environment and their internal structure. Beyond the "representative agent" used in many economic models, they can learn by generating, testing and evolving models of their environments and of other agents, converting these models into rules of stylised behaviour (Pahl-Wostl 1995). They use these abilities to change their environment, interact with other agents and solve group problems. Agents need 'sensors' to perceive their local neighbourhood and receive or send messages (Gilbert and Troitzsch 2000). ABMs are used to generate macro structures from local micro mechanisms.

Depending on the agents' number, their attributes and behavioral rules in their respective environments, ABM's can be of great variety and complexity, making them hard to analyse or predict. Since they are problem-specific, they can include many details matching reality, with processes occurring at different spatial and temporal scales (see Billari et al. 2006). Simulations have the character of experiments in virtual worlds, often with demanding computational requirements. Key challenges are to calibrate the models with data and to integrate ABMs into real-world applications such as stakeholder dialogues and negotiations.

5.6.2 Simulation environments and environmental simulation

Agent-based modelling and simulation is being increasingly used in environmental management. It permits the coupling and embedding of social interaction into environmental models, taking into account the adaptive, disaggregated nature of human decision-making as well as collective responses to changing environments and management policies. Special modelling–simulation environments or toolkits of various kinds are available for performing experiments, which abstract from the details and can be duplicated by other researchers. One example is the TRANSIMS model which attempts to capture the details of traffic flow and its consequences, for instance for the planning of road additions in Albuquerque. SWARM is a public domain software developed at the Santa Fe Institute, simulating collections of concurrently interacting agents and offering a wide spectrum of tools used for modelling experiments in many application areas, including economics, ecosystems, anthropology.[13] Vensim and Stella provide a software environment for developing, analysing, and packaging high quality dynamic feedback models.[14] Sensitivity testing allows to change assumptions about input values, to examine the uncertainty in selected output variables, and automatic calibration to fit historical data series. Among various applications, Vensim has been used to model the allocation of demands among competing suppliers and to track the efficiency of teamwork, based on information about skill and experience of the team member.

Hare and Deadman (2004) develop a taxonomy for environmental agent-based systems that could also serve as educational tools in environmental management. Six modelling requirements (coupling social and environmental models; micro-level decision-making; social interaction; intrinsic adaptation of decision-making and behaviour; population level adaptation and multiple scale level decision-making) are linked to a set of eleven case studies, listed in Table 5.1 (Hare and Deadman 2004). The case studies highlight that:

- social interaction tends to be implemented in algorithms imitating the behaviour of neighbours and friends;
- most popular are decision-making models based on simple heuristic rules; and
- the modelling of multiple scale decision-making is still in its infancy and needs to be further developed.

[13] Various materials and programmes can be found at website www.swarm.org.
[14] www.vensim.com/software.html; Hannon and Ruth (1999).

Table 5.1 Agent-based models in environmental assessment (Hare and Deadman 2004)

Name and type of model	Description	Type & number of agents
Bali model Rural water resource management	Investigates whether a specific Balinese system of water temple networks managing irrigation practices could have self-organised. A simulation is used to test the theory	Subaks (groups of farmers) (172)
SHADOC Rural water resource management	Investigates the viability of current irrigation practices in the Senegal river valley through development and interactive use of an ABM	Individual farmers, pumping station manager, water course manager (40–60)
CATCHSCAPE Rural water resource management	Investigates the viability of irrigation practices in Thailand with respect to future changes in drought conditions, changes in commodity prices, and farmer behaviour	Individual farmers, water manager (327)
Lake model Rural water resource management	Assesses farmers' adaptive responses to policy measures (i.e. taxation) for reducing phosphorus levels in a hypothetical lake	Individual farmers (100)
Thames model Urban water demand management	Investigates how social structure and learning affects the efficacy of a regulator's exhortations for consumers to save water as part of a drought management policy.	Households, policy agent (80–100)
MAGIC Flood mitigation decision support	Various "expert agents" cooperate with each other to come up with decision support advice for human flood catastrophe response teams.	Individual expert decision agents (<10)
Biomas Animal waste management	Explores possible negotiating strategies and outcomes used by simulated actors in managing the removal, transportation and processing of animal wastes.	Eleveur, Cultivateur, Transporteur, Transformateur (<10)

Name and type of model	Description	Type & number of agents
Rangeland model Rangeland resources management	Explores the range of collective responses of hypothetical pastoralists to regulators' policies for sustainability	Pastoralists, regulator (100)
FEARLUS Agricultural land use change	Investigates how well different social learning strategies employed by decision-makers compete in the face of a changing, heterogeneous environment	Farmer households (>40)
LUCITA Agricultural land use change	Explores how the characteristics of frontier families influence changing agricultural land use, and secondary succession, in Amazon rainforest near Altamira, Brazil	Farmer households (236)
Grand Canyon model Recreation management	Assesses impacts of river rafting trip management scenarios, where agents represent individual trips on the Colorado River through Grand Canyon National Park	Rafting trips (>50)

5.7 Stakeholders in Integrated Assessment

5.7.1 Participation and validation in Integrated Assessment modelling

Integrated Assessment models and tools are designed to aid the evaluation and decision-making process in complex nature-society environments. They provide decision-makers and stakeholders with a coherent framework to explore and reproduce future options and scenarios, addressing environmental, economic and social concerns in an integrated way. Users can interact and play with model units and other users, to better understand and anticipate opportunities and risks, facilitating decision-making under uncertainty. Integrating information from many different areas and knowledge across disciplines supports management and planning, and facilitates an informed debate of policy decisions and conflict resolution. An information management and decision support system, adapted to the user's needs, can provide scientific expertise to policy-makers quickly and improve transparency (Fordham et al. 1997). For longer time frames and at a larger scales, model-based scenarios can be used to increase understanding of the possible consequences of actions.

Integrated Assessment models range from highly aggregated models such as the DICE climate model of Nordhaus (1994), using dynamic optimisation to estimate optimal emission reductions, to process-based models, addressing details from climate to ecosystem change and human responses (Rotmans and Dowlatabati 1998). DICE has been criticised because of unrealistic assumptions based in standard economic theory (e.g. single optimiser, discount rate, simple damage function) which are hard to validate. Extensions and improvements have been suggested, including multiple actors with behavioral patterns, adaptive management and climate change mitigation (Hasselmann 1998, Weber et al. 2003).

To develop a new anticipatory IAM framework, tools are required to handle the complexity of the interactions between natural and social systems on a global scale (Moss et al. 2001). While models of some sub-systems (or modules) exist, there is no predictive theory or model on how those systems act together. However, software tools have been developed which enable the coupling of modules written in different languages.[15] Models of social systems are much less reliable and harder to integrate. Decision-making procedures, such as cost-benefit analyses, assume the ability to identify possible future outcomes and predict their values and probability of occurrence. Some social structures reduce uncertainty and mitigate the unpredictability of processes and events, such as stock changes, disasters or terror attacks. Norms and institutions increase societal efficiency and stability but also restrain the freedom of individual action and the pace of change. Predictive social theories would be rather prescriptive as by themselves they could change the structure of the system.

A modelling approach for policy advice not only legitimises from scientific criteria, but rather from its plausibility for the non-expert audience. Thus, integrated assessment needs to be embedded into a social process based on a dialogue with decision-makers and other stakeholders as well as the general public (cf. ULYSSES-project). A participatory integrated assessment approach directly involves stakeholders in the

[15] See the Community Integrated Assessment Module (CIAM) developed at the Potsdam Institute for Climate Impact Research. For reference see the website of the European Climate Forum (http://www.european-climate-forum.net).

specification and evaluation of social simulation models which help to identify the socially constructed reality, and the mental models of individual actors. They can specify the behavioural patterns and identify misspecification or implausibilities in the model.

Stakeholders can also be involved in the empirical validation of the model and its output. A program is said to be validated if it demonstrates to do in practice what it is designed to do, producing outputs that correspond to observable properties of real social systems. Given the size, complexity and diversity of integrated assessment models, none of these models will be strictly correct and complete or predict the future. Models of social systems cannot be fully validated experimentally and don't have to replicate in detail every property and event of real social systems but should capture and reproduce some essential qualitative features and statistical descriptions of the phenomena. Here the link between integrated assessment and qualitative modelling becomes obvious.

Validation of participatory agent-based social simulation is a reflexive process involving both modellers and stakeholders (Moss et al. 2001). The models should be convincing to stakeholders to understand the outcomes on different levels, coarse grain as well as fine grain. A user should be able to set the level of detail and to explore the implications of parameter choices and uncertainties. A practical goal is to understand the phenomena of interest sufficiently to mitigate adverse outcomes and to enhance positive outcomes. There are very few models which give an outsider an opportunity to comprehend the model structure and dynamics with reasonable effort (see the concept of mediated modelling, van den Belt 2004).

The combination of a physical model of the environment (e.g. climate) and a socio-economic model raises issues of scale in both time and space. While decision-making is local and short term, climate change, for instance, is essentially a long term and global phenomenon. Bringing these two scales together requires an iterative learning process which matches the diffusion and aggregation of local actions to global levels and the implementation of global decisions down to regional and local levels. Understanding the mutual relationship between global negotiation and local decision-making processes across hierarchies is one of the key challenges of Integrated Assessment modelling.

5.7.2 Examples of Integrated Assessment models

An early example for integrated environmental assessment is the project ICRA (Integrated Climate Risk Assessment) which is a core component of

CLEAR (Climate and Environment in Alpine Regions), developed at EAWAG in Zürich.[16] Seeking to span the interfaces between the physical, ecological, and social sciences, CLEAR combines two tools: computer models for organising scientific knowledge and focus groups for organising a social process involving scientists and non-scientists.

The EU funded project ULYSSES has used existing IA Models, such as TARGET and IMAGE, developed by the Dutch research institute RIVM, as an input for participatory integrated assessment. In addition, the PoleStar model was used, based on regional data about urban lifestyles. ULYSSES enhances the static framework of NAIADE, the Novel Approach to Imprecise Assessment and Decision Environments which is a multi-criteria decision analysis tool and allows policy-makers to seek "defendable" decisions that reduce the degree of conflict or that lead to greater equity in their impacts.[17]

The European project FIRMA (Freshwater Integrated Resource Management with Agents) aims to integrate physical, hydrological, social and economic aspects of water resource planning. Participation of stakeholders, like planners, decision makers, consumers, suppliers and environmental NGOs, is incorporated to improve the quality of the model, to raise the interest of stakeholders and to increase their confidence in the model results (Downing et al. 2000). The model framework combines agent-based modelling and Integrated Assessment.

- The *agent-based model* is based upon complex or cognitive agents, represented by independent sub programmes capable of reflecting on their goals and beliefs, and (re)acting to a changing (model) world, including the perceived behaviour of other agents. An update of beliefs, resulting from new information and changing perceptions, may lead to a change in goals and expectations about the impact of a particular strategy on the environment. Strategies are implemented and changed according to a pressure-state -response-impact (PSRI) concept, initiating a state changing process, triggered by threshold values. Behavioral changes result from rules assigned to each agent by declarative statements, written in declarative programming languages like SDML (strictly declarative macro language), MIMOSE, PartNet or others.
- The *Participatory Integrated Assessment* model uses mental models of organisations and institutions as input to the agent-based model, based

[16] Cebon et al. (1998). Project descriptions of CLEAR can be found at http://CLEAR.eawag.ch and http://hdgc.epp.cmu.edu/projects/abstracts/clear-icra.html.

[17] ULYSSES and other models are described at http://zit1.zit.tu-darmstadt.de/ulysses/models.htm.

on the most relevant actors and their interaction within the target system. Creating an interface between the participative process and the agent-based model allows stakeholders to develop and validate models, to learn, communicate and negotiate about planning strategies and policy measures (interviews, dialog methods, focus groups).

The integrated approach seems to be particularly suited to model the interaction between the natural environment and human actors, incorporating social dynamics from the early stages of planning activities. The combination of models and the simulation setup is a step forward to a decision support system (DSS) that enables modellers and planners, decision-makers and stakeholders to deal with a complex setting of interrelated issues. The consequences of human activities can be studied in an interactive, consistent and dynamic way. It allows explaining reasons for particular actions and their consequences on the environment as well as on other actors within a specific target system like a river basin. In applications such as climate change adaptation and natural resource management, physical parameters are taken into consideration, like river bed geometry, land use change or flood probability as well as external and internal pressures like climate change or demand for housing, and surprises. Agents can shape their landscape and react to events like floods, droughts or pollution. Here threshold values are significant, such as the height of dykes, carbon emission or fish harvested. The model also deals with agent–agent interaction, including communication and negotiation about planned activities, conflicts and coalition formation. Communication between stakeholders and modellers is an essential part of the entire modelling process. Increasingly the Internet is part of participatory group learning (Hare et al. 2001).

5.8 Integration and Outlook

As was discussed in this article, there are various tools and resources applicable to stakeholder assessment and interaction. Increasingly these tools become integrated in a manner that is useful for real-world decision-making. To achieve full integration is however a demanding task, given the complexity of the underlying problems. An integrated approach would seek to match the complex relation between environmental and socio-economic systems and the diversity of agents. An issue that is missing in most of the studies is the transition between heterogenous individual actors and collective units that emerge from the dynamic interaction, in the form of coalitions, networks and institutions. Understandig the conditions for a self-organised sustainability transition among multiple heterogeneous actors is an issue of great importance for successful stakeholder interaction.

To structure the analysis and identify the usefulness of the various tools, it is adequate to describe the policy decision-making and management process involving stakeholders as a multi-step process, including the following phases:

- Situational analysis and problem structuring
- Option identification and scenario modelling
- Concept development and criteria-based evaluation
- Decision-making and negotiation
- Planning and action
- Monitoring and learning

These phases form a repeated cycle with connecting processes such as evaluation, communication, capability building, information, simulation, and validation. In each of these phases, some of the tools mentioned in this article are particularly useful to support multi-stakeholder policy assessment (see the link between phases and tools in Figure 5.1). Analysing a particular situation, drawn from information about the environment, is important to understand the problem that stakeholders are dealing with. Here all kinds of sensors, measuring the key variables that characterise a particular problem, and other methods for data gathering and analysis are at hand, including interviews, questionnaires, polls and statistical methods using the data as input, including Bayesian Learning. To work on a problem, it is essential to understand the dynamic interactions among systems variables, elements and actors, using a wide range of methods from system dynamics and complexity science as well as agent-based and spatial modelling to analyse phenomena in time and space. Computer simulation has developed into the dominant tool to explore selected scenarios and identify the most relevant options to choose from. Using computer-based visualisation techniques, complex phenomena can be presented to stakeholders.

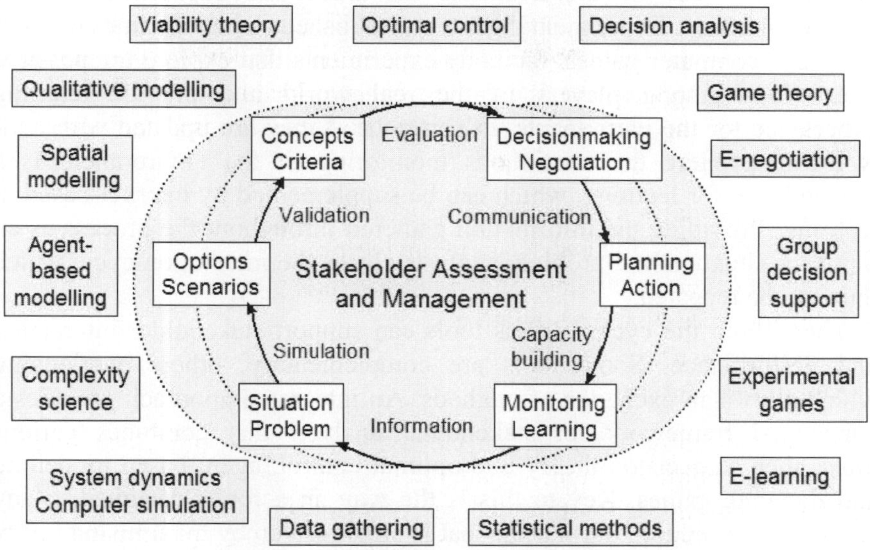

Fig. 5.1 The Stakeholder cycle and tools for stakeholder assessment and management

To select from the menu of validated options, stakeholders need to develop concepts for the future they want to design, and criteria to evaluate these concepts as well as the actions to take. By abstracting from unnecessary details which often cannot be validated, qualitative modelling tools support the reduction in complexity required for the design of concepts that can be understood and implemented. Viability theory translates value judgements into system states and trajectories that are to be achieved or avoided. The established approach of optimal control seeks to find those solutions that are perceived as the best to realise the concept according to the criteria chosen.

The highest burden of responsibility occurs during the decision-making and negotiation phase as well as during implementation because there the transition from concept to reality is actually made. Here the variety of tools developed in decision theory is relevant, including multi-criteria methods. Game theory is appropriate either for a few numbers of players or for coalition formation in negotiations. Increasingly game theory deals with dynamical problems. The Internet can be used for e-negotiations among distributed parties. Group-decision support is relevant in the whole process of decision-making and negotiations among several actors as well as setting up of work plans and their implementation in joint actions taken, making best use of the available resources and capabilities of the individual actors. Experimental games, established as board games or more advanced computer games, facilitate experiments that explore the possible interactions among players in the real world and provide learning experience for the participants, in particular if they are updated with real-world data. Here the continuous monitoring of the environment is a precondition for learning, which can be supplemented by internet-based e-learning. Providing the information gathered throughout the process as an input for situation and problem analysis closes the interaction cycle which thus can be repeated.

Throughout the cycle various tools can support stakeholder interaction and performance. Some tools are complementary, others overlapping which allows an exchange of methods. An integrated approach provides a conceptual framework for stakeholder analysis that combines various tools, such as decision theory and optimal control, agent-based modelling and dynamic games. Key to this is the way an actor is designed. Many models use a reduced-form actor that is either driven by maximising utility and minimising costs, following given behavioral rules and adapting to others, or seeking positions within viable boundaries. Real human beings however show each of these patterns, with a different emphasis in different contexts. Understanding the conditions and transitions between contexts and model settings in a more integrated model framework is a challenge which is also relevant for stakeholder assessment and interaction.[18]

[18] For an approach to design a more comprehensive framework of action and interaction of multiple agents see Scheffran (2000, 2001); Scheffran and Stoll-Kleemann (2003).

References

Ackoff, R. (1974). Redesigning the Future: A Systems Approach to Societal Problems. John Wiley

Aubin, J.P. (1991). Viability Theory. Birkhäuser

Aubin, J.-P., Saint-Pierre, P. (2004). An introduction to viability theory and management of renewable resources. In: Kropp, J., Scheffran, J. (eds). Decision Making and Risk Management in Sustainability Science. Nova Sci. Publ, New York, forthcoming

Axelrod, R. (1984). The Evolution of Cooperation. Basic Books, New York

Bächler, G., Spillmann, K. (1996). Kriegsursache Umweltzerstörung Vol. I – III. Rüegger, Chur, Zürich

Billari, F., Fent, T., Fürnkranz-Prskawetz, A., Scheffran, J. (2006). Agent Based Computational Modelling in Demography, Economic and Environmental Sciences. Springer-Physica

Bruckner, T., Petschel-Held, G., Toth, F.L., Füssel, H.-M., Leimbach, M., Schellnhuber, H.-J. (1999). Climate Change Decision-Support and the Tolerable Windows Approach. Environmental Modeling and Assessment 4, 217-234
See also the contributions by the same authors in the special issue of Climatic Change 56 (2003)

Carius, A., Lietzmann, K.M. (1999). Environmental Change and Security. Springer

Carraro, C., Filar, J. (1995). Control and game-theoretic models of the environment. Birkhäuser, Boston

Cebon, P., Dahinden, U., Davies, H.C., Imboden, D.M., Jaeger, C.C. (1998). Views from the Alps. Towards Regional Assessments of Climate Change. MIT-Press, Cambridge, Massachusets

Coleman, J.S. (1990). Foundations of Social Theory. Harvard University Press, Boston

Conte, R., Castelfranchi, C. (1995). Cognitive and social action. UCL Press limited, London

Diehl, P.F., Gleditsch, N.P. (2001). Environmental Conflict. Westview Press, New York

Dockner, E., Jorgensen, S., Van Long, N., Sorger, G. (2000). Differential games in economics and management science. Cambridge University Press

Downing, T.E., Moss, S., Pahl-Wostl, C. (2001). Understanding Climate policy Using participatory Agent-based Social Simulation. In: Moss, S., Davidsson, P. (eds). Multi-agent-based Simulation: Second international Workshop, MABS 2000. Revised Springer 2001, Boston, MA Lecture notes in computer science. Vol. 1979: Lecture notes in Artificial Intelligence

Ehtamo, H., Verkama, M., Hämälainen, R.P. (1999). How to select fair improving directions in a negotiation model over continuous issues. IEEE Transactions on Systems, Man and Cybernetics, Part C, 29 (1), 26–33

Ehtamo, H., Hämäläinen, R.P., Koskinen, V. (2003). An E-learning Approach for Teaching Mathematical Models of Negotiation Analysis. Helsinki University of Technology, Systems Analysis Laboratory

Eisenack, K., Kropp, J. (2001). Assessment of management options in marine fisheries by qualitative modelling techniques. Marine Pollution Bulletin 43 (7-12), 215-224

Eisenack, K., Scheffran, J., Kropp, J. (2006). Viability analysis of management frameworks for fisheries. Environmental Modeling and Assessment 11, 69–79

Elias, A.A., Cavana, R.Y., Jackson, L.A. (2000). Linking stakeholder literature and system dynamics: Opportunities for research, Proceedings of the international conference on systems thinking in management. Geelong, Australia, 174-179 http://www.esc.auckland.ac.nz/Organisations/ORSNZ/conf35/papers/BobCavana.pdf

Epstein, J.M. (1997). Nonlinear Dynamics, Mathematical Biology, and Social Science. Santa Fe Institute, Addison Wesley

Epstein, J.M., Axtell, R. (1997). Growing Artificial Societies. MIT-Press, Cambridge

Feichtinger G., Hartl R.F. (1986). Optimale Kontrolle ökonomischer Prozesse. De Gruyter, Berlin

Fisher, R., Ury, W. (1981). Getting to Yes: Negotiating Agreement Without Giving In. Houghton Mifflin, Boston

Fordham, D.P., Hood, L.M., Malafant, K.W.J. (1997). Complexity in landscapes and resource planning: Packaging science for decision makers. In: Frontiers in Ecology: Building the Links. Elsevier Science

Freeman, R.E. (1984). Strategic Management: A Stakeholder Approach. Pitman, Boston

Gandolfo, G. (1997). Economic Dynamics. Springer, Heidelberg

Gaylord, R.J., D'Andria, L.J. (1998). Simulating Society - A Mathematica Toolkit for Modeling Socioeconomic Behavior. Springer/Telos, Berlin et al.

Gilbert, N., Troitzsch, K.G. (2000). Simulation for the Social Scientist. Open University Press, Buckingham, PA

Grebogi, C., Yorke, J.A. (1997). The Impact of Chaos on Science and Society. University Press, UN

Haken, H. (1977). Synergetics. Springer, Berlin/Heidelberg/New York

Hannon, B., Ruth, M. (1999). Dynamic Modeling. 2nd edition. Springer-Verlag, Heidelberg

Hardin, G. (1968). The tragedy of the commons. Science 168, 1243-1248

Hare, M., Deadman, P. (2004). Further towards a taxonomy of agent-based simulation models in environmental management. Mathematics and Computers in Simulation 64, 25–40

Hare, M., Gilbert, N., Medugno, D., Asakawa, T., Heeb, J., Pahl-Wostl, C. (2001). The development of an internet forum for long-term participatory group learning about problems and solutions to sustainable urban water supply management. In: Hilty, L.M., Gilgen, P.W. (eds). Sustainability in the Information Society, 15th International Symposium Informatics for Environmental Protection, Part 2: Methods/Workshop Papers. Metropolis Verlag, Marburg, 743-750

Hasselmann, K., Hasselmann, S. (1998). Multi-actor opimization of greenhouse gas emission paths using coupled integral climate response and economic models. In: Schellnhuber, H.J., Wenzel, V. (eds). Earth System Analysis – Integrating Science for Sustainability. Springer, Berlin

Heinelt, H. et al. (2002). Participatory Governance in Mult-Level Context. Leske and Budrich, Opladen

Helbing, D. (1995). Quantitative Sociodynamics – Stochastic Methods and Models of Social Interaction Processes. Kluwer. Boston/London/Dordrecht

Hofbauer, J., Sigmund, K. (1995). Evolutionary Games and Population Dynamics. University Press, Cambridge

Holzinger, K. (1997). Evaluating Environmental Mediation. Results of a Participant Survey. Schriften zu Mediationsverfahren im Umweltschutz Nr. 18. Discussion Paper FS II 97–304. Wissenschaftszentrum Berlin für Sozialforschung

Homer-Dixon, T.F. (1991). On the Threshold - Environmental Changes as Causes of Acute Conflict. International Security 16 (2), 76-116

Hood, L. (2003). Tools for an Uncertain World: Agent Based Modelling, Greenhouse Beyond Kyoto-Conference www.affa.gov.au/corporate_docs/publications/word/rural_science/lms/greenhouse/agent_based.doc

Intriligator, M.D. (1971). Mathematical Optimization and Economic Theory. Prentice Hall, Englewood Cliffs

Jaeger, C.C., Renn, O., Rosa, E.A., Webler, T. (2001). Risk, Uncertainty, and Rational Action. Earthscan, London

Jelassi, M.T., Kersten, G.E., Zionts, S. (1990). An Introduction to Group Decision and Negotiation Support. In: Bana e Costa, C. A. (ed). Readings in Multiple Criteria Decision Aid. Springer-Verlag, Berlin, 537-568

Kersten, G.E., Noronha, J. (1998). Rational Agents, Contract Curves, and Inefficient Compromises. IEEE Transactionas on Systens, Man, and Cybernetics, part A: Systems and Humans 28, 326-338

Kettunen, E. (1999). Interactive Methods for Group Decision and Negotiation Support, Licentiate's Thesis. Helsinki University of Technology, Systems Analysis Laboratory

Köszegi, S., Kersten, G. (2003). On-line/Off-line: Joint Negotiation Teaching in Montreal and Vienna. Group Decision and Negotiation 12, 337-345

Kropp, J., Scheffran, J. (2006). Advanced Methods for Decision Making and Risk Management in Sustainability Science. Nova Science, New York, forthcoming

Kropp, J., Zickfeld, K., Eisenack, K. (2002). Assessment and management of critical events: The breakdown of marine fisheries and the north atlantic thermohaline circulation. In: Bunde, A., Kropp, J., Schellnhuber, H.J. (eds).

The Science of Disasters: Climate Disruptions, Heart Attacks, and Market Crashes. Springer Verlag, Berlin, 192–216

Kuipers, B. (1994). Qualitative Reasoning: Modelling and Simulation with Incomplete Knowledge. MIT-Press. Cambridge

Mitchell, R., Agle, B., Wood, D. (1997). Towards a theory of stakeholder identification and salience: defining the principle of who and what really counts. Academy of Management Review 22 (4), 853-886

Moss, S., Pahl-Wostl, C., Downing, T. (2001). Agent based integrated assessment modelling: the example of climate change. Integrated Assessment 2, 17-34

Mumpower, J.L. (1991). The Judgment Policies of Negotiators and the Structure of Negotiation Problems. Management Science 37, 1304-1324

Nash, J.F. (1950). The Bargaining Problem. Econometrica 18, 155-162

Nordhaus, W. (1994). Managing the Global Commons. The Economics of Climate Change. MIT-Press, Cambridge, MA

O'Riordan, T., Stoll-Kleemann, S. (2002). Biodiversity, Sustainability and Human Communities. Protecting beyond the protected. Cambridge University Press, Cambridge

Ostrom, E. (1990). Governing the commons. The evolution of institutions for collective action. Cambridge University Press

Owen, G. (1982). Game Theory. Springer

Pahl-Wostl, C. (1995). The Dynamic Nature of Ecosystems: Chaos and Order. Entwined Wiley, Chichester, 288

Pahl-Wostl, C. (2001). A New Understanding of Human Agency. Working paper, EAWAG
http://firma.cfpm.org/Documents/JASSforum2.htm

Peterson, S., (2002). Entscheidungstheoretische Analyse und Strukturierung von Mediationsverfahren zur Lösung von Umweltkonflikten. Dissertation, Agrar- und Ernährungswissenschaftlichen Fakultät der Christian-Albrechts-Universität zu Kiel. Internet Verlag www.dissertation.de

Petschel-Held, G., Schellnhuber, H.-J., Bruckner, T., Toth, F.L., Hasselmann, K. (1999). The Tolerable Windows Approach: Theoretical and Methodological Foundations. Climatic Change 41, 303-331

Priscoli, J.D. (1989). Public Involvement, Conflict Management: Means to EQ and Social Objectives. ASCE Jour. Water Res. Planning and Management 115 (1), 31-42

Pruitt, D.G., Carnevale, P.J. (1993). Negotiation in Social Conflict. Mc Graw Hill

Putnam, R.D. (1995). Bowling alone: America's declining social capital. Journal of Democracy 6 (1), 65-78

Raiffa, H. (1982). The Art and Science of Negotiation. University Press Cambridge, Belknap/Harvard , MA

Rotmans, J., Dowlatabadi, H. (1998). Integrated assessment of climate change: evaluation of methods and strategies. In: Rayner, S., Malone, L. (eds). Human Choice and Climate Change. Battelle Press, Columbus, Ohio

Scheffran, J. (2000). The Dynamic Interaction Between Economy and Ecology. Mathematics and Computers in Simulation (53), 371-380

Scheffran, J. (2001). Stability and control of value-cost dynamic games. Central European Journal of Operations Research 9 (7), 197-225

Scheffran, J., Stoll-Kleemann, S. (2003). Participatory Governance in Environmental Conflict Resolution. In: Deb, K., Srivastava, L. (eds). Transition Towards Sustainable Development in South Asia, New Delhi. The Energy and Resources Institute, 307-327

Schweitzer, F. (1997). Self-Organization of Complex Structures: From Individual to Collective Dynamics Vol. II. Gordon and Breach, London

Sebenius, J.K. (1992). Negotiation Analysis: A Characterization and Review. Management Science 38, 18-38

Sprinz, D.F., Wolinsky-Nahmias, Y. (2004). Models, Numbers, and Cases. Methods for Studying International Relations. University of Michigan Press

Teich, J. E., Wallenius, H., Wallenius, J., Zionts, S. (1996). Identifying Pareto-Optimal Settlements for Two-Party Resource Allocation Negotiations. European Journal of Operational Research 93, 536-549

Tuinstra, J. (2000). A discrete and symmetric price adjustment process on the simplex. Journal of Economic Dynamics and Control 24 (5-7), 881-907

van den Belt, M. (2004). Mediated Modeling. A System Dynamics Approach to Environmental Consensus Building. Island Press, Washington, Covelo, London.

van der Veen, A. (1999). Interactive decision-making: a review Report to the FIRMA project. ICIS http//firma.cfpm.org/documents/interactdecmaking.htm

von Neumann, J., Morgenstern, O. (1944). Theory of Games and Economic Behavior. Princeton University Press, Princeton N.J.

Weber, M., Barth, V., Hasselmann, K. (2003). A Multi-Actor Dynamic Integrated Assessment Model (MADIAM) of Induced Technological Change and Sustainable Change Economic Growth. Paper presented at EGS-AGU-EUG Joint Assembly. Nice, France, 06.-11. April

Weidlich, W. (2000). Sociodynamics - A Systematic Approach to Mathematical Modelling in the Social Sciences. Harwood Academic Publishers

Weidner, H. 1998). Alternative Dispute Resolution in Environmental Conflicts-Experiences in 12 Countries. Edition Sigma , Berlin

Welp, M., de la Vega-Leinert, A., Stoll-Kleemann, S., Jaeger, C.C. (2006). Science-based stakeholder dialogues: Theories and tools. Global Environmental Change 16, 170-181

6 To See or not to See, that is the Question: Geo-information Visualisation Tools as a Means to Facilitate Stakeholder Dialogues in Land and Water Management Planning

Marleen Maarleveld[1], Rob van de Velde[2,3], Joost van Uum[2], Irene Pleisier[3]

1 ARCADIS, the Netherlands
2 Service for Land & Water Management, Dutch Ministry of Agriculture, Nature and Food Quality, the Netherlands
3 Vrije University Amsterdam, the Netherlands

6.1 The stakeholder dialogue context addressed

As discussed in the chapters of this book, coping with complex natural resource management problems calls for an approach that involves stakeholder dialogues. Stakeholder participation is essential because of the character of the natural resource management problem addressed. Many of the natural resource problems faced may be characterized as a social dilemma. A social dilemma occurs when people find themselves in a situation in which their individual interests appear to conflict with their collective interests. In such situations people will often choose in favour of their individual interests though this choice is both individually and collectively disadvantageous. Well-known examples of such dilemma situations are the prisoner's dilemma and Hardin's tragedy of the commons (Hardin 1968, Ostrom 1990). In particular the latter example has shown how people in social dilemma situations will make choices that are not sustainable for the natural resources involved.

However, research and practice in natural resource management have pointed out that people are able to make more sustainable choices in social dilemma situations (Ostrom 1990, Ostrom et al. 1995, Gunderson et al. 1995). Essential in coping with such situations is that stakeholders understand the problem situation and the manner in which their decisions and actions interact and have consequences for the natural environment. Thus, to resolve such dilemma situations, the stakeholders involved will need to collectively reflect on and act in the situation they are facing. Stakeholders need to develop a shared understanding of the problem at hand, explore alternatives, take action, and evaluate outcomes. Stakeholder dialogues may contribute to understanding the different perspectives, insights and actions that lead to the problem situation and its resolution.

In this chapter, we will focus on experiences in facilitating stakeholder dialogues for sustainable land and water management planning. In land and water management, planning different social dilemma situations may be encountered. For example, rice farmers in the Philippines may choose to forego collective maintenance of the terrace system for short-term livelihood decisions. Or in the Netherlands, scarcity of space requires people to choose between natural environment interests and housing development interests when towns want to expand and develop houses in river plains. In addition, the growing number of actors involved, an increasing amount of information to be processed, and uncertainties involved contribute to the complexity of land and water management planning.

In a number of cases, geo-visualisation tools have proven to contribute to collective reflection and action processes in stakeholder dialogues. Cartographic and dynamic geo-visualisation of problem situations and possible action alternatives may help stakeholders to better understand the situation they face, to develop alternatives and to undertake collective decision-making and action. Accordingly, geo-information visualisation tools may provide a means to facilitate stakeholder dialogues for more sustainable land and water management planning. In this chapter we will further discuss and illustrate the value of geo- visualisation tools in this context.

With regard to theory, we draw on the perspective of planning as a learning process (Friedmann 1987, De Geus 1988, Van der Vlist 1998, Maarleveld 2003). This type of planning perspective helps to provide insights into collective reflection and action processes in land and water management planning. To better understand the possibilities of geo-based visualisation tools to facilitate stakeholder dialogue and decision-making, we draw on insights from geo-information science and cognitive science (Van Lammeren and Hoogerwerf 2003, Batty et al. 2002, Bill 1999, Weinman 1988). The main theoretical bases are discussed in Section 6.2. In Section 6.3, three case examples are presented and discussed in light of these insights. In conclusion, we draw a number of lessons in terms of facilitating stakeholder dialogues and the quality of geo-visualisation tools.

6.2 Theoretical perspectives for facilitating stakeholder dialogues through geo-information visualisation tools

In this section, insights from planning and learning theory are discussed to gain a better understanding of collective reflection and action processes that may play a role in stakeholder dialogues in land and water management planning. In addition, the potential of geo-information visualisation tools in such stakeholder dialogues is discussed in terms of developments in geo-information science and cognitive science.

6.2.1 Planning as learning

There are many different ways to regard planning undertaken in land and water management. In his review of major planning traditions, Friedmann (1987) distinguishes planning as social reform, policy analysis, social learning, and social mobilisation. Each of these perspectives has its own strengths and weaknesses. In this article we will focus on planning as learning in order to gain insight into the collective reflection and action processes to be facilitated in stakeholder dialogues. Studies on regional planning involving spatial, environmental and water management policy support such a focus (Van der Vlist 1998). In various fields directly or indirectly related to spatial planning, a learning viewpoint has also been found to act as a potential perspective for bringing about change for sustainable development. For example, in development practice a learning approach has been found conducive to developing sustainable community and farmer practices (Korten 1980,1984). Organisational and management practice and theory have turned to learning as a means to effectively cope with a more interconnected world and, as a consequence, with a more complex and dynamic business environment. In order to cope with such complexity, collective learning and organisational learning have been put forward (De Geus 1988, 1997, Senge 1990, Argyris and Schön 1996). In the field of policy analysis, learning has been used as a factor to explain and improve policy development (Glasbergen 1996, Eberg et al. 1997). For researchers and practitioners in natural resource management, learning has provided a means of capturing and managing sustainable development as an ongoing process versus a stable end state (Lee 1993, Finger and Verlaan 1995, Gunderson et al. 1995).

The notion of learning captures the link between understanding and action necessary to develop knowledge continuously and the ability to use it. This is illustrated in Kolb's learning cycle in Figure 6.1. Concrete experiences may be reason for reflection. One's window on the world (normative cognitive frame) determines which issues are viewed as problematic and which are not. Abstract conceptualisation may lead to the development of new ideas. These ideas need to be tested in practice, which leads to new concrete experiences.

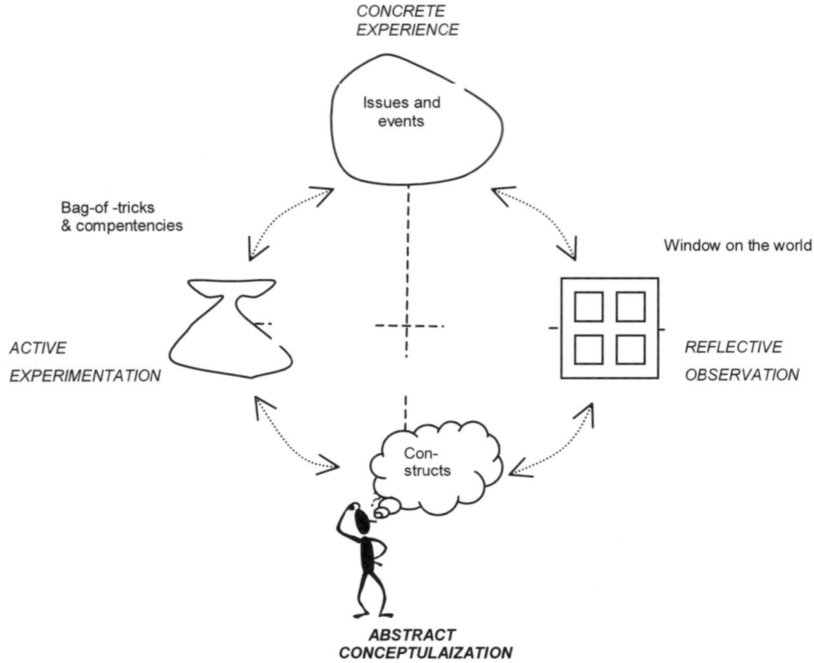

Fig. 6.1 Kolb's learning cycle (adapted from Kolb 1984).

Planning may be viewed as a learning process, as illustrated in Figure 6.2. People involved need to become aware of a problem issue, analyze the problem more closely, explore options, implement the option of choice, and monitor and evaluate whether actions have the desired result. As most planning processes involve a number of people, there is dialogue and decision-making throughout the process.

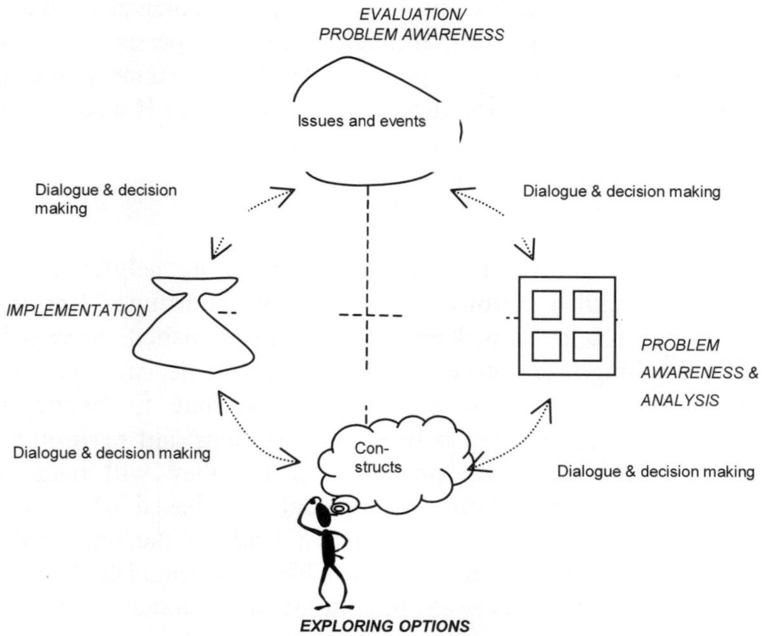

Fig. 6.2 Planning as learning.

Moreover, the constituting components of a planning cycle also entail learning cycles. Problem awareness and analysis, exploring options, implementation, and evaluation all involve cycles of concrete experience, reflection, abstraction and experimentation. As such, planning may be viewed as a complex of multiple learning cycles.

Viewing planning from a learning perspective draws attention to a number of insights to take into account when facilitating stakeholder dialogues. These insights may be summarized in terms of four questions: Who is learning? What is learned? How is it learned? and Why is learning taking place?

Who is learning?

In terms of the question of who is learning, the stakeholders involved in the learning process are an important point of focus (Lee 1993, Röling 1994). Systemic change, i.e., change in the normative frames that guide people's behaviour, has been found to occur primarily when all parts of the system learn to understand how the system works (Weisbord and Janoff

2000). Thus, to realize sustainable land and water management planning, the whole range of stakeholders, all having their own perspective, need to be involved in the learning process. Collective understanding by stakeholders of how the system they are a part of works is a starting point for learning to renew (see "What is learned?").

What is learned?

The learning loops of Argyris and Schön (1996) are helpful diagnostic concepts to distinguish various aspects of what constitutes learning. In Figure 6.3, the three levels of learning loops distinguished are visualized. Single-loop learning takes place when the results of decision-making and action are evaluated in terms of the way they contribute to realizing goals and expectations. A mismatch between expectations and performance is resolved by improving actual practices so that they will better meet existing goals and expectations. These goals are based on underlying values and assumptions. When a mismatch leads to the questioning of existing goals and expectations, it is possible to distinguish double-loop learning. Such learning leads to a reframing of values and assumptions that underlie behaviour. The cognitive frames questioned may be individual windows on the world as well as collective ones embedded in organisations and institutions. In this questioning process, people may learn that common underlying values and assumptions underlie the contradictions and dilemmas they are facing. Such shared values and assumptions may be the basis for new joint goals. Such deep, systemic change enables a collective to renew.

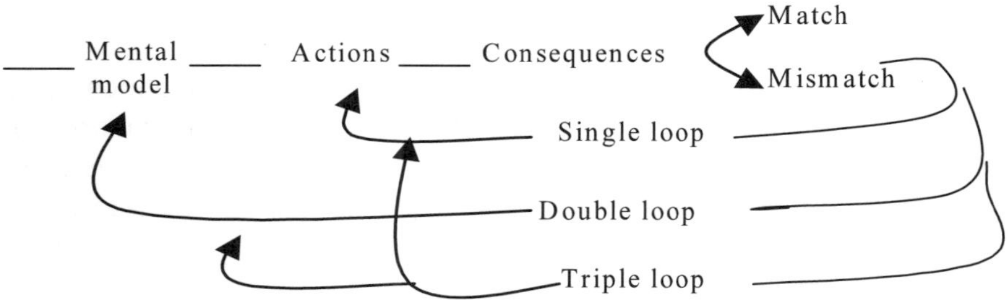

Fig. 6.3 Single, double, and triple loop learning.

Triple-loop learning may be viewed as a specific type of double-loop learning, namely, when such learning concerns the way learning itself takes place.

How is it learned?

Kolb's learning cycle in Figure 6.1. illustrates how learning may take place. Different people have been found to have different biases for ways of learning. In such learning biases, different aspects of the learning cycle dominate. Three different learning modes may be distinguished. In learning through direct experience, concrete experiences and active experimentation form the basis of learning. A bias for reflective observation characterizes learning through observation. Furthermore, abstract conceptualisation is characteristic of learning through abstraction, i.e., extracting common features from seemingly diverse responses and formulating rules of behaviour that go beyond what has been experienced or observed.

Why is learning taking place?

Learning may be triggered by both proactive and reactive motivations for change. On the one hand, some people have an innate desire to create and develop. Such proactive motivations also provide triggers for learning and change. On the other hand, some people will react negatively and even resist change when mismatches between expectations and performance make clear that a previously set goal will need to be adjusted. In other words, maintenance of existing cognitive frames may trigger people's learning.

6.2.2 Geo-information visualisation tools as a means to facilitate stakeholder dialogues and decision-making

Geo-visualisation techniques that support communication on spatial conflicts, challenges and future scenarios have been in the spatial planner's toolkits for decades (Van Lammeren 2003). Graphical presentation of information has a long history. Maps are some of the earliest existing geo-information visualisation tools. Cartography has had, and continues to have, an important role in the graphical presentation of geospatial information (Fairbairn 2001). Fairbairn (2001) defined cartographic representation "as the transformation that takes place when information is depicted in a way that can be perceived, encouraging the senses to exploit the spatial structure of the portrayal as it is interpreted." It is very hard to imagine stakeholder dialogues in planning without maps playing a significant role to inform, communicate, and design.

Geographical information systems (GIS) date back to the early seventies. First applications of GIS originated in landscape architecture and physical planning. GIS have become a useful tool in visualising complex computer-based data for spatial land and water management plans. In a field where transparency is of key importance and where many public and private stakeholders are involved, high quality mapping of current and future situations is needed. Developments in information technology have made it possible to efficiently store, manipulate, and visually present complex and large amounts of data. Moreover, developments in information technology enable the development of geo-information infrastructure, which supports participation of stakeholders in planning processes. For example, interactive technology developed for computer games may be adapted for geo-visualisation tools. Geographic information systems have already been widely used in stakeholder dialogues in third world countries (Harris and Weiner 2003) as well as in the developed world (Harrison and Hacklay 2002). In both cases, GIS has proven to be a good tool to support an interactive planning process. GIS allows instant interactivity, can visualise the plans in 2D at various locations, and can be instantly altered to process the comments and suggestions from all stakeholders involved.

The presentation of spatial plans to stakeholders and participants in planning processes mostly makes use of 2D visualisations. Gradually this method of presentation has been extended with presentations that make use of computerised 3D visualisations (see Batty et al. 2002). 3D visualisation provides an effective way of presenting large amounts of complex information to a wide audience. 3D visualisations help to give a more realistic picture of future changes in landscapes and allows the user to relate information and reality more easily.

A combination of scale models and GIS seems to be an ideal basis for 3D presentation and development of spatial plans. For many years, planners have combined real world representations with virtual/future objects in scale models to represent future changes in the landscape. Such scale models have been used to present detailed spatial plans to the public. This type of representation has been found to be easy to comprehend and to give a good overview of the plans. However, a scale model also has numerous disadvantages. A scale model might be a large, rigid, solid thing that can only be kept at a specific location. In the field of spatial planning, this location is usually a project office or an information centre. Moreover, interaction with a scale model is difficult. Background information cannot be offered on the fly, and new ideas cannot be visualised instantly.

In combining the scale model and GIS approaches, the 3D effect of a scale model for visualising the future situation and the interactivity and adaptability the GIS component are brought together. This combination is called virtual reality or VRGIS (Hacklay 2001). A 3D computer model is generated that shows the current or future situation (Dias et al. 2003, Verbree 1998). Users can explore the model by simply navigating through the virtual reality environment. Virtual reality can be very useful for presenting large amounts of information effectively to participants within spatial planning. Participants without any planning experience can effortlessly relate the visualized information to the real world. Virtual reality is described by Fisher and Unwin (2002), as "the ability of the user of a constructed view of a limited digitally-encoded information domain to change their view in three dimensions causing update of the view presented to any viewer, especially the user".

There are important cognitive aspects related to visualisation, perception and understanding of spatial information to take into account when developing geo-visualisation tools. Understanding the different information dimensions and media types and how these relate to different senses is useful for understanding how to develop geo-visualisation tools for spatial objects and spatial planning processes (Bill 1999). The media used for visualisation of spatial information may have four types of functions according to cognition science (Weinman 1988):

- the function of demonstration,
- the function of putting into context,
- the function of construction,
- the function of motivation.

The *function of demonstration* is achieved by using media to give a realistic picture (demonstrate the idea, object or landscape). This can be achieved with the support of photos, videos or virtual reality. The media with the *function of putting into context* should help the user put the detailed information into a bigger context, like an overview of the area (for spatial context), or sounds that are related to a particular area may help the user to identify and position the given information. The *function of construction* is related to the creation of complex mental models by the user (mental models are constructions of knowledge about information units and relationships). Abstract media of pre-prepared information is best suited for this function, such as graphs, diagrams or abstract layers. Finally, the media can have the *function of motivation*. Media with this function intend to arouse the user's interest and attention. This can be achieved with animations, interactive objects, e.g. interactive flyovers are a typical example of this function (Bill 1999).

Besides the media functions, when developing a system to visualize and perceive spatial information, one should pay attention to cognitive processes of:

- Short-term memory's limited cognitive capacity;
- Increasing important information;
- Avoidance of overloading a single sense;
- Supporting double encoding of information.

Because *human short-term memory* is only able to process seven information units at the same time, the spatial information system should not provide too much information simultaneously. Multiple representations can overcharge the human cognitive capacity, but they can also emphasize important information and improve information processing, if used in the right way. (Maps, pictures, sounds, and videos can be used in combination to *increase important information*). Also, a combination of visual and sound information helps the user's perception by avoiding the *overload of a single sense*. The human memory can store information in pictorial and textual formats (*double encoding*), so pictures in combination with written or spoken text should be used to describe information (Bill 1999).

6.3 Geo-visualisation practice in the facilitation of stakeholder dialogues and decision-making in land and water management planning

In this section three cases are discussed in which geo-visualisation tools have contributed to facilitating stakeholder dialogues in land and water management planning. For each case a brief overview of the case context, the geo-visualisation tool used and its effect, and the lessons learned are presented. The different cases have been chosen because they illustrate how geo-visualisation tools may play a role at various phases of a planning cycle. The cases discussed are: joint learning for water management in the Ifugao, Philippines (planning cycle: realizing the problem); visualizing consequences of flood management choices in the EU (planning cycle: exploring alternatives); and flying through planned urban expansion in Groningen, the Netherlands (planning cycle: abstraction/experiencing the future).

6.3.1 Realizing the problem: Joint learning for watershed management in the Ifugao, Philippines[1]

The case context

The landscape of the Ifugao consists of rugged mountains, low-lying hills, and an alluvial area along the Magat River. The province is located about 320 kilometres north of the Philippines' capital city Manila. The Ifugao's centuries-old rice terraces are world famous for their ingenuous engineering in extreme environmental conditions. The terraces reach the highest altitude (1600m) found in the Asia-Pacific region. They are a well-built, extensive engineering and hydraulic system, using traditional skills. Research has shown that this traditional agro-ecological system has been able to support a relatively high population density for many centuries without depleting its natural resources. As such, the Ifugao terraces have been added to the UNESCO World Heritage List of cultural and natural properties considered to be of outstanding value.

However, closer inspection makes visible the crumbling walls of the terrace system. Abandoned and broken-down terrace walls, thinning forests, landslides, erosion, slash-and-burn farming, extremely high poverty of inhabitants, loss of traditional knowledge, irresponsible tourism, and dependency on government and project support are just a number of the problem issues facing the Ifugao people and the terraces. So while ecologists and conservationists regard the terraces as one of the soundest soil and water conservation structures ever built by people, Ifugao has the largest area affected by moderate to severe erosion in the region.

Local, regional and national public and private efforts have been undertaken to reverse the situation. However, divergent views, goals and working methods of the different actors involved have led to clashes as organisations involved try to work according to their own development paths. As a result, not only has the erosion of the environment not been stopped, but the interacting network of external and local stakeholders has created a situation in which, at the time of the research of Gonzalez (2000), the divergent views and goals clash rather than converge toward a collective understanding of the problem situation and a working strategy.

The GEO-visualisation used and its effect

In order to better understand spatial dimension of the Ifugao's problem situation, Gonzalez combined aerial photographs and remote-sensing data

[1] Gonzalez (2000)

and discussed them with local inhabitants. The GIS visualisations were also used in discussions with provincial board and other stakeholders involved.

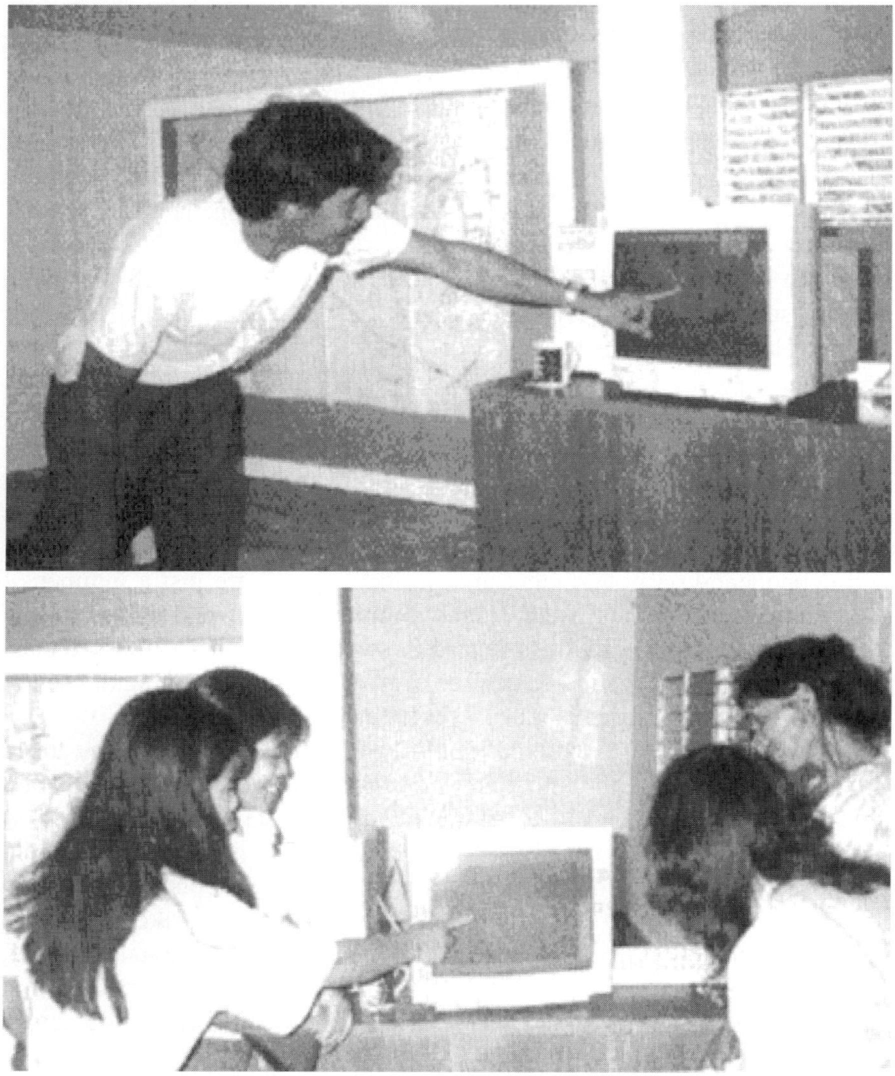

Fig. 6.4 Combining maps, aerial photos, and GIS to discuss and understand the watershed management problems.

Because the Ifugao is a rather remote area, some creativity on the part of the researcher was necessary. The battery of the "jeepney", the local transportation, provided power for the computer laptop. The locals proved capable of greatly improving the data, as traditional terrace management revolved around geographical agricultural dimensions. In the discussions of the maps, people (locals and outsiders) became more aware of boundary issues and degradation problems. Locals were so enthusiastic that they decided to start watershed monitoring with the help of the GIS visualisation tool. Overall, the geo-visualisation tool helped with the functions of putting into context, of construction, and of motivation.

Lesson learned

Direct involvement of local Ifugao stakeholders in developing GIS-based watershed management data and visualisation has made a twofold contributed to learning. On the one hand, the approach taken has improved integration of quantitative and qualitative spatial information available from the local level up to international levels. On the other hand, anchoring the development of GIS and its outcomes in the experience of local stakeholders has created a tool for facilitating a dialogue of ideas about the space that the Ifugaos are managing with others (see Figure 6.5).

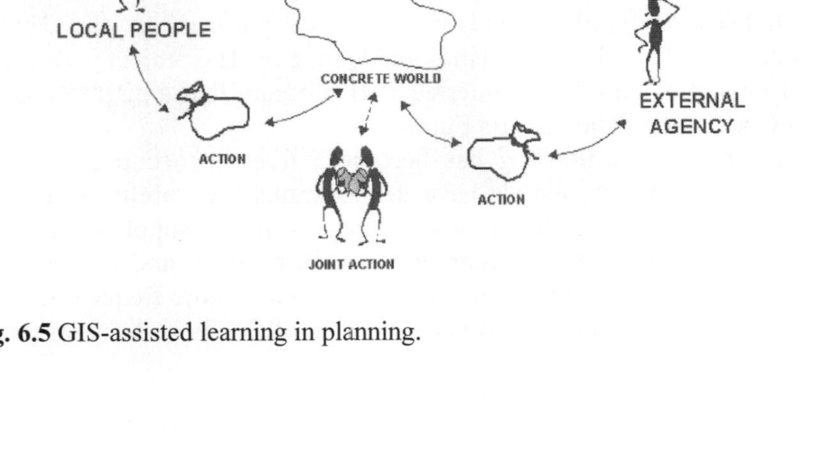

Fig. 6.5 GIS-assisted learning in planning.

With GIS at their disposal, stakeholders were able to construct alternative perspectives about their environment (e.g. as map layers) and discuss or negotiate them (e.g. as screen displays, overlays, aggregation) in order to arrive at shared knowledge, and hopefully the wisdom to act accordingly. The researcher gained new insights with regard to the meanings people ascribe to and agree upon regarding their environment. Seeing together the area they live in on a regional level, local people realized how they fared compared to others. They also learned new GIS techniques and thought of new, local applications. At the provincial level, stakeholders recognized the potential of the use of GIS tools in management activities, such as settling boundary disputes, monitoring reforestation projects, and getting an overview of terrace conditions. Overall, local, provincial, and outside stakeholders were involved in the learning process that involved single, double, and even triple (learning how to learn) loop learning. Learning took place by direct experience but also by observing how others used the GIS tool. Moreover, successfully using the GIS visualisation tool involved learning by abstraction. The learning process was at first triggered by reactive motivations for change, but slowly more proactive triggers started to play a role.

6.3.2 Exploring alternatives: Visualizing consequences of flood management choices in the EU[2]

The case context

Floodscape is a project that aims to demonstrate that flood management can be achieved by making space for water during flood events while maintaining normal use of the land. In addition, the project aims to involve local stakeholders in this new approach to flood management. It is a four-year transnational, EU-funded project with partners in the United Kingdom, Belgium, the Netherlands and Germany. The project is financed and promoted as part of the Interreg IIIB program that aims to generate interregional cooperation across Europe.

River and coastal flooding has become a frequent occurrence across many parts of Europe. Floods have demonstrated their ability to cripple cities and towns, destroy homes and businesses, power supplies, transport infrastructure, and communication systems. In recent years, flooding has increased noticeably. Climate change has resulted in more frequent rainfall and an increasing number of storms, all causes of flooding. As cities grow,

---

[2] www.floodscape.net

many natural flood plains have been built upon in response to high demand for housing. Such housing development reduces space available to rivers for floodwater and means that flood defences have to be built to protect properties.

In the past, floodwater has been controlled by building walls, embankments, gates, and barriers. As climate changes and its consequences have become more unpredictable, new solutions need to be found. Higher flood defences are no guarantee for flood damage prevention. Building higher defences is also expensive and has major impact on the landscape, wildlife, and people's enjoyment of river spaces. In this project, a new approach has been developed to manage flooding that will benefit people by:

- restoring wetlands and river habitats, making space for nature;
- providing open and aesthetic riversides;
- enabling easy access to and egress from the river.

The Floodscape project aims to gain experience with this approach in seven pilot actions in the participating countries.

One of the pilot actions is taking place in the *Hurwenense Uiterwaard* in the Netherlands. The *Hurwenense Uiterwaard* is part of the Rhine flood plain of the river *Waal*, a tributary of the Rhine. The foreland of this tributary is expected to be able to contribute to the flood-risk reduction plan for the Rhine flood plain. The challenge is to meet targets for flood-risk reduction through means other than raising existing dykes. In this light, possibilities are explored for large-scale nature-development programs in flood plain areas, for example, lowering the winter bed, creating parallel side channels to the river channel, and creating large-scale pools. For the *Hurwenense Uiterwaard*, such an approach provides opportunities to explore the development of new natural habitats such as marsh vegetation, rough grasslands, and possible river bound forest.

Pollution of sediments as a result of former industrial activity is a problem for large parts of the *Hurwenen* floodplain. Large quantities of polluted soil need to be extracted and disposed of in an environmentally sound way. The *Hurwenense Uiterwaard* pilot action, therefore, seeks to develop more space for water to manage flood risk, create sustainable nature conservation, and maximize agricultural and recreational opportunities as well as to address the problem of soil pollution. The pilot action was further developed during 2003 – 4 in an environmental impact assessment process (EIA) that includes:

- formulation of an area development plan for the floodplain;
- assessment of the plan in terms of different options for its development;

- undertaking of specific research, e.g. ecological and archaeological surveys;
- community consultation and involvement in preparation of the plan;
- relationship with the EU Habitat Directive: *Hurwenen* is to be appointed as a Habitat Directive protected area.

The combination of the area development plan and the environmental impact assessment has made the planning process and finding of solutions more complex because competing national objectives (flood relief and nature development) need to be balanced.

The development of the area plan and the EIA is expected to be an iterative process. The planning process involves different planning cycles in which area consultations and a local advisory group participate. The advisory group includes representatives of local community organisations, local business, representation the local branch of the provincial agricultural organisation, local fishing and hunting organisations, and local nature, environment and landscape organisations. The Advisory Group constantly re-evaluates results of the planning process and where necessary call for further research, analysis of additional options, and re-evaluation of existing alternatives.

The GEO-visualisation used and its effect

Fig. 6.6 Geo-information based visualisation of water retention effects in *Hurwenense Uiterwaard.*

The Service for Land and Water Management, a national government organisation responsible for the implementation of rural development, has developed a geo-visualisation tool in which geo-information data (geomorphologic data, contour maps) is used in combination with aerial photographs of the *Hurwenense Uiterwaard* to visualize effects of different choices to allow plain flooding. In Figure 6.6., both the current situation and two alternatives in the first planning document are visualized as they appear in the GEO-visualisation tool. This tool is interactive. In the right hand corner of the computer screen, users can manipulate the water level and see the water-retention effects accordingly. The geo-visualisation tool is used in the discussion with experts, decision-makers, and will - in the next phase of the tool's development – also be used with people in the area. In a later stage in the planning process, alternatives and their effects will also be visualized in greater detail. Overall, the geo-visualisation tool has aided with the functions of putting into context, of construction, and of motivation.

Lessons learned

The geo-visualisation tool contributes to an increased understanding of the effects of the different alternatives. The visualisations have helped to make clear what the different experts involved mean with their sometimes complex jargon. The proverb a picture says more than a 1000 words has been proven once again. Such understanding improves the quality of the problem analysis, discussion, and decision-making regarding the possible alternatives. The quality of the discussions and decision-making will increase commitment of stakeholders and contribute to the EIA. Improved communication:

- increases transparency of the planning process;
- speeds up decision-making;
- increases possibilities for stakeholder participation.

In terms of learning, at this point in the planning process, experts are primarily involved. By direct experience, they are learning about the effects of their communication and how to be more effective communicators. The EU project contexts provide a proactive frame for the development of the GIS visualisation tool.

6.3.3 Experiencing the future: Flying through planned urban expansion in Groningen, the Netherlands[3]

The case context

Groningen Meerstad is the name of a housing development project on the eastern side of the city of Groningen (see Figure 6.7), which is located in the northern part of the Netherlands and has 175,000 inhabitants.

[3] www.meerstad-groningen.com

Fig. 6.7 Bird's-eye view of the urban housing development project *Groningen Meerstad*, the Netherlands

Meerstad is a complex project with different, at first sight contradictory, goals and many stakeholders. The city of Groningen has been searching for new locations to construct housing developments for a growing population and its demand for quality housing. In this respect, Groningen is not different from many other cities in the Netherlands or for that matter in the world. Initially, administrators, civil servants and project developers expected that the eastern part of the city would not need to be developed for housing until after 2010. However, successful economic development has made shorter-term housing developments at this potential location necessary. In 1998 plans to this end gained momentum when a design for multi-functional development of housing, landscape, and water management was nominated by STIR (*Stimulerings fonds Intensief Ruimtegebruik*), a fund set up by the Dutch Ministry of Housing, Spatial Planning and Environment that aims to stimulate multiple land use. The Groningen project aims to combine housing development with water and nature management. A creative spatial design and the choice to "invest a priori in landscape" make this project an innovative example of the implementation of the "red for green" policy principle. The current situation is an agricultural area of 4,000 hectares. A new urban settlement with 10,000 dwellings in a landscape of high quality (environmentally friendly light industry and sufficient room for recreation and leisure) will be developed. A lake of 650 hectares with recreational, water buffering, and storage purposes is central in the landscape development. The name *Meerstad* is a word play in Dutch, meaning both 'more city' as well as 'lake city'.

From the beginning, *Meerstad* started as an open planning process. Starting point for the development of *Meerstad* is an open communication process in which participants work together to create a final master plan in 2006. In 2004 and 2005 several sub-plans were scheduled for completion, resulting in a final master plan in 2006. The actual implementation of the master plan will start in 2006 with an end in 2020. Together, the governing bodies, private companies, local citizens, and societal organisations have created a concept master plan in 2003. The main objective of this open planning process is to take into account the wishes, ideas and thoughts of all stakeholders: in other words, to compare all the needs and demands of the different groups of stakeholders, resulting in a master plan. The challenge in this process is to create a high-quality land use in a balanced way for all stakeholders and to enable all land-use functions in one area by using a multi-sectoral approach.

The GEO-visualisation used and its effect

The Service for Land and Water Management has developed a virtual landscape for the Groningen *Meerstad* project, and a prototype has been "filled" with local Groningen data to allow stakeholders (and others interested) to fly through the newly planned urban development. The virtual landscape viewer integrates different geospatial datasets into a 3D landscape through which stakeholders are able to fly over to "zoom in", for which more detailed and different geo-referenced data is used. The user interaction is made possible through the keyboard, mouse movements and clicks, or any other computer interaction hardware like a joystick, similar to that used in computer gaming. The interaction hardware allows the user to move in the landscape freely. Moreover, a user can increase and decrease speed and click on objects to retrieve extra multimedia information. Other features of the virtual landscape include the possibility to follow predefined paths and move to relevant predefined positions.

Orientation and navigation concepts are related to the "travel metaphor". A user is able to:

- identify the current position through an orientation map of the overall landscape;
- reconstruct the route that leads to that position;

- distinguish the different options for moving on from the current position via the menu buttons;
- distinguish direction movements; a compass indicates the direction the user is facing.

Recognisable landmarks act as orientation points for relevant places and also contain multimedia links to extra information. Links can be pictures, videos, web pages, sound, and messages, and the virtual landscape tool generally contributes to the functions of demonstration, construction, and motivation.

Lesson learned

In the Groningen *Meerstad* project, the stakeholders group is numerous and heterogeneous, with different sensibilities and varying interests and concerns about the project. It is fundamental to communicate the complex information involved in an understandable manner. In this way, stakeholders will have a same understanding of the goals and consequences of the project. A 3D geo-visualisation tool such as the virtual landscape provides an effective way of presenting large amounts of complex information to a wide audience, including those with no Geographic Information Systems (GIS) or mapping experience. The system design has taken into consideration cognitive principles and is able to integrate high-quality mapping of the current situation, 3D representations of the future, and (geo) multimedia (regarding real world information). Stakeholders have indicated that the virtual landscape tool helps them to understand the proposed plans and proposed changes. Moreover, stakeholders have fun using the virtual landscape to fly through the newly planned urban development. Such a mood motivates participants for discussions and commitment to the project.

6.4 Conclusion: Seeing is believing

Facilitating stakeholder dialogues

The theoretical insights with regard to planning as learning and the use of GIS visualisation tools, as well as the cases discussed, make clear that facilitating stakeholder dialogues plays an important role in sustainable land and water management planning. The problems faced are complex in nature. This complexity makes it impossible for any single individual to resolve them all. In other words, different individuals will need to work together in order to gain an understanding of the problems faced and their resolution. For example, understanding the erosion problems faced in the Ifugaos, exploring alternatives for flood management in the *Hurwenense Uiterwaard*, and planning a new urban housing development project in Groningen *Meerstad* all require the involvement of various people with diverse interests and knowledge. The problems faced are also complex in nature in the sense that no one single, objective solution exists. Often new knowledge needs to be developed and new combinations of existing knowledge established; solutions are the result of learning and negotiation processes among stakeholders. The three cases presented illustrate how different stakeholders with different perspectives interact in terms of individual and collective learning. In creating opportunities for interaction, stakeholder dialogues can facilitate linking learning cycles of people involved in land and water management planning.

Facilitating stakeholder dialogues with geo-information visualisation tools

Both the theory and cases discussed underscore the value of geo-visualisation tools to facilitate stakeholder dialogues for sustainable land and water management planning. Geo-visualisation tools stimulate functions of demonstration, of putting into context, of construction, and of motivation. Through these functions, geo-visualisation tools contribute to important aspects of stakeholder dialogues. The cases presented show how different 2D and 3D visualisations help stakeholders get a realistic picture of the situation (demonstration); understand how their situation fits in a larger picture (putting into context); give meaning to the planning process (construction); and arouse participants' interests and attention (motivation). Depending on the use of the geo-visualisation tool in the planning process, these functions may be triggered at the individual and/or collective level.

Quality of geo-information visualisation tools

The value of geo-information visualisation tools in stakeholder dialogues not only depends on the manner in which the tool is used in land and water management planning but also on the quality of the tool. This means taking into account quality standards with regard to visual materials used and the user friendliness of the tool. The quality of the tool is dependent on the degree to which stakeholders are able to recognise their environment and relate it to the changes occurring or desired. Geo-visualisations need to take into account the possibilities and limits of people's cognitive capacity.

This often means limiting both the quantity and complexity of information. It can also mean trying to involve different senses to avoid the overload of a single sense. The case studies and the learning perspective also point out that involving stakeholders in the tool design and analysis, i.e., participatory technology design and participatory planning processes, empowers the individual and collective reflection and the action taking place.

Overall, it may be concluded that geo-visualisation tools are useful to involve people in stakeholder dialogues for land and water management planning. Visualisation is powerful in different ways. More than fifty percent of the neurons in the brain are used in vision. In addition, visualisations helps to make visible collectively what may be hidden in the thinking and action of individuals. Thus geo-visualisations tools provide a means to facilitate and improve the quality of stakeholder dialogues.

References

Argyris, C., Schön, D.A. (1996). Organizational learning II: Theory, method, and practice. Addison-Wesley, Reading

Batty, M., Fairbairn, D., Ogleby, C., Moore, K.E., Taylor, G. (2002). Introduction Virtual cities. In: Fisher, P., Unwin, D. (eds). Virtual reality in geography. Taylor and Francis, New York

Bill, R., Dransch, D., Voigt, C. (1999). Multimedia GIS: concepts, cognitive aspects and applications in an urban environment. In: Camara, A., Raper, J. (eds). Spatial Multimedia and Virtual Reality. Taylor and Francis, New York

De Geus, A. (1997). The Living Company. Growth, learning, and longevity in business. Brealy, London

De Geus, A. (1988). Planning as learning. Harvard Business Review March-April, 70-74

Dias, E., Van de Velde, R., Nobre, E., Estêvão, S., Scholten, H. (2003). Bridging the gap between spatial perception and spatial information. Proceedings of the 21st International Cartographic Conference (ICC), Durban, South Africa, 10.-16. August 2003. Cartographic Renaissance, 626-636

Eberg, J., van Est, R., van der Graaf, H. (1996). Leren met beleid. Beleidsverandering en beleidsgericht leren bij Nimby-, milieu- en technologiebeleid. Spinhuis, Amsterdam

Fairbairn, D., Andrienko, G., Andrienko, N., Buziek, G., Dykes, J. (2001). Representation and its relationship with cartographic visualisation: a research agenda. CaGIS 28 (1)

Finger, M., Verlaan, P. (1995). Learning our way out: A conceptual framework for social-environmental learning. World development 23, 503-513

Fisher, P., Unwin, D. (2002). Virtual reality in geography. Taylor and Francis, New York

Friedmann, J. (1989). Planning in the public domain. From knowledge to action. Princeton, Jersey

Gemeente Groningen, dienst Ruimtelijke Ordening en Economische Zaken. Groningen (1998). Maximum laadvermogen, voorbeeldproject intensief ruimtegebruik, Groningen

Glasbergen, P. (1996). Learning to manage the environment. In: Lafferhand, W.M., Meadowcraft, J. (eds). Democracy and the environment. Problems and prospects. Edward Elgar, Cheltenham

Gonzalez, R.M. (2000). Platforms and terraces. Bridging participation and GIS in joint-learning for watershed management with the Ifugaos of the Philippines. Wageningen, Published dissertation

Gunderson, L.H., Holling, C.S., Light, S.S. (1995). Barriers and bridges to the renewal of ecosystems and institutions. Columbia University Press, New York

Haklay, M. (2001). Virtual Reality and Geographical Information Systems: Analysis and trends. In: Fisher, P., Unwin, D. (eds). Virtual Reality and Geography. Taylor and Francis, London

Hardin, G. (1968). Tragedy of the commons. Science 162, 1243-1248

Harris, T., Weiner, D. (2003). Linking community participation with geospatial technologies. Aridlands newsletter 53

Harrison, C., Haklay, M. (2002). The potential of public participation GIS in UK environmental planning: appraisals by active publics. International Journal of Environmental Planning and Management (JEPM) 45 (6), 841-863

Kolb, D. (1984). Experiential learning: Experiences as the source of learning and development. Prentice-Hall, Englewood Cliffs

Korten, D.C. (1980). Community organization and rural development: A learning process approach. Ford Foundation, New York

Korten, D.C. (1984). People-centered development: Contributions toward theory and planning frameworks. Kumarian, West Hartford

Lammeren, R. van, Hoogerwerf, T. (2003). Geo-Virtual reality and participatory planning. CGI-report 2003-07,Wageningen

Lee, K. (1993). Compass and Gyroscope. Integrating science and politics for the environment. Island Press, Washington D.C.

Maarleveld, M. (2004). Facilitating social-environmental learning for sustainable natural resource management. Published Doctorate Thesis, Wageningen

Ostrom, E. (1990). Governing the Commons. The evolution of institutions for collective action. Cambridge University Press, Cambridge

Roling, N. (1994). Platforms for decison-making about ecosystems. In: Fresco, L., Stroosnijder, J., Bouma, J., Van Keulen, H. (eds). The future of the land: Mobilizing and integrating knowledge for land use options. John Wiley and Sons, Chichester

Senge, P.M. (1990). The fifth discipline: The art and practice of the learning organization. Century Business Press, London

Van der Vlist, M. (1998). Sustainaibility as a task of planning. Coordination and integration at the regional level among spatial planning, environmental planning and water management planning in the rural areas of the Netherlands. Published Dissertation, Wageningen

Verbree, E., van Maren, G., Germs, R., Jansen, F., Kraak, M.-J. (1999). Interaction in virtual world views – linking 3D GIS with VR. International Journal of Geographical Information Science 13 (4), 385-96

Weinman, B., (1988). Psychische Prozesse beim Verstehen von Bildern. In: Bill et al. (1999)

Weisbord, M., Janoff, S. (2000). Future search: An action guide to finding common ground in organizations and communities. Berret-Koehler, San Francisco

Part III
Case Studies in Environmental Policy, Management and Science

7 Science-based Stakeholder Dialogues in Climate Change Research

Martin Welp[1], Anne C. de la Vega-Leinert[2], Susanne Stoll-Kleemann[3], Cornelia Fürstenau[2]

1 University of Applied Sciences Eberswalde, Germany
2 Potsdam Institute for Climate Impact Research, Germany
3 Humboldt University of Berlin, Germany

7.1 Introduction

Science-based stakeholder dialogues are structured communication processes linking scientists with societal actors that are relevant for the research problem at hand. Rather than being objects of research, the stakeholders are partners in dialogues, in which the exchange of arguments is the distinguishing feature. The richness and relevance of such dialogues usually increases if there is a safe space in which a broad range of viewpoints can be freely expressed. Scientists have started to create forums which provide a platform for such interaction and consciously seek dialogues by organising workshops or by launching joint research projects. Science-based stakeholder dialogues can be regarded as a distinct approach to knowledge creation, in which researchers actively seek to incorporate non-scientific knowledge in the research process.

Different streams of literature implicitly or explicitly deal with science-based stakeholder dialogues. Post-normal science, transdisciplinary research and 'Mode 2' knowledge production are the most prominent approaches addressing the need for more stakeholder involvement, each emphasising different aspects of dialogues. The concept of post-normal science developed by Funtowicz and Ravetz (1993) can be characterised as science where the traditional fact/value dichotomy can not be maintained. This line of literature is therefore relevant for what will be outlined in this chapter. We speak of post-normal-science when "facts are uncertain, values in dispute, stakes high and decisions are urgent" (Funtowicz and Ravetz 1993: 744). Under the conditions of 'soft' facts, hard value-related decisions must be made. This requires the involvement of non-scientific knowledge. In this context Hage et al. (2005) see quality assurance as one of the major challenges of post-normal science. To address this challenge Funtowicz and Ravetz suggest 'extended peer communities', who "deploy 'extended facts' and take an active part in the solution of their problems" (Ravetz 1999: 647). Participants of these 'extended peer communities' can be all kind of stakeholders from the business, policy or NGO world. Each group can enrich the research process with their local, environmental or sectoral knowledge.

In transdisciplinary research issues are addressed from more than one viewpoint simultaneously (Pohl 2005). To solve complex problems, such as biodiversity loss or climate change a traditional disciplinary approach is not enough. But the line of argumentation goes beyond this: besides different disciplines researching together, research needs to take into account the knowledge outside the scientific sphere. Research will be socially relevant only if the traditional ways knowledge is produced and organised change. One line of interdisciplinary research for example concerns the collaboration between research institutes and industry/the private sector - an issue which is of high relevance for science-based stakeholder dialogues.

Mode 2 knowledge production described by Gibbons et al. (1994) also emphasises transdisciplinarity. The authors observe a shift in those organisations that produce knowledge away from established institutions towards more heterogeneity and reflexivity. Mode 2 knowledge production implies that dialogues play an increasingly important role in critically scrutinising arguments presented by different organisations. A common nominator for the three above-mentioned lines of literature is viewing research as a process of mutual learning. Science-based stakeholder dialogues are part of the practices that have been described as transdisciplinarity, Mode 2 knowledge production or post-normal science.

From the scientists' point of view relevant stakeholders in a research process may include representatives of the private sector, NGOs, governments, citizen groups or lay persons (Welp et al. 2006). The main difference between using traditional social science approaches (such as interviews or questionnaires) and facilitating science-based stakeholder dialogues is that the latter fosters participatory and collaborative research and promotes mutual learning between all actors involved.

The objectives of science-based stakeholder dialogues may include some of the following: identifying socially relevant research questions, providing a 'reality check', incorporating ethical and value considerations in assessments, and accessing stakeholders' knowledge.

A research process should ideally include several iterations of dialogues, which take place over a long period of time. Different stages may have different objectives. Cycles of stakeholder dialogues may start with identifying relevant research questions and move on to phases of consultation, developing models, reviewing and modifying these models and coming to new conclusions.

There are few recipes to guarantee a successful stakeholder dialogue. The degree of stakeholder involvement, its timing and iteration, and the methods to collect and analyse knowledge uncovered and produced during dialogues are critical aspects to consider. Each research project is designed and run according to its specific research objectives, participants, available resources, etc. Science-based stakeholder dialogues are as diverse as the research questions they explore. Each dialogue is thus a unique process, which will yield unique results. This uniqueness however does not mean that valuable scientific insights and useable qualitative and quantitative knowledge cannot be systematically produced, discussed and tested in stakeholder dialogues. On the contrary, the authors strongly believe that such dialogues substantially increase the social relevance of research and improve the quality of results, provided they are adequately thought out and conducted.

Participation in decision-making has been hailed as one of the pillars of sustainable development and integrated resources management. It has thus been advocated as a means to improve the relevance, legitimacy and implementation of decisions taken, as well as the credibility and accountability of decision makers with regard to civil society. The same principles are increasingly applied to climate change research, which is to a large extent funded by tax-payers via government bodies under the understanding that science has a role to play in informing and guiding society along the path of sustainability transition.

The present paper reflects on stakeholder dialogues and experiences made at the Potsdam Institute for Climate Impact Research (PIK, Germany[1]). PIK conducts integrated assessment projects in the field of adaptation and mitigation of climate change. Most projects have a strong focus on computer modelling of global change, on its potential impacts

---

1 See the PIK web site: www.pik-potsdam.de

and possible adaptation. We have selected three initiatives and projects with a particularly strong stakeholder component for further analysis.

In more general terms this paper explores how science-based stakeholder dialogue can play an important role in the generation of knowledge and what the relevance of such dialogues is for the wider society. To this end two objectives are set. First, key aspects relevant to stakeholder dialogues are discussed in the light of the examples and lessons are drawn from an evaluation of PIK's stakeholder experience. Second, theoretical considerations introduced in Chapter 2 and extended in Chapter 12 of this book (Welp and Stoll-Kleemann 2006, Stoll-Kleemann and Welp 2006) are revisited in the light of the practice in science-stakeholder dialogues commented below.

7.2 Stakeholder dialogues in climate change research

7.2.1 Experiences at PIK

PIK has in the last decade played a significant role in climate change research, particularly in model-based integrated assessment studies. It has sought to develop a holistic approach for climate change and climate impact studies, with horizontal integration (via interdisciplinary staff and projects), and vertical integration (via the consideration of all major research aspects from problem formulation to recommendations to policy-makers). At the core of PIK's mission is the wish to produce meaningful insights and to encourage a transition to sustainability.

PIK's mission, research focus and structure have constituted a suitable environment within which science-based stakeholder dialogues have found a natural place. The authors of the present paper have all been involved in innovative participatory environmental research, in particular via stakeholder dialogues. The dialogue initiatives considered here range from the creation of platforms for dialogues, such as associations and forums, to individual projects funded by different sources (EU, national research funding, private companies).

The three selected examples are the European Climate Forum[2] (ECF), ATEAM[3], and SilviStrat[4] (see Table 7.1). ECF is a platform for the exchange of arguments regarding long-term climate policy and other controversial issues related to climate mitigation and adaptation. ATEAM (2001-2003) was concerned with ecosystem service provision and European vulnerability to global change and SilviStrat (2001-2004) with local forest management responses to global change (in the state of Brandenburg, Germany). The above examples are representative of the diversity of stakeholder initiatives at PIK (for further examples see de la Vega-Leinert et al. 2006, in review). While ECF is consolidating a long-term stakeholder process, ATEAM and many other projects run over 3-5 years only. While the research agenda in ECF is responsive to stakeholders' expectations, ATEAM and SilviStrat are product-oriented projects which thus have a well-set agenda defined to a large extent at the project proposal stage.

The stakeholders involved in PIK's stakeholder activities have been diverse, ranging from interested individuals to international corporations. Creating bridges for long-term collaboration between scientists and stakeholders requires intensive attention. Researchers need to be aware that stakeholders may become weary of being approached repeatedly for different activities. To avoid overlaps a PIK stakeholder database was created. This improved communication with various stakeholder groups and facilitated synergies between different projects. In the following the objectives, main issues and involved stakeholders of each case are described.

---

[2] See the ECF web site: www.european-climate-forum.net

[3] The full name of the project is Advanced Terrestrial Ecosystem Assessment and Modeling, Project number: EVK2-2000-00075, funded by the 5th Framework Programme of the European Commission under the topic "Energy, Environment and Sustainable Development". See the web site: http://www.pik-potsdam.de/ateam/

[4] The full name of the project is Silvicultural Response Strategies to Climate Change in Management of European Forests. See the web site: http://www.efi.fi/projects/silvistrat/

Table 7.1 Project description

Project/ initiative	**European Climate Forum ECF**	**ATEAM**	**SilviStrat**
Research focus	– long-term climate policies – energy systems – what is dangerous climate change?	– ecosystem service modelling – European vulnerability assessment	– forest management strategies to mitigate the impacts of climate change – management strategies to enhance carbon sequestration, – multi-functional forest management
Dialogue objectives	– new research questions – joint projects	– evaluation of modelling components, scenarios and results – evaluation of sectoral adaptive capacity	– new research questions – forest functions and their ranking
Types of stakeholders	– private companies – NGOs – Policy-makers – scientists	– private land and forest owners – sectoral representatives – private and public environmental resource managers and consultants – climate and environmental policy advisers – NGOs – scientists	– private forest owners – employees of national forest services – wood industry – scientists – NGO's – forest related business sectors (e.g. tourism, water)

Project/ initiative	**European Climate Forum ECF**	**ATEAM**	**SilviStrat**
Methods	– joint studies – conferences – workshops – working groups – teleconferences	– Workshops – Questionnaires – interviews	– Workshops – Questionnaires – interviews
Types of involvement	– exchange of arguments – stakeholders influence the research focus	– influencing case studies and indicators modelled – influencing the presentation of results – help shape future research agenda	– defining forest management objectives – ranking the objectives according to stakeholders' preferences
Deliverables	– discussion papers – conference reports – climate games – model coupling tools	Evaluation of: – modelling indicators – sectoral driving forces – land use scenarios – vulnerability mapping methods and results – dialogue activities	– modelling the impact of different management and climate scenarios on forest functions – evaluating trade-off effects between conflicting management objectives

7.2.2 European Climate Forum (ECF)

The European Climate Forum (ECF) is a non-profit organisation established in 2001 by seven leading research institutions in the field of climate research, energy research and integrated assessment as well as diverse members, which include traditional and renewable energy industries and companies, major energy users, insurance and finance, policy-makers, environmental NGOs, and scientists. Strategic decisions are made in monthly telephone conferences of the board, which includes both scientists and stakeholders.

ECF provides a platform for discussions on controversial climate change issues. The objectives of stakeholder dialogues have ranged from identifying new research questions to combining ethical and factual arguments and accessing stakeholders' local knowledge with respect to impacts of climate change. ECF has focused on issues, for which there exists at present strong disagreement and controversy. Examples of such controversial issues include carbon capturing and storage (CCS), long-term climate policies (Hasselmann et al. 2003), the role of biofuels in the transport sector (European Climate Forum 2003), and the question "What is dangerous climate change?"[5] (European Climate Forum 2004). Such questions have typically been discussed in the annual ECF events or a thematic workshop.

A further way of cooperating with stakeholders is to initiate and carry out joint studies. An example of close collaboration between researchers and stakeholders was a project on videoconferencing with the Deutsche Telekom. The project, which was carried out jointly by PIK and the Deutsche Telekom AG under the umbrella of ECF assessed the potentials of information and communication technologies (ICT) to contribute to a more sustainable development of the transport sector. The focus was on the potential of substituting business travel by using video- or teleconferencing (Runge 2004).

Stakeholders who are involved in ECF activities include the founding members, members who have joined the forum later and invited guests. The stakeholders typically include representatives from corporations and companies operating in sectors such as insurance, car manufacturing, and energy. Also small and medium-sized enterprises (SMEs) and non-governmental organisations (NGOs) such as WWF are involved.

---

[5] According to the UN Framework Convention on Climate Change the ultimate objective of climate policy is to "stabilise greenhouse gas concentrations in the atmosphere at a level that would prevent dangerous anthropogenic interference with the climate system". There is however no agreement what warming levels qualify as dangerous according to this piece of international law. Some stakeholders consider the already observed changes of 0.8° Celsius above pre-industrial levels as dangerous, while others suggest that a 2° Celsius increase in global mean temperature over a long period of time would have dangerous impacts on ecosystems and human livelihoods (ECF 2004).

Since the involved individuals and institutions have very different perceptions on the climate change problem, there are differences in the preferred course of action to mitigate and to adapt to climate change. The members have different areas of competence too. For example some have technical know-how, others are strong in economic analysis, while others have data which is of high relevance for integrated assessment studies. These particular competencies combined with the richness of perspectives make the dialogues attractive for both researchers and stakeholders.

The selection of participants has been based on personal contacts rather than on a systematic approach. The group of potential participants to be approached for specific events depends on the issues to be discussed. For example, for the event focusing on biofuels in road transport both energy suppliers, energy distributors and policy-makers were taken into account. The involved NGOs were mainly environmental organisations. Other groups representing sections of civil society, such as consumer and car drivers' associations are interesting as potential future members. It should be noted that policy-makers have been invited to ECF events as guests. Membership is not open to political or government organisations. This is a conscious choice and reflects the view that ECF should be independent from political bodies. Some argue that funding from private businesses may corrode scientific independence as well. Experience has however shown that in such funding arrangements scientific freedom has not been more limited than with public funding sources, in particular since NGOs play a balancing role in the dialogues.

7.2.3 ATEAM

ATEAM was a large interdisciplinary (> 60 scientists) project funded under the 5th EU Framework Programme (Schröter et al. 2005). Its focus was on ecosystem modelling and vulnerability assessment to global change at European scale. The research in ATEAM was not stakeholder-centred, in that it was neither initiated nor run in close collaboration with stakeholders. However, the research process and content throughout the project continuously focused on producing results which could be of use to the stakeholder community targeted by ATEAM. Stakeholders could significantly influence, but not fundamentally change the research work plan, the modelling framework or methodologies developed within the project without stakeholder consultation. The direct role of stakeholders within the project was punctual, mostly feedback orientated, and focused on evaluating specific modelling components as well as the overall scientific results. Stakeholders were involved at key points during the research process, in between which scientists improved and tested their models with consideration of stakeholders' suggestions. The research agenda in ATEAM was further centred on top-down quantification in a natural science context, rather than exploratory bottom-up qualitative research, as for example in the scenario development part of the Millennium Ecosystem Assessment[6]. It was comparable in its aims and process to the dialogue developed within the Delft process (van Daalen et al. 1996).

The goals of the ATEAM stakeholder dialogues were manifold. In particular the project aimed at: 1) opening up the ecological modelling world to a wider audience, 2) fostering greater knowledge integration through inter- and transdisciplinarity, 3) learning from stakeholders what scientific information is meaningful for natural resource managers and decision-makers, 4) improving and evaluating ecosystem service modelling and vulnerability assessment, and 5) raising awareness on global change issues.

The issues ATEAM focused on were: 1) improving modelling of ecosystem service provision under global change, 2) developing a multi-scenario approach (including climate, nitrogen and land use scenarios), 3) producing a preliminary aggregated indicator for adaptive capacity at subnational level, and 4) combining the above elements innovatively to produce maps of European vulnerability to global change over 4 time slices till 2100. These maps, as well as their interpretation were ATEAM's main deliverables. ATEAM's strong points were thus: its state-of-the-art natural science and terrestrial ecological modelling, for which it has obtained high scientific recognition and credibility, and its willingness to create bridges of collaboration and meaning to social sciences modelling and to a wide range of stakeholders.

ATEAM's stakeholders included public and private sector consultancies (e.g. DHI Water and Environment or Associazione Cultura Turismo Ambiente), sectoral representatives (e.g. cereal growers, paper-agro industries), private businesses (e.g. land and forest owners), public organisations which act as advisers (e.g. European Environmental Agency), managers (e.g. forest, water and or natural park management),

[6] See: www.millenniumassessment.org.

non governmental organisations (e.g. Royal Society for the Protection of Birds), independent umbrella organisations (e.g. Comité International pour la Protection des Alpes) and other scientific institutions not involved in the ATEAM consortium.

ATEAM aimed at developing participative methods and activities, and to obtain feedback to improve the usability of its model outputs. It was thus important to collaborate with stakeholders who understood sectoral as well as scientific issues related to global change. Stakeholders' scientific affinity and competence eased the discussion with ATEAM scientists, who especially at the beginning of the project were relatively inexperienced with, and critical of, the dialogue concept. Often stakeholders had different hats (e.g. private sector consultant and academic), and participated as individuals, rather than official representatives of specific organisations. Initially stakeholders from personal networks were invited. Then a matrix was designed with relevant sectors vs. geographical focuses/scales and organisation type. The aim was to systematically produce a sample/database of stakeholders to contact and hopefully involve during the course of the project. Few stakeholders with a purely local focus (e.g. regional nature park managers) were approached since the project was producing limited meaningful results at this scale.

Despite the participation of many stakeholders who represented private sector activities, ATEAM did not sample comprehensively purely commercial interests. ATEAM often targeted environmental managers and consultants who had a green bias, despite keeping in mind at all times the requirements for sectoral competition and market viability. They provided a fresh, albeit controversial (from the point of view of ATEAM scientists) view of the realities of many businesses. Noticeably absent from the spectrum of ATEAM participants were representatives from the transport or financial/insurance sectors, farmers, consumer and/or citizen associations, or downstream manufacturing or distributing activities such as the agro-chemical industry and food processing (except for paper, and energy and water distribution). Contacts were made to fill some of these gaps (e.g. IKEA, Gerling Reinsurance, local farmer(s) and national farmer's association). Non-attendance was then mostly due to the lack of relevance of ATEAM's results for specific commercial activities, of time, or of remuneration (i.e. the project could not cover all expenses, e.g. consultants' fees).

7.2.4 SilviStrat

SilviStrat was also an EU 5th Framework Programme funded research project with 7 partners spread over the different forest regions in Europe from the Mediterranean to the Boreal region. The main focus of the project was to investigate adaptive management strategies to enhance carbon sequestration in European forests and to find ways to mitigate adverse impacts of global climate change. The assessment at a scale of forest management units (Badeck 2005) was based on simulations performed with two forest growth models, 4C (Lasch et al. 2002) and FINNFOR (Kellomäki 1993). Furthermore a wood product model (Eggers 2002) was used to calculate/simulate the fate of carbon in wood products and landfills. The stakeholder dialogue was part of a subproject – a regional study, which evaluated the effects of forest management on forest functions in Brandenburg, Germany.

The overall objective of the stakeholder dialogues within SilviStrat was to investigate forest management objectives and management preferences assigned by single stakeholders. A further objective was to present first relevant findings about the impact of management and climate change on carbon storage and timber production. The effects of global climate change to forest ecosystems and their functions, and the awareness to and relevance of these impacts were discussed and possibilities to react to changes were compiled. The workshop had the aim to bring together scientists and stakeholders to exchange knowledge and experience, discuss the problems of climate change for the forestry sector and define new research questions.

SilviStrat analysed direct and indirect impacts of present forest management operations and climate on carbon sequestration, timber production, biodiversity, and groundwater recharge in European forests . The aim was to develop a better understanding of how management could be improved to maintain sustainable forest production and increase carbon sequestration capacity, and sustain multi-functionality of forest ecosystems under current and changing climate conditions. Additionally, costs and benefits of adaptive management strategies were assessed at the management unit level. In representative management units of major east German forest types the impact of forest operations were simulated with the goal of increasing carbon sequestration, maintaining sustainable forest production and other forest functions under changing climatic conditions. Furthermore the impact of forest management on reduction of drought risk and the potential of forest management for carbon sequestration and mitigation of climate-change-induced impacts was estimated at the European scale.

SilviStrat focused on local stakeholders using goods and services provided by forest ecosystems or managing forest ecosystems, including private forest owners, local environmental organisations, federal forest services (from local forester to higher administrative levels), the wood industry, scientists as well as representatives of forest-related business sectors (water management, tourism).

The regional study within SilviStrat tried to involve mainly representatives of stakeholder groups closely related to forest ecosystems, such as forest managers and businesses, which directly or indirectly market forest products (timber and non-timber products). Excluded were the local population, tourists and policy-makers. The focus was on local stakeholders who were from the region covered by the study area, as well as regional representatives of national or international organisations and scientists. The selection of potential participants was partly based on personal contacts and partly on a systematic approach. Not all invited groups were represented in the workshop and questionnaires due to lack of relevance of the research topic or time. To secure a larger, more representative group, stakeholders from outside Brandenburg were asked to participate. Nevertheless, some gaps still remained.

7.3 Methods applied in the dialogues

Methods used in stakeholder dialogues need to be tailored so that they fit the objectives. Two kinds of tools for stakeholder dialogues can be distinguished: tools for facilitating communication (communication tools) and tools for formalising actors' mental models, assessments, etc. (analytical tools) (see Chapter 2 in this book). Analytical tools can be applied to complement and support communication tools. Approaches such as Bayesian Belief Networks can be used to formalise stakeholders' assessments (Welp et al. 2006). Structuring and framing the research problem at hand can benefit from conducting group model building exercises (Vennix 1999) or by eliciting mental models of stakeholders (Morgan et al. 2002), or by combining both methods.

Workshops, brainstorming sessions, and regular teleconferences have been used by ECF in a search for new socially relevant and intellectually challenging research questions. Various controversial themes, such as the potential of biofuels in the transport sector, have been discussed in some technical and economic detail and documented in proceedings. A way to approach the broad public and experts in the academic, NGO and business world has been to draft discussion papers. Such papers, for example on long-term climate policies, have been published in refereed journals while press releases targeted the general public (Hasselmann et al. 2003). The writing of such discussion papers in small writing teams has provided an opportunity for mutual learning. For example, the ECF discussion paper "The Challenge of Long-Term Climate Change" was drafted by a group which was formed at the annual ECF conference 2002. This paper did however not represent an ECF consensus view, since the forum does not endorse specific views expressed by its members.

ECF has promoted and been closely involved in the development of communication tools, most notably of climate computer and board games. A computer game for two players was launched in November 2002 at a climate exhibition in the 'Deutsches Museum', Munich. In the game, players control future climate policy by adopting the role of either the government, a Chief Executive Officer of a global company or a typical private household from an industrialised country. The players endeavour to maintain a sustainable climate in the future while pursuing their own individual welfare goals. According to a survey among visitors to the Museum, the game was evaluated as being interesting and fun to play. Usually visitors played the game for 15 minutes.

Finally the internet has also been one important communication tool: documents, software and a climate game have been made available to interested people around the world. Email circulars have been effective in informing members of new events, publications or opportunities for cooperation. The public outreach activities of ECF have included besides discussion papers (Hasselmann et al. 2003) also press releases.

A board game (Winds of Change) was developed in close collaboration with the Munich Reinsurance Company. This game depicts challenges in technological learning, investments and keeping climate warming beyond dangerous levels and has been applied with stakeholders from the business world as well as students. ECF also supported the development of a further board game (Keep Cool), where players play a decision-maker of a region (such as Europe, USA, developing countries, etc.). The ECF family of climate games[7] thus comprises at the moment one computer game and two board games, which can be used in fostering team learning on climate change.

Within ATEAM, networking and contacting stakeholders has been a major activity throughout the project. Communication and dissemination material included flyers that have been produced in different languages, posters, executive summaries and full reports of meetings, a webpage[8] and facilitation/moderation during events. Activities included stakeholder workshops, questionnaires and interviews. The main dialogue focus of ATEAM has been in presenting its research, obtaining feedback from stakeholders, and seeking ways of accommodating stakeholders' suggestions within the pre-defined ATEAM framework. The specific objectives of each event were shaped to evaluate the progress of the research either in plenary or in dedicated sectoral working groups. Additionally stakeholder questionnaires and interviews of ATEAM scientists were carried out as part of the evaluation of the dialogue's outcome. Finally, independent observers participated in each general stakeholder meeting and provided the stakeholder dialogue coordination team with critical feedback, which helps to improve the following events.

Half way into the project it became clear that a digital compilation of the project's most salient results would be a useful communication tool for interested stakeholders. This led to the development of the ATEAM Atlas of European Vulnerability[9] (Metzger et al. 2004 and http://www.pik-potsdam.de/ateam/). The tool allows users to select indicators of impact and vulnerability, using the socio-economic, climate and land use scenarios they are most interested in. The maps are placed in a fact sheet, which provides succinct information on the models and scenarios used, the main assumptions made, the indicators themselves and additional references. Whenever aggregated or relative indicators are shown, users can decompose the results into their components or choose to view absolute data. Furthermore, users can perform simple queries, as well as focus in on specific environmental regions or countries. During final

[7] Further information about the ECF family of climate games can be found on the web site: www.european-climate-forum.net/games

[8] Some of this material is available at: //www.pik-potsdam.de/ateam/stakeholderweb/ateam_stakeholder_material.html

[9] The ATEAM Atlas of European Vulnerability is available to download at: www.pik-potsdam.de/ateam/

dialogue activities, stakeholders viewed early versions of the tool and commented on ways to improve it.

The SilviStrat project used a combination of communication and analytical tools. A workshop was organised in cooperation between PIK and the Landesforstanstalt Eberswalde (Brandenburg) to identify forest management objectives and the preferences of 23 stakeholders. The workshop started with a brainstorming session in which important forest functions were identified. Later on the group elaborated these main functions. The relevance of different forest management objectives was evaluated with the help of a questionnaire, which each stakeholder was asked to fill in. A summary of the presentations, discussion and findings of the workshop was sent to all interested parties. Stakeholders who could not participate in the event were also asked to fill in the questionnaire.

Forest management objectives and preferences of stakeholders were investigated using Saaty's Analytical Hierarchy Process (AHP) and Saaty's eigenvalue method (Saaty 1990). The AHP is a mathematical method for analysing multi-criteria problems. The forest management objectives are ranked by pair-wise comparisons where stakeholders have the option to express their preference between two functions on a rating scale from equally important to absolute priority. The ratings are arranged in a symmetric matrix and the local priorities of the elements in the matrix are calculated by the normalised right eigenvector. The expected utility of alternative management options was calculated by means of a multi-criteria analysis method based on an additive utility theory (Kangas 2002, Lexer 2000), which incorporated results from the stakeholder dialogue. The potential success of simulated forest management plans were analysed and trade-off effects between conflicting objectives were discussed.

The three examples thus applied very different tools ranging from games to computer models. The originality of SilviStrat's and ATEAM's stakeholder dialogue exercises was that the results derived from the dialogue were directly used in model development. SilviStrat used multi-criteria analysis to reflect with stakeholders on management alternatives in the forest sector. ATEAM developed innovative land use scenarios and an interactive interface for integration of its main results: the ATEAM mapping tool and Vulnerability Atlas (Metzger et al. 2004). ECF created different communication tools including board and computer climate games. All projects developed diverse and lively stakeholder networks and created situations in which stakeholders were confronted with state-of-the-art science on climate change.

7.4 Reflections

7.4.1 How can we evaluate science-based stakeholder dialogues?

What are adequate and useful evaluation[10] criteria for science-based stakeholder dialogues? Chapter 4 in this book gave an overview of some approaches to evaluation. In the following we will expand on this and focus on criteria that are especially relevant for science-based dialogues. Criteria that can be used for other participatory processes, such as city or road planning, do not necessarily apply to scientific dialogues. The main reason for this is that in planning and decision-making a consensus or clear majority view regarding a decision or action is striven for. In scientific dialogues on the other hand, a consensus view may emerge, but it is not the primary aim. Disagreement may prevail, as it often does, and shape future research. Nevertheless, evaluating stakeholder processes faces similar difficulties as when evaluating other participatory processes.

Evaluation of science-based stakeholder dialogues helps to adjust the course of the exercise and improve it gradually. There are few papers exploring the theoretical underpinning and practical steps of evaluating science-based stakeholder dialogues. However, relevant literature can be found in adjacent areas of evaluation, such as critical theory (Webler 1995), risk communication (Rowe 2000), public participation (e.g. Webler 1995, Rowe 2000) and democratic theory (Fiorino 1990). A distinction can be made between evaluations conducted by outsiders and participatory evaluations. While outsider evaluations are said to be independent and less biased it is important to have, in addition, evaluations from the participants themselves. The latter is viewed as credible and useful because the diverse needs of participants are more likely to be fulfilled (Chess 2000).

Important criteria to evaluate stakeholder dialogues are accountability, performance and direction (Abrams et al. 2003). Accountability means that scientists are accountable to the invited stakeholders and focus on transparency (free flow and access to all relevant information). Performance includes responsiveness to stakeholder concerns, effectiveness and efficiency (making the best use of resources) as well as

[10] Chelimsky and Shadish (1997) define evaluation as „determining merit or worth".

using adaptive approaches. Direction finally focuses on strategic vision and effective leadership (how new ideas are generated and innovative processes to address and resolve difficult issues launched) as well as the use of collaborative learning in various forums.

A main distinction can be made on the object of the evaluation. This can explore how stakeholder dialogues take place (process evaluation) or assess the results themselves (outcome evaluation). Both can be performed during and/or after stakeholder dialogue efforts. It is useful anyway to reach consensus between the participants in advance on which goals to evaluate. For science-based stakeholder dialogues outcome-related evaluation is likely to be the most relevant one (e.g. was a stakeholder dialogue beneficial in identifying faults and gaps in the research strategy?). The output of the stakeholder dialogue should have a genuine impact on the research carried out (criterion of influence). One danger of science-based stakeholder dialogues is that some projects often only want to fulfil the conditions of research funding agencies, which increasingly require stakeholder dialogues as components of research projects without there being any intent of considering the knowledge of stakeholders formulated in the science-based dialogue. One approach that might lead to fulfilling this criterion is to ensure that there is a clear acceptance from all participants beforehand as to how the output will be used and how it might direct research. A more process related criterion is that of transparency. It means that the stakeholder dialogue should be transparent so that the stakeholders can see what is going on and how they influence the research process. The nature and scope of the stakeholder dialogues should thus be clearly defined. It is important to ensure that there is reflection regarding the scope of a stakeholder dialogue and its expected output,. The effectiveness of a procedure, as well as its credibility, is likely to be influenced by any dispute caused through misunderstandings. Documenting the process of reaching a shared view (as well as the outcome) will increase transparency, and hence the credibility of the exercise. Furthermore it will increase the efficiency of the process.

7.4.2 Achievements

ECF has consolidated a rich and dynamic network of stakeholders. The main difference to other networks, such as the MIT Energy Modelling Forum, PEW Foundation, Climate Strategies (RIIA) and the Electric

Power Research Institute (EPPRI), is the focus on joint studies and exchange of arguments among members with very different interests. In contrast to the ATEAM and Silvistrat projects which lasted only some years, the ECF was created as a permanent structure. ECF has for example positively contributed to structuring the debate on "What is dangerous climate change?" As a result of an international symposium in Beijing scientists and stakeholders came to the conclusion that a 2°Celsius increase in global mean temperature over a long period of time would be dangerous for ecosystems and humans (European Climate Forum 2004). This message was conveyed and well received at a side event of the United Nations Framework Convention 10th Conference of Parties in Buenos Aires, Argentina. The development of climate games has been an activity that has given ECF public visibility. Games can be used as communication tools to engage people in thinking about and discussing climate change in an entertaining way. Coupled climate-economy models have served as a point of reference for game development, in particular in the development of the computer game.

ATEAM's achievements through stakeholder involvement are significant. Firstly, a group of leading natural science modellers opened up to stakeholder interactions and more generally to the need of integrating social sciences in ecosystem modelling. The stakeholder participation has in itself been a powerful driver for more interdisciplinarity and a continuous help to focus and prioritise research efforts and resources to better address stakeholders' needs. Secondly, scientists were instead led to question the basic assumptions, methodologies and indicators used in their scenarios and models, the meaningfulness of the models themselves and of their results (incl. specific temporal and geographical scales) for stakeholders. Thirdly, efforts were mobilised to address stakeholders' suggestions when time and resources allowed this. For example additional case studies were carried out, the focus of one PhD thesis was significantly changed and specific modelling indicators were adapted to better suit issues raised by stakeholders. When stakeholders' recommendations could not be catered for within the scope and time-horizon of the project, they contributed to drawing a future research agenda, which fed ongoing research. That the ATEAM modelling and assessment approaches achieve clear scientific credibility was critical for stakeholders, who also placed a high value on the transparency of the methods used for aggregation and integration of the results. This has influenced vastly the ATEAM research work plan. Indeed this led to the development of innovative analytical and communication tools to promote better understanding of potential global change impacts on ecosystem service provision. In particular, the integrated vulnerability mapping methodology and mapping tool, as well as the concept of summary map information sheets were designed to address this very need. Consequently, scientists participating in the process have gained a more open attitude to participatory research since they have had direct, positive experience that this can be stimulating and fruitful, despite being resource and time consuming (See de la Vega-Leinert et al. 2006, in review for a more detailed evaluation of the ATEAM stakeholder dialogue).

In the SilviStrat project stakeholders played a key role in the research process. The results of the stakeholder dialogue were essential in particular for the regional study within SilviStrat. The multi-criteria analysis method that was applied in the study of forest management needed the inputs of stakeholders. These rankings were in past research projects provided by scientists/experts rather than by professionals in the forest sector. Stakeholders' assessments have now been integrated into the process and therefore the study is more closely linked to local knowledge on the management level. The dialogue helped to understand stakeholders' concerns, problems, restrictions and uncertainties in current forest management and under the aspect of changing climate. The project results provided an overview of differences and similarities of stakeholder interests in relation to forest ecosystem management. Stakeholders identified new research questions during the research process. These were collected and either addressed in the ongoing SilviStrat project, or if this was not feasible, used in drafts of future research projects.

By and large PIK has established itself as an institute which is interested in, and has gained the capacity to conduct stakeholder dialogues, which are of interest for both scientists and people outside the scientific community. It is also important to note that PIK is performing this together with other European and international research institutes. Since the time stakeholders can dedicate to such activities is limited, a coordinated effort among researchers and research institutes creates synergies and increases the efficiency of such dialogues.

7.4.3 Dealing with different expectations

Stakeholders may have different views on what the objectives and outcomes of a dialogue may be. It is important to reflect on the expectations and to develop a shared view on what stakeholders and scientists may gain by engaging in the time-consuming effort of stakeholder dialogues. Reflecting on these expectations should be part of the evaluation scheme of scientific stakeholder dialogues.

Within ECF the agenda is set jointly by scientists and stakeholders. Joint studies can be created flexibly if the provided resources are available. In other types of research projects the possibility to change research strategies is more limited since they are funded projects, which have an agreed research programme and products to deliver to the funding agencies. Thus for example ATEAM and SilviStrat have made successful and useful incursions into participatory integrated assessments. Although the full diversity of stakeholders' needs and preferences could not be catered for, flexibility and adaptation has been developed to explore new research directions when stakeholders' suggestions suited the interests, expertise and willingness of the involved scientists.

Although stakeholders' and scientists' interests sometimes differ, strong efforts have been made to listen carefully to, and accommodate the expectations and research needs of stakeholders. Within ECF new research areas have been intensively searched for and debated. Research projects were mostly easily agreed upon and initiated (e.g. project on the role of telecommunication in CO_2 reduction). However in one instance a proposed project on carbon capturing and sequestration (CSS) was subject to heavy debate. Some stakeholders considered carbon sequestration as an unacceptable technical fix, and were strictly against such a project and considered the issues as a no-go area. Others saw it as a potentially low cost technology for climate mitigation. Embedded in a broader assessment of technological options, carbon capturing and storage were eventually accepted for a project proposal. The framing of the research question is thus of great importance: many climate-related issues can be framed either as narrow technological questions which put aside for example the question of societal acceptability, or as a broader set of social-scientific and natural-scientific sub-questions.

Researchers are not always ready to engage in dialogues with stakeholders. Dialogues in natural sciences are a very recent development and many natural scientists have hardly dealt with dialogue methods (which have more affinities to social sciences). At the beginning of the ATEAM project, some scientists in the consortium were uneasy about the decision to engage in dialogue activities. The project had chosen to step out of the known paths of fundamental ecological modelling research and there was some uncertainty on whether this was a valid choice from the scientific point of view, and on how to perform this well. In the peer community some viewed this initiative 'at best' as a marketing trick to attract funding or 'at worst' as a 'non scientific' goal, which would discredit the project's overall scientific credibility. The project leadership thus took a significant risk and had to dedicate much time to convincing some project members and peers that it would be worth the effort. The latter was achieved by not compromising in core parts of the research plan (e.g. the detailed modelling developments and the benchmarking exercise, see Morales et al. 2005), which were not presented to stakeholders. These formed the main scientific achievements per se of the project, and guaranteed scientific credibility in the ecological modelling peer community. As consensus was forged on the originality and feasibility of the overall methodology, including the generic adaptive capacity index, and of the importance of the stakeholder dialogue component, the project achieved scientific recognition also in the interdisciplinary global change assessment community. Explicitly, the dialogue has aimed at elucidating ATEAM's work to raise stakeholder interest on the future results, and awareness on potential impact of global change. Implicitly however ATEAM scientists aimed at obtaining an overall consensus on the validity of ATEAM's approach through plenty of room for discussion on the limits of the research and needed future improvements.

Within SilviStrat the aim was a wide-spread understanding of forest services and functions which are required by various groups. One particular problem was to secure the participation of stakeholders from the wood industry, tourism and water management. Due to lack of time or the low priority they gave the workshop these groups were not represented. Furthermore representatives of environmental organisations had a forest-related background and therefore did not present a strict nature conservation point of view. Thus the dialogue between stakeholders was not as controversial as expected.

7.5 Conclusions: dialogue practice in view of the Integrative Theory of Reflexive Dialogues

The selection of stakeholders contacted for a dialogue exercise is, consciously or not, biased towards some specific actors rather than others. As discussed by Stoll-Kleemann and Welp in Chapter 2 and Chapter 12 (Welp and Stoll-Kleemann 2006, Stoll-Kleemann and Welp 2006), stakeholder dialogues are distinct from public participation exercises in that they do not aim at achieving a representative sample of the population, but rather a wide range of different opinions on a specific topic. Before initiating a dialogue exercise the spectrum of interested parties should thus be identified, leaving aside those which do not seem relevant to the problem at hand. In this selection process the personal networks, preferences, interests and priorities of the researcher will induce some amount of bias towards specific actors. To minimise this, a systematic selection process can be developed to complement the often used 'snowball approach' (Biernacki and Waldorf 1981). The creation of a stakeholder database can play a critical role not only in storing and analysing contact information of individual/organisations approached, but also their background, expertise, level of interest for the research topic, as well as any further contacts they might have suggested. However, if biases may be restricted at the selection and invitation phases, others will appear as stakeholders accept or decline invitations. Stakeholders are often busy and need to be convinced that they will gain significant benefits before they commit time and effort to activities which are not the focus of their work. Communication skills and a strong feel for how to engage stakeholders and demonstrate the relevance of the dialogue process for their personal activities will certainly help in gaining stakeholder support. Nevertheless in some cases the research/dialogue topic is simply too disconnected from stakeholders' activities to secure their interest and participation. Since biases cannot be avoided, reflecting on these and on their influences on the dialogue process and outcomes is an important step in evaluating the dialogue's achievement as well as in planning future exercises.

Citizens, i.e. non-expert lay persons, were not in the focus of any of the above-mentioned case studies. This was however the targeted audience of an earlier project carried partly out at PIK. The ULYSSES project engaged 600 citizens in structured Integrated Assessment Focus Group sessions to discuss climate change impacts and possible solutions (Stoll-Kleemann et al. 2001, Kasemir et al. 2003). Participants were confronted with the latest knowledge on climate change and synthesised their newly gained understanding in citizen assessments of the causes and impacts of climate change. These included suggestions on mitigation and adaptation measures (e.g. within the transport, energy and household sectors) as well on who should act, where and when. Welp et al. (in press) have pointed out that such exercises should be linked to parliamentary decision-making more strongly than has been the case so far.

The integrative theory of reflexive dialogue as outlined in Chapter 2 highlights the need to incorporate both analytical and communication tools in stakeholder dialogues. In all three case studies both types of tools were used. The examples can be characterised as dialogues with a focus on expert stakeholders. Although ECF engaged also in dialogues with lay person and studied for example their perception of the movie "The day after tomorrow" (Reusswig et al. 2004), the vast majority of contacts were climate change experts, such as representatives of companies, NGOs, and government bodies.

Social psychological theories are highly relevant for science based stakeholder dialogues, since group processes, and prejudices often play an important role especially in the phase where relationships between researchers and stakeholders are consolidated (trust building). This usually takes some time and being aware of such process may help to design meetings and events in a manner where personal relationships can evolve. Linked to the aspect of trust is that that expectations of those involved in science-based stakeholder dialogues need to match reasonably well. It is important to be explicit about them and to make the rules of the game clear at an early phase of the dialogues. During the course of dialogues the expectations may change and it is important to be flexible in this respect too. Thus being explicit about the objectives is key requirement for a working relationship based on trust.

Theories of organisations learning are helpful in finding out how representatives of different organisations can jointly create shared meaning. The development of a language which is understandable for all participants is a key element of science-based stakeholder dialogues. In discussions and dialogues language is created and altered. Communication and analytical tools thus complement each other. Analytical tools help in structuring an issue and in finding the crucial differences between the assessments of different individuals.

Science-based stakeholder dialogues are structured communication exercises, which are directed by researchers. Although stakeholders' views are taken into account, the choices on the ultimate research direction remain the responsibility of the scientists. In some cases decisions are made jointly by scientists and stakeholders. Critics of the current practice of scientific-based stakeholder dialogues often claim that for scientists dialogues appear to be a substitute for 'real scientific inquiry'. They argue that stakeholders are consulted and asked to provide the important parameters, conceptualise problems and do the actual thinking. This view is based on a misconception of what scientific dialogues aim at. Good research increasingly takes place in small interdisciplinary teams, in which the individual scientists meet regularly to think together. Science-based stakeholder dialogues are an extension of this practice and an effort to bring together even more different knowledge domains than the different academic disciplines. Stakeholder dialogues are not a substitute for thinking but rather they foster the art and practice of thinking together.

References

Abrams, P., Borrini-Feyerabend, G., Gardner, J., Heylings, P. (2003). Evaluating Governance: A Handbook to Accompany a Participatory Process for a Protected Area. Draft, July. Parks Canada and TILCEPA

Badeck, F.W., Fürstenau, C., Lasch, P., Suckow, F., Peltola, H., Garcia-Gonzalo, J., Briceño-Elizondo, E., Kellomäki, S., Lexer, M.J., Jäger, D., Lindner, M., Thiel, D., Kaipainen, T., Lehikoinen, N., Junge, S., Feliu, J. (2005). Adaptive forest management at the scale of management units. In: Kellomäki, S., Leinonen, S. (eds). Management of European forests under changing climatic conditions. Research Notes 163. University of Joensuu, Faculty of Forestry, 315-382

Biernacki, P., Waldorf, D. (1981). Snowball Sampling. Problems and Techniques of Chain Referral Sampling. Sociological Methods and Research 10 (2), 141-163

Chess, C. (2000). Evaluating Environmental Public Participation: Methodological Questions. Journal of Environmental Planning and Management 43, 769-784

van Daalen, C. E., Thissen, W. A. H., Berk, M. M. (1996). The Delft process: experiences with a dialogue between policy-makers and global modellers. Global Environmental Change 6, 267-285

Eggers, T. (2002). The impacts of manufacturing and utilisation of wood products on the European carbon budget. Joensuu, European Forest Institute, 90

European Climate Forum (2003). The Biofuels Directive. Potential for climate protection? Conference Summary (Norwich, UK, 8th - 10th September, 2003). URL: http://european-climate-forum.net/events/norwich2003/pdf/ecf_conference_summary.pdf

European Climate Forum (2004). What is dangerous climate change? Initial results of a Symposium on Key Vulnerable Regions and Climate Change and Article 2 of the UNFCCC, Buenos Aires, 14 December 2004. URL: http://www.european-climate-forum.net/pdf/ECF_beijing_results.pdf

Fiorino, D.J. (1990). Citizen Participation and Environmental Risk: a survey of institutional mechnisms. Science, Technology and Human Values 15, 226-243

Funtowicz, S., J. Ravetz (1993). Science for the post-normal age. Futures 25, 739-755

Gibbons, M., Limoges, C., Nowotny, H., Schwartzman, S., Scott, P., Trow, M. (1994). The new production of knowledge. The dynamics of science and research in comtemporary societies. Sage Publications, Thousand Oaks, CA

Hage, M., Leroy, P., Willems, E. (2005). Participatory approaches in governance and in knowledge production: What makes the difference? Paper presented at the International Sociology Association Conference "Environment, knowledge and democracy", 6-7 July 2005, Marseille, France

Hasselmann, K., Latif, M., Hooss, G., Azar, C., Edenhofer, O., Jaeger, C. C., Johannessen, O. M., Kemfert, C., Welp, M., Wokaun, A. (2003). The Challenge of Long-term Climate Change. Science 302, 1923 - 1925

Kangas, J., Kangas, A. (2002). Multiple criteria decision support methods in forest management - An overview and comparative analyses. In: Pukkala, T. (ed). Multi-Objective Forest Planning. Kluwer Academic Publisher, Dordrecht, 37-70

Kellomäki, S., Väisänen, H., et al. (1993). FinnFor: a model for calculating the response of boreal forest ecosystem to climate change, University of Joensuu, Faculty of Forestry, 1-120

Lasch, P., Badeck, F. W., et al. (2002). Sensitivity of simulated forest growth to changes in climate and atmospheric CO2. Forstwissenschaftliches Centralblatt Supplement 1, 155-171

Lexer, M. J. (2000). Ein multi-attributives Nutzenmodell zur Unterstützung der waldbaulichen Entscheidungsfindung dargestellt am Beispiel sekundärer Nadelwälder. Forstwissenschaftliches Zentralblatt 119, 377-394

Metzger, M. J., Leemans, R., Schröter, D., Cramer, W., the ATEAM consortium (2004). The ATEAM vulnerability mapping tool. Office C.T. de Wit Graduate School for Production Ecology and Resource Conservation (PE&RC), Wageningen, The Netherlands. CD-ROM publication No. 27. URL www.pik-potsdam.de/ateam

Morales, P., Sykes, M.T., Prentice, C., Smith, P., Smith, B., Bugmann, H., Zierl, B., Friedlingstein, P., Viovy, N., Sabaté, S., Sánchez, A., Pla, E., Gracia, C.A., Sitch, S., Arneth, A., Ogee, J. (2005). Comparing and evaluating process-based ecosystem model predictions of carbon and water fluxes in major European forest biomes. Global Change Biology 11 (12), 2211

Morgan, M. G., B. Fischhoff, A. Bostrom, Atmanet, C.J. (2002). Risk Communication; A Mental Models Approach. Cambridge University Press, Cambridge

Pohl, C. (2005). Transdisciplinary collaboration in environmental research. Futures 37, 1159-1178

Ravetz, J. (1999). What is Post-Normal Science? Futures 31, 647-653

Rowe, G., Frewer, L. J. (2000). Public Participation Methods: A Framework for Evaluation. Science, Technology, & Human Values 25, 3-29

Runge, D., Reusswig, F. (2004). Substitution von Geschäftsreisen durch Videokonferenzen. Zusammenfassung des Endberichts vom Dezember 2003. Potsdam Insitute for Climate Impact Research, Germany. URL: http://www.european-climate-forum.net/pdf/Zusammenfassung%20ECF.pdf

Reusswig, F., Schwarzkopf, J., Pohlenz, P. (2004). The climate blockbuster 'the day after tomorrow' and its impact on the german cinema public. PIK Report No. 92.

http://www.pik-potsdam.de/pik_web/publications/pik_reports/reports/pr.92/pr92.pdf

Saaty, T. L. (1990). Multicriteria Decision Making - The Analytic Hierarchy Process, Planning, Priority Setting, Resource Allocation. RWS Publications, Pittsburgh

Schröter D., Cramer, W., Leemans, R., Prentice, I.C., Araújo, M.B., Arnell, N.W., Bondeau, A., Bugmann, H., Carter, T.R., Gracia, C.A., de la Vega-Leinert, A.C., Erhard, M., Ewert, F., Glendining, M., House, J.I., Kankaanpää, S., Klein, R.J.T., Lavorel, S., Lindner, M., Metzger, M.J., Meyer, J., Mitchell, T.D., Reginster, I., Rounsevell, M., Sabaté, S., Sitch, S., Smith, B., Smith, J., Smith, P., Sykes, M.T., Thonicke, K., Thuiller, W., Tuck, G., Zaehle, S., Zierl, B. (2005). Ecosystem Service Supply and Vulnerability to Global Change in Europe. Science 310 (5752), 1333-1337

Stoll-Kleemann, S., O`Riordan, T., Jaeger, C.C. (2001). The psychology of denial concerning climate mitigation measures: evidence from Swiss focus groups. Global Environmental Change 11 (2), 107-117

Stoll-Kleemann, S., Welp, M. (2006). Linking Case Studies to the Integrative Theory of Reflexive Dialogues. In: Stoll-Kleemann, S., Welp, M. (eds). Stakeholder Dialogues in Natural Resources Management. Springer-Verlag, Heidelberg

Vega-Leinert, de la, A.C., Schröter, D., Leemans, R., Frisch, U., Pluimers, J. (in review). A stakeholder dialogue on European vulnerability. A contribution to the ATEAM special issue submitted to Regional Environmental Change

Vennix, J.A.M. (1999). Group model-building: tackling messy problems. System Dynamics Review 15 (4), 379 - 401

Webler, T. (1995). Models for Environmental Discourse. In: Fairness and Competence in Citizen Participation. Renn, O., Webler, T., Wiedemann, P. (Eds). Kluwer Academic Publishers, London, 35-86

Welp, M., Stoll-Kleemann, S. (2006). Integrative Theory of Reflexive Dialogues. In: Stoll-Kleemann, S., Welp, M. (eds). Stakeholder Dialogues in Natural Resources Management. Springer-Verlag, Heidelberg

Welp, M., de la Vega-Leinert, A., Stoll-Kleemann, S., Jaeger, C.C. (2006). Science-based stakeholder dialogues: tools and theories. Global Environmental Change 16, 170-181

Welp, M., Kasemir, B., Jaeger, C.C. (in press). Citizens' Voices in Environmental Policy: The Contribution of Integrated Assessment Focus Groups to Accountable Decision Making. In: Coenen, F.H.J.M., Paterson, R. (Eds). The Promise and Limits of Participatory Processes for the Quality of Environmentally Related Decision-making. Springer-Verlag, Heidelberg

8 Science in Support of the Forest Biodiversity Programme for Southern Finland

Working from the inside

Eeva Hellström[1]

Finnish Forest Association, Helsinki, Finland

8.1 Introduction

The scientific community has an important role in producing information for decision-making processes. Moreover, the effective and proper utilisation of scientific knowledge in support of policy-making is a profound goal of the scientific community.

In order to bring valuable information to the policy table, and help focus subsequent research studies on policy-relevant topics, Mills and Solberg (1998) emphasise the need to build a collaborative infrastructure and relations between science and policy-making. This can be accomplished, for example, through proactively conducting research on anticipated policy issues, regular conferences, joint research studies, adaptive management, and boundary spanners. All these approaches help to strengthen the scientific community's input to policy-making, while operating from "outside" the policy process itself.

However, as the complexity of issues increases, and as the evaluation or assessment of policy implications (e.g. strategic environmental assessment, SEA) is becoming an increasingly integral part of policy processes, the question has been raised, to what extent the scientific community may or should be involved "inside" the policy process.

This paper illustrates a forest policy process, where scientists and the scientific community have been involved in policy-making in a non-traditional way, by supporting the process not only from the "outside" but also from the "inside".

---

[1] The author was assigned as permanent expert to a commission set up to work out a proposal for a Forest Biodiversity Programme for Southern Finland, and as chair of the Commission's interim Working Group for Policy Means.

First, traditions of forest protection in Finland are described (Chapter 8.2). Finland has protected great parts of its northern forests. These areas are mostly publicly owned and scarcely populated. In recent years, nature protection in the more productive and predominately privately owned, Southern forests have been a heated issue of debate in Finnish forest policy. In particular, new instruments, which go beyond the traditional "strictly protected areas", have been called upon.

In Chapter 8.3, the process of compiling the Forest Biodiversity Programme for Southern Finland is described. The programme places particular emphasis on developing new, innovative means for nature conservation in private forests. In this paper, particular focus is placed on describing the conflicts and challenges related to the work and on the different ways in which scientists supported the process from the "inside". Finally, Chapter 8.4 reviews the lessons learned when involving scientists in policy processes not only from the "outside" but also from the "inside".

8.2 Traditions of Forest Protection in Finland

Conservation of untouched forests or forests, which have sustained the essential features of their original natural value despite slight human impact, was in the main focus of forest protection policy in Finland during recent decades. Establishing protected areas, where forest management and logging were no longer allowed, was the most natural way of securing the natural values of these forests, called "wilderness areas", "old growth forests", or "ancient forests" in the public debate.

The traditional way of maintaining natural values was to establish nature conservation areas and strict nature reserves. In Finland, this has been done since the 1930s. Later on, various national programmes were initiated to protect the main types of ecosystems. In addition to national parks and strict nature reserves, the Finnish government has approved national conservation programmes of e.g. mire conservation areas, herb-rich forest areas, old-growth natural forest areas, shoreline protection areas and wilderness areas.

In Finland, forest protection policies typically divided forests dualistically into either strictly protected areas excluding all forest management activities, or commercial forests managed according to overall environmental guidelines. Few forms of forest protection existed in addition to these two basic types. The dualistic approach to forest protection policy is also reflected in the everyday language. "Forest protection" is generally understood as the strict protection of a specific forest area drawn on the map and marked on the field, from all forest management and logging activities.

Owing to this setting, forest protection policy in Finland has traditionally emphasised regulatory control, particularly judicial control through legislation concerning threatened species and protected areas. In the 1990s, judicial control expanded to arise not only from environmental legislation but also from forestry legislation, and national control was supplemented by international control by the EU.

The 1990s also marked an era of expansion of non-regulatory means of control over the environmental values of forests (e.g. public ownership and planning, economic incentives, information and negotiation). For example, the Finnish Forest and Park Service was transformed from a state agency into a state enterprise, and new weight in the legislation controlling was given to preservation biological values. On state lands, ecological landscape planning and participatory planning were adopted, and the state forest enterprise has actively continued the voluntary establishment of forest areas designed for recreation or preservation purposes.

In private forestry, normative forest management guidelines were transformed into voluntary-based recommendations and forest certification. Furthermore environmental management systems were introduced and a series of environmental guidelines and biodiversity programs by various actors (e.g. forest owner and forest industry organisations and the State Forest Enterprise) were published. Moreover, economic incentives for the protection of natural values of forest were made available for nature management activities and environmental projects that fulfil the requirements set by the Act on Financing Sustainable Forest Management (1997).

These recent developments in the use of other policy means than regulatory ones did not remarkably reduce the tensions related to the traditional dualistic setting of forest protection policy. In the minds of many Finns, the majority of forests were still either "strictly protected" or "commercial forests" managed according to overall environmental guidelines. For example, although the Finnish government proposed to the EU Commission the inclusion of approximately 12 per cent of Finland's surface area into the Natura 2000 Network in 1998, the dualistic protection policy was not seriously challenged. The areas proposed mainly consisted of sites already included in existing conservation programmes.

This dualistic forest protection policy was a very fruitful setting for the emergence of a number of intense conflicts related to the protection of the last "old growth forests" or the "wilderness areas" of Finland in the 1980s and 1990s. Mostly, these conflicts focused on remote state-owned forests in Northern and Eastern Finland. Although the conflicts led to important policy reforms, the culture of environmental forest conflicts in Finland in 1984-95 can be characterised as including a tendency for strong value clash, intense struggles, and poor relations between different actors (Hellström 2001). However, it can also be argued that many of the conflicts were rooted in the intensification of forest management, and a subsequent belief that establishing strictly protected areas was the most effective way to combat the threats caused to biodiversity by forestry practices.

The dualistic setting of forest protection also led into a situation, where the percentage of strictly protected areas became a central indicator of the level of forest protection. Measured in such quantitative terms, the share of strictly protected forests of the total forest area is higher in Finland (6,6 %) than in any other European country. Most European countries (e.g. France, Germany, UK, Austria, Switzerland, Belgium, Netherlands, Italy, Spain) have less than one percent of their forests strictly protected (Parviainen et al. 2000).

Typically, protected forests are usually concentrated in the most remote areas. In Finland, for example, only about one per cent of the forests in Southern Finland are strictly protected. This imbalance between forest protection levels in Northern and Southern Finland set the frame, within which the Forest Biodiversity Programme for Southern Finland was prepared in the beginning of the new Millennium.

8.3 Scientific involvement in compiling the Forest Biodiversity Programme for Southern Finland

8.3.1 From "outside" involvement to "inside" involvement

In Finland, debate on regional differences in the level of forest protection has long roots. However, a political break-through in the issue took place when the National Forest Programme 2010 (Ministry of Agriculture and Forestry 1999) recognised the imbalance between the level of forest protection in Northern and Southern Finland, and stated the need to assign a broad-based group of specialists to identify the potential needs for increased forest protection in Southern Finland. Developing forest protection in Southern Finland was taken into the programme of work of the Finnish government in 1999, as recommended by the National Forest Programme.

The National Forest Programme was above all a political programme, which did not fully utilise available scientific information. Several aspects of the scientific basis were strongly criticised in various environmental assessments of the programme (e.g. Hildén et al. 1999). This raised pressure for increasing the involvement of scientists in future processes. Later, the need to develop participation of the scientific community in national forest policy was also noted in the evaluation of the National Forest Programme (Kivinen and Paldanius 2002).

To begin with, the Ministry of Environment set up a working group comprising mainly of specialists in ecology and protection biology, to evaluate the status and needs of forest protection in Southern Finland. In September 2000, the working group reported the need for better protection of herb-rich forests, mineral-soil sites with abundant decayed wood, and spruce mires in Southern Finland. Moreover, commercial forests should contain more decaying and burnt wood, large aspens and other hardwood species (Ministry of Environment 2000). The recommendations were made on ecological scientific grounds only, not taking into consideration their potential social or economic impacts.

The idea behind the scientifically oriented working group was that compiling ecological information would facilitate further decision-making in a subsequent multi-stakeholder process. However, this did not prove to be the case. Although the contents or conclusions of the working group were only questioned to a minor degree, there was reluctance among many stakeholders to utilise the findings of the working group, because they were not involved in the process and in drawing up the conclusions.

This process would have been a typical example of instrumental utilisation of science (see Box 8.1) if it had led to a decision on the future of protection of forests in Southern Finland. Information needs on forest protection in Southern Finland were identified, scientific information was gathered, and to a minor extent also produced. Then, they were interpreted in the framework of the decision-making situation., Because the information was interpreted only in an ecological framework, however, the working group was never given the mandate to finalise the instrumental use of science by deciding on the choice of solution. Instead, this issue was left to an explicitly politically dominated policy process, which was to follow.

Box 8.1 Types of utilising scientific knowledge as identified by Lampinen (1985).

Instrumental utilisation has direct influence in decision-making. It is best described as problem solving. This process may be described through the following chain: analysis of decision-making situation – identification of information needs – production or gathering of scientific information – interpretation of the research results within the framework of the decision-making situation – choice of solution. In short, the decision-maker uses scientific evidence consciously in order to fill in gaps of knowledge that are strategic to his decision-making. At large, the instrumental utilisation of science in decision-making is open to many types of criticism.

In *conceptual* utilisation of science, research does not provide direct answers to predefined questions but has a more indirect influence on decision-making. Research helps to conceptualise the problem in question. Most often, research has more impact on problem formulation than problem resolution. In this approach, science has no monopoly on "correct" information. Decision-making is also based on previous experiences, and other non-scientific communication.

Political utilisation is another form of indirect influence of science to decision-making. Instead of using research to search for the best possible solution, science is used to support a specific policy. Often, in political utilisation, research results are harnessed to serve purposes for which they were not produced. However, researchers may also themselves offer decision-makers such results that they are themselves comfortable with. Their motivation may be increased research funding or to influence decision-making.

In December 2000 the Finnish Council of State appointed a commission to work out a proposal for a Forest Biodiversity Programme for Southern Finland (later referred to as the METSO Commission and the METSO Programme). The Commission was established to present the goals and a schedule of work for improving the protection status of forests in Southern Finland, and appoint necessary means and funding for the work. The commission also had to examine the readiness and possibilities of different actors in the forestry sector for promoting forest protection in Southern Finland. Finally, the impacts of the proposed actions on private economies and national economy, on employment and other social aspects were to be identified.

The METSO Commission had 25 members representing a broad variety of economic, social and environmental interests related to forests. Although this new process was based on interest group representation, also scientists were assigned several roles "inside" the process:

– Five permanent experts to the commission were appointed to support the Commission's work. These experts represented knowledge in policy processes and conflict management, resource management and environmental impact assessment, ecology, environmental economics, and forestry development. The experts had no right to vote in the Commission but this was of little consequence since the Commission aimed at consensus and did not vote on a single issue. Although the status of the experts was not made clear, it was expected that they act neutrally in relation to the different interest groups, and base their work on expertise only.

- Some of the experts were assigned to chair interim working groups of the Commission (e.g. Working Group for Policy Means, and Working Group for Assessment Criteria).
- The design of the process was constantly developed and evaluated by a small working committee, consisting of the chair, co-chair, secretaries, working group leaders and the experts to the Commission.

Although officially assigned as experts to the Commission, the experts were not the only scientists to participate in the process. For example, the Finnish Association for Nature Conservation appointed one of the best-known ecological scientists in Finland as their representative. That is, not all scientists operated from a neutral position in relation to the interests involved. Moreover, several of the people that were involved (e.g. the Secretary General and the vice-chairman of the Commission) had a scientific education and career prior to their present positions in administration.

The task of the Commission was challenging already because of its wide scope and large number of participants. However, perhaps even more challenges were related to the novel nature of the work, differences related to information production and use, trust, commitment, funding and innovativeness. In the following sections, these challenges are elaborated, and the roles of the scientists in meeting them are described.

8.3.2 Setting the stage for information-sharing and trust-building

Work in the METSO Commission began in a situation, where deep distrust existed between parts of the stakeholders involved. Resolving the forest protection issue in Southern Finland was made difficult by the burdens of old conflicts over forest protection, and the distrust between actors it had generated. Despite this distrust, all the stakeholders involved in the process appeared to be rather well committed to the basic idea of securing biological values in the forests of Southern Finland, according to the recommendations made in the National Forest Programme. However, there was no overall commitment to increase the use of any specific type of forest protection measure, nor the need to introduce a new protection means. In the beginning, the participants even had rather different views on the overall necessity of the work to be conducted. They also had different perceptions of the level of each other's commitment to the work.

Although a large amount of information was compiled by the groups of specialists assigned to evaluate the status and needs of forest protection in Southern Finland (Ministry of Environment 2000), the METSO Commission still faced the problems of availability and credibility of information, which did not ease the problems related to lack of trust and commitment. For example, in the 1990s, major changes took place in forest management, with some positive impact on the forest environment. However, the final impacts of changes in forest management practices are not easy to evaluate only within a decade, when the rotation period of the forest is ten times longer. Insufficient information also existed on the ecological values that were already protected in the existing national parks and nature reserves. Moreover, information on the economic impacts of forest protection was limited, and the contents of social sustainability were still being defined. There was also considerable concern about political utilisation of science (see Box 8.1), e.g. using science to support particular policies.

Different stakeholders reacted on the lack of information and fears of political utilisation of science in different ways. For some, it was a motivation to act rapidly ("avoidance of potential ecological threats"). For others, the level of information was too low to trigger any action ("waiting until there is sufficient evidence"). Such debate also increased tension among the members of the Commission.

Owing to the substantial lack of trust, the METSO Commission was not willing to divide into any smaller working groups, although it was obvious from the beginning that the meetings, which were usually participated by 30 people, could not work very effectively. Everyone wanted to be present in every meeting, and be able to safeguard his or her interests at every time. In the beginning, there was also an evident need to build at least some common knowledge base.

Subsequently, during the first months, the Commission's work focused only on hearing external experts, sharing information and discussion (box 1 in Figure 8.1). Although the establishment of a common knowledge base did not succeed in all necessary aspects, giving enough time for discussion helped in clarifying some of the concepts used, and in learning to communicate with each other. This builds enough seeds of trust and commitment, in order to be able to continue with other working methods.

Although hearing external specialists brought valuable information to the table, it did not resolve all the problems related to information.

Subsequently, in May 2000, the Commission appointed itself an interim-working group (the Working Group for Research, box 2 in Figure 8.1), for identifying research needs for the future. The working group also evaluated what information could be produced within the time span of the Commission's work. The working group was participated by representatives of all major interest groups (forest owners, forest industry, state forestry, forestry professionals and nature conservation). This is an indication of how politicised the issue of information was within the work of the Commission. In addition, all the permanent experts to the Commission had the possibility to participate in the work. This working group was active until the final stages of the Commission's work.

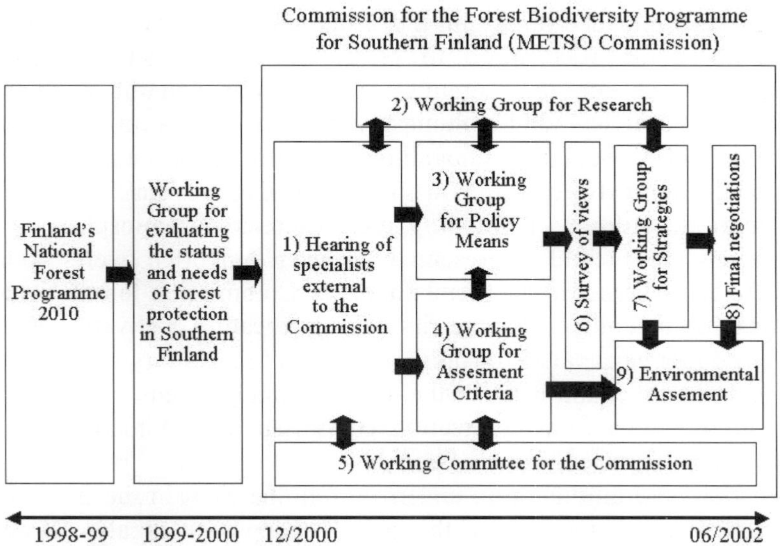

Fig. 8.1 The process of compiling the Forest Biodiversity Programme for Southern Finland (METSO Programme).

8.3.3 Conceptual work and process support

The amount of information gathered by the METSO Commission during the first six months was enormous, but it still did not fulfil all information needs. This led into some frustration on the progress of the Commission. Accordingly, the members of the Commission were finally ready to search for solutions by working in smaller dynamic groups that could work in more creative ways.

Innovative solutions were called upon for various reasons. Perhaps the most important practical barrier to the work was lack of funding. The programme was drafted in a situation, where the previous government of Finland had already decided upon using nearly € 600 million for implementation of old protection programmes during 1996-2007, with an additional € 250 million to be used in protecting state owned land. Because of this already ongoing significant input in nature protection, the government stated in appointing the METSO Commission, that no additional funding from the state budget could be assigned for forest protection in Southern Finland until after the year 2007.

Such tight financial frames required high innovativeness in designing new, cost-effective means of forest protection, and new models for funding such protection. Accordingly, in the assignment letter by the government, the Commission was urged to search for cost-effective, innovative solutions. Moreover, the Commission itself was commited to develop ecologically efficient and voluntary approaches. Innovativeness was a challenge also because innovations require good and confidential communication between different stakeholders, and overall motivation for the work. Accordingly, it was important to create an atmosphere of trust, where new ideas could be presented and even supported over stakeholder borders, and where also the members of the Commission felt motivated by the work to be conducted. The first step in creating motivation and trust was clarifying what was to be done.

The assignment and work of the Commission differed in many respects from all previous forest protection commissions in Finland. The most important differences were related to how the words "protection" and "protection programme" were understood. In the work of the Commission, "protection" no longer meant the strict conservation of areas drawn on maps and marked in the forest. Instead, it meant securing biological values of forest by using a variety of means both in conservation areas and in managed forests. Accordingly, the "protection programme" that the committee was drafting, was not a traditional map of areas to be protected, but rather a comprehensive framework policy for a variety of protection measures for the future.

This shift of focus brought about the need for conceptual work, which was strengthened when the Commission was finally ready to appoint additional working groups in June 2001. The Working Group for Protection Means (box 3 in Figure 8.1) was given the task to evaluate present means of forest protection, discuss their further development, and suggest potential new means of forest protection in Southern Finland. To begin with, the Working Group listed all means that were used to preserve forest biodiversity today. This was essential in order to increase understanding that the Commission was to deal with the whole spectrum of policy means and not only with traditional designation of lands for protection purposes. For classification of the policy means, a framework of both policy means and protection strategies was utilised. The idea was to illustrate how a certain protection strategy could be implemented through the use of several alternative policy means, and how one type of policy means could be used to fulfil several types of protection strategies. Finally, a SWOT analysis was conducted on the various policy means and protection strategies. This systematic, conceptual approach ensured that the search for solutions also focused on such new possibilities of forest protection, which were not in use in Finland yet. Accordingly, it broadened the scope of solutions to be considered by the Commission as a whole.

Simultaneously with the Working Group for Policy Means, also a Working Group for Environmental Assessment (box 4 in Figure 8.1) was assigned. Its task was to suggest methods and criteria for evaluation of the ecological, economic, and social impacts of the programme. The fact that the assessment criteria were designed simultaneously but within a different group that designed the new protection means, increased potential for creativeness. Those responsible for designing new policy means did not have to care for the consequences, but could rely on the fact that each suggestion would eventually be evaluated by using jointly agreed criteria. In the work of the Working Group for Impact Assessment, the overall concept of sustainable development had to be conceptually opened and defined in such a practical way that it could guide decision-making on the final programme.

Both working groups conducted a significant amount of conceptual work. Scientists who had been appointed as permanent experts to the METSO Commission chaired both working groups. Accordingly, these two processes formed a phase of predominately conceptual utilisation of science. As described in the Box 8.1, conceptual utilisation of science does not provide direct answers to predefined questions. Instead, research helps to conceptualise the problems in question.

These two working groups worked in close interaction during their whole existence. Moreover, the progress of work was regularly discussed and further developed in the meetings of the working committee of the Commission, participated by the chair, co-chair, secretariat, working group leaders, and other permanent experts to the Commission (box 5 in Figure 8.1). Accordingly, the work conducted at the working committee formed an important further channel for the participation of scientists in the policy process. However, this input was essentially neither instrumental nor conceptual. Instead of producing or disseminating, or sharing information related to the substance of the work, the scientists offered procedural support to the process.

8.3.4 Strategies and outcomes

Although the work of the Working Groups for Policy Means and Impact Assessment were completed in January 2002, procedural support from scientists continued. The list of potential means for future forest protection in Southern Finland was used as a basis for a survey among the members of the METSO Commission, in order to find out if there were any means that the members of the Commission could agree upon (box 6 in Figure 8.1). The survey was conducted anonymously so that the members of the Commission did not know which interest groups supported which means. On the basis of the survey, the suggested means were divided into two groups: those where some common interest existed, and those where significant disagreement existed.

In the following strategy work, the Commission decided to first find out, what could be done in relation to the means where most agreement existed, and only then consider whether there is need to supplement the selection of means with some more disputed ones. This helped focus the strategy work more on common than contrasting interests.

Next, a Working Group for Strategies was appointed (box 7 in Figure 8.1). Although it was suggested by the working committee that the experts to the Commission continue to lead this process, the members of the Commission disagreed. At this time, the process had come to a point where the cards had been dealt, and it was time to play them. This called for a chairman who was in a position to be able to carry political responsibility for the decisions to be made. Accordingly, this task was assigned to the chairman of the Commission.

After the strategy work that set the frame for final solutions, the final decisions on what means to include in the final programme, and to what extent they should be used, were made in negotiations participated by all members of the Commission (box 8 in Figure 8.1). Despite an extremely challenging process, the Commission for the Protection of Forests in Southern Finland was able to hand over their proposal to the Council of State of Finland in July 2002, by the time of the deadline that was given to the committee. Only one dissenting opinion and three supplementary statements were annexed to the report of the Commission.

As part of the final report, economic, social and ecological impacts of the proposed programme were evaluated (box 9 in Figure 8.1). Although the evaluation used the ecological, social and economic criteria designed by the Working Group for Assessment Criteria, the evaluation itself was not made in name of the Commission, but as an expert opinion of one of the experts to the Commission, which was annexed to the report. This was justified by the argument that, although the choice of assessment criteria was of political nature, the evaluation of impacts based on these criteria was predominantly a scientific effort. The practical reason was that the METSO Commission simply ran out of time. It would not have been possible to reach an agreement on the evaluations. Many of the disagreements concerning the validity of data that hampered the work in the early phases of the Commission's work would have been resurfaced.

Accordingly, the METSO Commission did not propose the immediate drafting of a traditional forest protection programme, in which strictly protected areas would be created by acquisition to the state. The strategy of the proposal is to first investigate the potentials related to the new voluntary means, and only then decide on the need of increased use of more traditional protection measures.

Five percent of the funding reserved for the pilot projects on new voluntary protection measures is to be allocated to research. Accordingly, at the same time as the piloting of the new voluntary protection means was started, vigorous and multi-disciplinary research has been launched to evaluate the economic, social and ecological impacts of the new means, in order to provide information for further decisions to be made on the protection of forests in Southern Finland. By 2006, an assessment of the impact of the measures taken will provide the basis for further decision-making. Accordingly, science also plays an important role in the implementation and evaluation of the programme.

Box 8.2 Forest protection measures of the Forest Biodiversity Programme for Southern Finland (METSO Programme)

The METSO Programme includes a variety of measures to improve forest protection in Southern Finland. Part of the means were based on previous protection means, whereas part of the means are new, thus, requiring pilot projects before use in full scale. In total, the programme includes 17 actions to be taken during 2003-07. All these actions were approved by the government of Finland, which has included the METSO programme in its new programme of work. Below, only the most important measures suggested are described.

During the first stage, the focus is on the *restoration* and *management* of the present nature conservation areas on public lands. Restoration aims at making areas, which have been changed by human activities, revert to as close to their natural state as possible. For this purpose, the Programme also calls for organised compilation of data from existing protected areas. The Programme also proposes that the Finnish State Forest Enterprise takes natural values into account more effectively in state-owned forests, and that areas valuable from a nature conservation perspective are inventoried and protected in special areas and in areas adjacent to nature conservation areas.

On private lands, the Programme launches four new protection instruments that operate on a voluntary basis. In *competitive bidding*, the authorities ask the landowners to offer areas to be designated as protection areas, after which the best offers are selected for implementation. The *trade with natural values* is a system where the landowner, under a special contract, maintains or adds to the natural values in his forests, and is compensated with an income from the buyer of natural values, such as the state or a foundation. In *biodiversity networks*, natural values in more extensive areas are safeguarded through local co-operation between forest owners and other local bodies. Finally, a *nature management area* could be established on the application of the landowner, so that besides commercial use there would be efficient protection of biodiversity. The landowner would be fully compensated for the economic loss ensuing from limitations in the use of these areas.

The Programme also proposes increased funds for the enhancement of biodiversity and a correction of the drawbacks of the present support system, and improved *information means*, e.g. education, extension, monitoring and research, in support of forest protection. For example, the final report of the committee includes a list of research needs and funding to start a new research programme on forest biodiversity.

Finally, the Programme includes increased *financing* external to the state budged (e.g. establishment of a foundation for financing forest protection) so as to make it possible to respond to the willingness of landowners to protect forests. Despite these sources of financing, the Programme also proposes a total of about € 60 million of state financing for the implementation of the action plan during 2003-2007.

The METSO-process was successful in the sense that it completed its task in time and on a rather high level of consensus. However, consensus was reached at the expense of not being able to define accurate, measurable goals for the level of forest protection (e.g. hectares, percentages, species, biotopes). Yet, the majority of the Commission also felt that the programme had made important contributions to the development of forestry practices in Finland in line with ideas of sustainable development. In particular, in the traditions of Finnish forest protection, this solution is a major step from a dualistic strategy based on regulatory means (see Chapter 8.2), towards a more pluralistic strategy using also voluntary means. The environmental groups that submitted the dissenting opinion and supplementary statements, however, saw that the process simply postponed necessary decisions and thus represented a "victory" for those opposing "proper" biodiversity protection through the establishment of more traditional protection areas. This tension is the background against which future decisions on the protection of forests in Southern Finland are to be made.

8.4 Lessons Learned

8.4.1 New perspectives on the utilisation of science

The traditional role of scientists in policy processes is instrumental or conceptual (see Box 8.1), taking place mostly "outside", and in the best case in close interaction with the actual policy process.

In the beginning of the work of the METSO Commission, the environmental community expected and the forestry community feared that ecological science would be used instrumentally in decision-making. Strong expectations existed particularly among the environmental community that the process would continue in a linear way, basing final decisions primarily on ecological facts compiled by the working group that was set up prior to the METSO Commission to evaluate the status and needs of forest protection in Southern Finland.

However, instead of a linear process, where goals are first defined, and means are then selected for reaching the goals, an iterative process occurred. The final goals of the METSO Programme were actually defined after agreement on the use of different means was reached. In fact, for many interest groups represented in the Commission, agreeing upon the means of forest protection (e.g. how protection would be implemented and who would pay for it) was a more important decision than agreeing upon the amount of forest protection. Such a setting challenged the basic ideology behind instrumental utilisation of science, and induced some feeling of betrayal among those, who had expected the whole process to be based on instrumental utilisation of science.

As described in Sections 8.3.1 and 8.3.3, the process involved expectations on instrumental utilisation of science, and eventual utilisation of conceptual utilisation of science. However, ultimately the appointed experts also played a very different role: that of developing procedures and of facilitating discussions within the process. Scientific knowledge and experiences of the experts were used to support the designing and selection of working methods of the process. Here, the task of scientists was to support the policy process in such a way, that the task of the METSO Commission would be accomplished and that agreement could be reached within the time frame set for the work. This kind of use of scientists and science does not fit into the categories of Lampinen (1985, see Box 8.1).

In the process of preparing the METSO Programme, this setting was not always understood. Reaching agreement and completing the task that was assigned was not in the interest of all parties involved. Accordingly, even process support from the scientists was sometimes understood as a political statement in favour of a specific policy action.

In many policy processes facilitators are used to guide the process through conflicting situations. In addition to facilitation skills, facilitators are usually expected to be neutral in relation to the issues to be resolved and to have sufficient expertise in the field of question. The demands for neutrality and expertise easily lead to temptations to turn to the scientific community. As pointed out in this section, this has advantages and disadvantages. A question that remains unanswered here is, in which ways scientist facilitators differ from pure process facilitators, and how these differences should be taken into consideration in policy processes.

Environmental impact assessment is another form of utilisation of science, which challenges the typology of utilisation of science presented in Box 8.1. The environmental assessment was in this case an integral part of the preparation of the programme. The criteria that were developed contributed to the development of means and the development of means to be included in the programme developed the criteria. This kind of reflexive use of science has features that can be said to reflect conceptual use, but may also provide instrumental basis for the choice between alternatives. It is also political in the sense that the discussions on criteria are colored by political considerations even when they are based on e.g. quantitative calculations of monetary costs of ecological benefits.

Another example of reflexive use of science in the METSO process, which goes beyond instrumental and conceptual utilisation of science, is related to the innovative solutions that were called upon. Dialogue between scientific knowledge and practical experience is a major source of innovation, and as the METSO process illustrates, the innovative nature of policy processes may be greatly increased, if the scientific community is able to participate also from the "inside".

8.4.2 Challenges in working from the "inside"

The preparation of the METSO Programme is an example of a policy process, where scientists were involved not only from the "outside" but also from "inside". A common fear is that deep involvement in the actual policy process may easily lead to political utilisation of science (see Box 8.1), which is discussed in the following.

Efficient utilisation of science within the policy process requires understanding from those leading the process, and also courage from the scientists themselves. Any statement, even though made on scientific grounds, and expressed only in order to support the process as a whole, may be interpreted as political utilisation of science by those with an opposite interest to the proposal made. In fact, one factor which has been preventing the participation of scientists in policy processes is the scientists' own fears of being labelled to support specific interests, which may endanger their integrity as researchers. In such situations, it is important that the leader of the process may stand in support of the integrity of the scientists.

When involving scientists "inside" policy processes, it is also essential that the role of the scientists is clearly defined and also communicated to all parties involved. At best, the involvement of the scientific community is planned already simultaneously with the assignment of the task and nomination of the committee. This would give the scientists a firm background to stand upon, in such conflict situations where members of the policy process have problems in differentiating between political and other forms of utilisation of science.

Traditionally, it is advised that the role of scientists is to produce and transmit information to policy processes, and not to participate in value-based decision-making. A similar fear has also been expressed by many forest research administrators, emphasising that the importance of scientists in support of policy-making is based on credibility, which should not be compromised (Mills and Solberg 1998, Lewis and Koch 1999, Guldin et. al 2005).

Involving scientists in policy processes from the "inside" does not need to challenge this view by introducing political utilisation of science (see

Box 8.1). In fact, the close involvement of scientists in decision-making may even increase the legitimacy of the policy outcomes. In modern pluralistic societies, there is growing call for participation of various interests in policy processes. Why should the scientific community make an exception? Even if we accept that the scientific community as a whole should not take political stand in relation to policy outcomes, we could still acknowledge that the scientific community has a legitimate interest in policy processes.

Jaatinen (1999: 22) defines lobbying as influencing political decision-making in the interest of a group by communicating with publics relevant to the political process in a certain issue. Accordingly, central aspects of lobbying include communication, influence and interest. The science community is not exempt from these aspects. In fact, the science community is expected to disseminate and communicate information provided by scientific methods in an open manner. Inevitably, one important motivation for communication is to influence decision-making. Moreover, it is in the interest of the science community that decision-making is based on sound scientific information as a foundation for reasonable and accepted decisions. This overall goal of the scientific community in policy processes should be acknowledged and separated from the types of interests involved in political utilisation of science.

If the participation of scientists is seen as an asset for policy processes, the question remains who should be involved and how. Experiences from the preparation of the METSO Programme suggest, that scientists being able to disseminate both instrumental and conceptual knowledge are necessary. Scientists being able to give procedural advice may also be essential for the outcome, particularly when conflicting issues are involved.

At best, individual scientists working within the policy processes may act as "bridges" to the rest of the scientific community. However, constructing a firm bridge requires considerable activity from the scientists themselves. In an ideal case, the scientist involved in a policy process may identify information needs along the process, use his expertise and contacts to gather such information from the rest of the scientific community, and disseminate it back to the committee. In practice, however, scientists often tend to act as individuals, without a mandate from the whole scientific community. This means that no matter who are involved, there usually exist some disagreements within the scientific community on which information to use and how. However, internal disagreement does not only feature the scientific community. It is present in almost all interest parties involved in policy-making, and thus, it should not discourage the closer participation of scientists in policy processes.

References

Guldin R.W., Parrotta, J.A., Hellström, E. (2005). Working Effectively at the Interface of Forest Science and Forest Polic. Guidance for Scientists and Research Organizations. IUFRO Task Force on the Forest Science-Policy Interface. IUFRO Occasional Paper 17

Hellström, E. (2001). Conflict Cultures – Qualitative Comparative Analysis of Environmental Conflicts in Forestry. Silva Fennica Monographs 2

Hildén, M., Kuuluvainen, J., Ollikainen, M., Primmer, E. (1999). Kansallisen metsänohjelman ympäristövaikutusten arviointi (Environmental Assessment of the national forestry programme. Ministry of Agriculture and Forestry

Jaatinen, M. (1999). Lobbying Political Issues. A Contingency Model of Effective Lobbying Strategies. Inforviestintä Oy, Yliopistopaino, Helsinki

Kivinen, K., Paldanius, J. (2002). Kansallinen metsäohjelma 2010:n evaluointi. Diskurssi Oy & Maa- ja metsätalousministeriö. 16.9.2002

Lampinen, O. (1985). Yhteiskuntatieteellisen tutkimuksen hyödyntäminen poliittis-hallinnolliessa päätöksenteossa. Suomen Akatemian julkaisuja 6/1985. Helsinki

Lewis, R., Koch N.E. (1999). Forest Science-Policy Interface. IUFRO Task Force Science-Policy Interface. Papers presented at a Side Event of the Third Session of the Intergovernmental Forum on Forests (IFF 3), Geneva, Switzerland 3-14 May, 1999. IUFRO Occasional Paper 13, 1-5

Mills, T.J., Solberg, B. (1998). Research Contributions to Forest Policy Formulation: Opportunities and Cautions. International Consultation on Research and Information Systems in Forestry. Proceedings. An Austrian and Indonesian Initiative in Support of the Programme of Work of the Intergovernmental Forum on Forests. September 1998, Gmunden, Austria. Edited by the International Union of Forestry Research Organisations (IUFRO) and printed by the Federal Ministry of Agriculture and Forestry of Austria, 19-31

Ministry of Agriculture and Forestry (1999). Finland's National Forest Programme 2010. Publications 2/1999

Ministry of Environment (2000). Metsien suojelun tarve Etelä-Suomessa ja Pohjanmaalla. Suomen ympäristö 437

Ministry of Environment (2002a). Etelä-Suomen, Oulun läänin länsiosan ja Lapin läänin lounaisosan metsien monimuotoisuuden turvaamisen toimintaohjelma. Suomen ympäristö 583

Ministry of Environment (2002b). Finnish Government Commission on Forest Protection Proposes Action Plan. Press Release of the Ministry of the Environment, 2 July 2002

Parviainen, J. et al. (2000). Forest Reserves Research Network in Europe. Mission, Goals, Outputs, Linkages, Recommendations and Partners. Final

Report. COST Action E4. The Finnish Forest Research Institute, Joensuu Research Station.

9 Public Participation during Site Selections for Natura 2000 in Germany: The Bavarian Case

Melanie Eben

German Development Service, Brazil

9.1 Introduction

The Bavarian case study presented here will explore the participatory process applied for site selections for the European-wide Natura 2000 network aimed at protecting Europe's biodiversity and analyse how successful participation really was. For this purpose, representatives of key groupings such as the Environment and Agricultural Ministries; government agencies such as the Bavarian State Agency for Environmental Protection, the District Council, and the Bavarian Forestry Agency, landowner associations such as the Bavarian Farmers' Union, the Bavarian Landowner and Forest Owner Associations, as well as two nature conservation NGOs had been selected for interviews. The interviews with the representatives of each stakeholder group were open-ended to ensure an unlimited expression of opinion in order to gain in-depth knowledge about attitudes, to reveal existing views on protected area management, to explore how rules and regulations are interpreted, and to explore opinions and suggestions regarding the future of biodiversity.

9.2 Public participation – just a new buzz word?

Participation can express itself in many ways, but regardless what shape it may take, it presents a powerful tool for the public to make their voice heard. It plays an important role in a democratic system where political decision-making involves - or at least theoretically ought to involve - integrating the public's voice. However, its success will be determined by the institutional framework that in turn will make effective public participation possible and by maintaining stakeholder involvement over time (see also Chapter 1). Stakeholders can be individuals or groups involved or affected by a development or conservation project, or who hold influence or affect the project or decision in some way or other, such as government agencies.

Six different types and degrees of participation can be distinguished: 1. passive participation, 2. participation by consultation, 3. "bought" participation, 4. functional participation, 5. interactive participation 6. independently taking action and self-mobilising stakeholder groups (cf. Pretty et al. 1999). Participation by stakeholder consultation is the way participation may take place in Bavaria. However, we shall see that it is not as powerful as other types.

Looking at public participation in environmental decision-making is interesting because nature-conservation-related matters are a relatively recent addition to the political agenda. It is no surprise, therefore, that integrating the public into decision-making processes is still not commonly practiced by the various levels of government.

Nature conservation should work with people and not against them and should apply tools such as public participation. This approach was acknowledged in Principle 10 of the 1992 Rio Declaration: *"Environmental issues are best handled with the participation of all concerned citizens, at the relevant level. At the national level, each individual shall have appropriate access to information concerning the environment that is held by public authorities, including information on hazardous materials and activities in their communities, and the opportunity to participate in decision-making processes. States shall facilitate and encourage public awareness and participation by making information widely available. Effective access to judicial and administrative proceedings, including redress and remedy, shall be provided".*

This principle was then reaffirmed by the World Summit on Sustainable Development in Johannesburg in 2002. Equally important is the Aarhus Convention, adopted in 1998, which constitutes the first international legally binding instrument for access to information, public participation in decision-making, and access to justice in environmental matters.

9.3 The Biodiversity Strategy of the European Union: the Natura 2000 network

Biodiversity in Europe is distinct in the sense that most habitats have been modified over many centuries by farmers, foresters, fishermen, and hunters, leaving behind semi-natural habitats rich in biodiversity with special habitat-species relationships. This stands in contrast to many tropical countries, which are very rich in terms of biodiversity but mostly left untouched and unmodified by human beings (such as the so-called *hot spots* – areas characterised by great biodiversity).

This biodiversity, as we know it today, still very much depends on being maintained by traditional low-intensity agricultural or silvicultural practices that have also shaped and influenced the European landscape in the past. In Germany, the state of the environment continues to be a reason for concern and is far from being managed on a sustainable basis. This has been repeatedly pointed out by the German Advisory Council on the Environment (SRU 2001): 69% of the existing 500 biotope types in Germany are threatened, while one third of them face serious threats; 36% of the fauna and 26.8% of the flora are endangered, while 90% of these threatened plants and animals are found in remaining areas of natural biotopes or sites under extensive use.

The evidence suggests that the rate of loss of habitats and species will not slow down in the near future (BfN 2000). There are manifold reasons for this, including habitat fragmentation, intensive agriculture, an ever-expanding infrastructure, and the traditional nature conservation approach. The last-mentioned focuses on setting aside a small percentage of land as protected areas but ignores the integration of areas outside these designated sites into the approach, thereby omitting to protect biodiversity on a large scale (SRU 2000). A new approach for safeguarding biodiversity, not only in Germany, but on the European level, is therefore needed to prevent the continuing deterioration of habitats and the loss of species.

The European Union's reply to international attempts to protect biological diversity (as acknowledged in the international Convention on Biological Diversity of 1992, a consequence of the Rio Summit) first came in 1979. It was embodied in the Birds Directive, which primarily sought to protect wild bird species by designating *Special Protected Areas* (SPAs) (Directive on the Conservation of Wild Birds, 70/409/EEC) with Article 3 recognising the importance of protecting habitats as an essential pre-requisite for the survival of birds. EU members' implementation of the directive has been very slow, and site designations are still incomplete today. Thirteen years later, in 1992, this idea was to be followed by the Habitats Directive (Directive on the Conservation of Wild Fauna and Flora, 92/43/EEC) which was to set up *Special Areas of Conservation* (SACs). The Directive's principle aim, as stated in Article 2(1), is to *'contribute towards ensuring biodiversity through the conservation of natural habitats and of wild fauna and flora in the European territory of the Member States to which the Treaty applies'*. At the same time, it seeks to '*take account of economic, social and cultural requirements and regional and local characteristics*' (Article 2(3)).

The approach adopted by the European Union was subsequently to combine protected sites under both the Birds and Habitats Directives to create the European-wide Natura 2000 network with what are called "Sites of Community Interest" (SCIs). The significance of the Natura 2000 network lies in its presenting a legally binding document created and ratified by all European Union members that establishes a common basis for a coherent ecological network. This network grants the European Commission substantial power to oblige EU member states to adhere to this agreement by, for example, withholding structural funds in case of non-compliance: in Bavaria this would have resulted in the loss of almost 1 billion German Marks (BayStMLU 2000d).

However, recent research by the WWF (2001) seems to indicate that despite an initial reluctance amongst member states to implement the Habitats Directive, Natura 2000 is now gaining in political importance, and it is increasingly considered as an influential driving force for nature conservation (SRU 2000). Although participatory measures during the implementation phase are not required by the Habitats Directive and are left at the discretion of member states, some EU countries such as Finland, Austria, and most German federal states have initiated consultative procedures, as was also the case in Bavaria.

As already hinted at in the introductory paragraph, it is important to distinguish between various levels of participation: consultation can be one type of participation – it serves to ask people for their opinion on a certain subject, but decision-makers can refuse to adopt any of the ideas expressed during the actual decision-making process. Consequently, consultative procedures are a first positive step forward but often lack the power to influence decision-making actively and independently, something that can only be achieved by participation through self-mobilisation and connectedness (Pretty et al. 1999), where it is the people - rather than decision-making organisms - who take action to initiate changes to a proposed decision or regulation.

9.4 Implementation Procedures of Natura 2000

The federal structure of Germany (see Figure 9.1), composed of sixteen *Bundesländer* (federal states; sing. *Bundesland*), functions in such a way that the federal government in many cases only provides framework legislation, leaving each *Bundesland* responsible for its own implementation of the federal laws. This is the case for legislation related to nature conservation. The Federal Nature Conservation Act provides the legal basis, i.e. the framework, for the nature conservation acts of all *Bundesländer* and, therefore, plays an important role for nature conservation in Germany. This often results in variations in the implementation of national policies.

Fig. 9.1 Map of all 16 German Bundesländer. Source: www.bundesrat.de

This has also been the case with the Natura 2000 network, where each *Bundesland* is responsible for ratifying the Directive in its own legislation, a result of which has been delays in adopting appropriate measures. Some of these delays can be linked to political, economic, and social factors (see WWF 1999, 2000, 2001, Dieterich 1999), but opposition to the designation of SCIs is one of the major issues and will be examined further in this chapter.

9.5 Opposition to Protected Areas in Bavaria Due to a Lack of Participation

Although renowned for its environmental awareness and general support for nature conservation issues, Germany is plagued by strong opposition to designating areas for nature conservation purposes from a range of stakeholders such as farmers, private landowners, forest owners, and sometimes even entire communities or key political players (see also Stoll-Kleemann 2001a).

There is a variety of reasons for certain social stakeholder groups to oppose designated protected, and the phenomenon is observed in many countries (e.g. McNeely 1992). One common reason is the lack of participation during the designation process, whilst another is the fear that nature conservation measures are to be carried out without any compensation for landowners. Whatever lies behind the opposition, however, the result is problematic because it has been clearly demonstrated that acceptance forms the broad basis for successful protected area management - an institutional framework alone being insufficient to protect nature effectively if the people involved are not prepared to contribute actively to the process of implementing the political measures.

An unfortunate reality in Germany is that nature conservation officials often display a patronising attitude and are known to ignore local land users. Not surprisingly, this has resulted in conflicts amongst various interests. It is slowly being recognised that problems have to do with people and not so much with environmental issues and that, as a consequence, society on the whole has to be integrated into nature conservation processes (Stoll-Kleemann 2001c). This would imply introducing other types of participation rather than mere consultations with stakeholder groups.

The situation in Bavaria is interesting in that a strong sense of independence - based undertaking actions things voluntarily - and a deep-rooted opposition to interference are traits of Bavarian culture (see BayStMLU 2000b), and these appear to be at the base of the recalcitrance against and the resistance to selecting Sites of Community Interest for the Natura 2000 network. As a gauge for detecting signs of opposition in Bavaria, the following indicators have been used:

- a slow, delayed and incomplete implementation process (e.g. BN 1999, EU Commission 2001, LBV 2001);
- the priority given to economic interest that resulted in generating an incomplete list of proposed Sites of Community Interest (pSCIs) (e.g. WWF 2000);

- written protest (e.g. BBV 1999);
- the rejection of designated sites that became obvious during a consultation procedure (e.g. BayStMLU 2000a).

9.6 Reasons for Opposition

When nature conservation measures lead to actual or perceived loss of or limitations in personal freedom or rights related to one's personal property, which can have major emotional repercussions, they are likely to produce negative sentiments toward establishing areas for nature conservation (Stoll-Kleemann 2001a-c), whereby the extent of non-acceptance will depend upon the subjective importance of the perceived "losses" involved. It has been demonstrated that the higher the degree or stakeholder participation in deciding and enacting these measures, the more likely there will be a higher degree of acceptance. Landowner representatives in Bavaria unanimously considered limitations in their property rights and resulting losses in the value of their land caused by the Natura 2000 regulations as unacceptable. They also felt threatened and provoked by "greens" trying to re-educate them, fearing that they would end up as landless farmers whose rights had been abrogated (e.g. BBV 2001).

Opposition on principle to any changes related to the implementation of nature conservation means that landowners react with scepticism, and it often proves difficult to change this negative attitude. A particular manifestation of this widespread feeling of mistrust and resentment of insufficient information was the public request by the Bavarian Farmers' Union to withdraw all designations of private land in order to avoid any legal disputes that could potentially arise through a designation (BBV 2000). It was reasoned that if designations were vetoed or rejected, nothing could happen to them nor would their land be designated as a Site of Community Interest. This ploy was propagated in weekly news bulletins and farmers' magazines. The lack of information about the Habitats Directive deeply affected farmers; they felt ignored and enraged at being confronted with a new regulation. No information or accompanying explanations had been provided, nor were they able to seek advice regarding details about the implications of the site designations.

Limitations in personal decision taking and regulations or controls being imposed upon property through a top-down approach also hinders acceptance. Landowners fear that their interests are not adequately heeded and that new rules and regulations (particularly when they are obligatory) will have negative implications for the economic development of their property (e.g. BBV 1999). This not only created mistrust and reluctance to accept new regulations in the Bavarian case study, but also in most other EU member states, according to a study by WWF (2000), a situation mainly attributed to the insufficient creation of awareness amongst stakeholders (such as farmers and landowners) throughout the European Union.

The top-down approach pursued by former Federal Environment Minister Jürgen Trittin is especially disliked throughout Germany, as the agricultural lobby feels that their interests are being ignored and their land turned into conservation areas, preventing them from engaging in their agricultural activities in these areas and resulting in a loss of value of their land. The lobby also regards site designations as "interventions on private property" that would involve expropriation and/or hinder economic development.

How members of certain social groups perceive their role or identity is determined by traditional values, beliefs, or emotions. Cultural values, for example, determine the relationship between 'nature' and 'man'. Landowners regard themselves as keepers of the countryside who have always 'looked after' the land (e.g. Verband der Bayerischen Grundbesitzer 2001b). They now, however, see their traditional role being questioned and are thus opposed to shifting emphasis to a relationship where 'man' plays an 'inferior' role to nature. In Germany, landowners perceive this practice as resulting from an overly zealous attitude on the part of the country's previous red-green coalition government.

Differences or ambiguities in the interpretation or regulations on the national level amongst the *Bundesländer*, but also between stakeholders such as ministries, government agencies or NGOs within the *Bundesländer*, have also led to a number of hindrances to nature conservation. These include lack of cooperation, co-ordination and/ or disagreements in how best to implement Natura 2000, incomplete site designations, and the failure to implement management plans and provisions for monitoring. The interpretation of the implications of Natura 2000 have often caused confusion, not only amongst landholders, but also amongst "experts" in the nature conservation scene, as it remained unclear (as has been unclear right from the beginning in 1992 when agreeing to adopt the Habitats Directive) what exactly had to be done to implement the Directive. There was also a lack of co-ordination on the higher administrative level amongst various government agencies, some of whom were not willing or did not have the capacity to cooperate.

Inadequate financial provisions (e.g. lack of compensation payments for farmers) and lack of staff at the national and *Bundesländer* level have also contributed to delays, and the insufficient allocation of necessary resources in Germany was confirmed by a study by WWF (2001). The Advisory Council of Experts on the Environment (SRU 2000) stresses that adequate funding and personnel have to be provided to guarantee prompt implementation of the EU directives.

The lack of political will to respond to the implementation requirements of the Habitats Directive resulted in the *Bundesländer* often feeling restricted in their planning authority. On the other hand, the resistance to implementing the Directive resulted in their producing very little publicity about Natura 2000 and what it implies: landowners were informed very late about the designation process and its consequences.

9.7 The Participatory Process in Bavaria

In 1998 the Bavarian Nature Conservation Act (Bayerisches Naturschutzgesetz) was amended to implement the Habitats Directive into Länder legislation. An area of 120,000 ha (1.8% of the Bavarian territory) had initially been proposed as SCIs (Sites of Community Interest). Although one of the first *Bundesländer* that provided a list with SCIs, Bavaria's proposed sites only presented a fraction of the valuable habitats and ignored the protection of other already designated sites. For instance, in 1985 a Bavarian biotope network *Bayerischer Biotopverbund* had been established; it is maintained by means of nature conservation contracts and currently comprises 8.5% of valuable habitats in Bavaria.

However, none of the sites contained in the Bavarian biotope network was proposed as a Site of Community Interest, as strong emphasis is placed on acceptance and voluntary commitment by farmers: designating them as SCIs would consequently be against the principle of acting voluntarily. This in turn prompted criticism by the non-governmental organisation *Bund Naturschutz* since these areas represent important habitats that would fulfil the scientific criteria of Natura 2000. As a consequence WWF-Europe and *Bund Naturschutz* declared the proposed designations as incomplete and not fully representative; they subsequently produced so-called 'shadow lists' to complement gaps in the official site designation lists (*Bund Naturschutz* 1999, 2001, WWF 2000).

In order to overcome strong resistance to site designations, i.e. to increase the acceptance of site designations by making the selection process more participatory, a three-month public consultation procedure ('Dialogverfahren') was initiated by the Bavarian Environment Ministry in February 2000 (BayStMLU 2000d). This was done in accordance with the European Commission, which had approved the "Bavarian Solution" (c.f.

BayStMLU 2000e), and was the first dialogue procedure of its kind to take place in Bavaria. The principal reason for this new approach was the strong criticism by environmental organisations as well as the reluctance of landowner groupings to agree on site proposals.

In various districts, public meetings were held beforehand (organised by the Bavarian Environment Ministry and the Bavarian State Agency for Environmental Protection) whose intention was to inform communities, private property holders, farmers, citizens, organisations and associations such as farmers' unions, as well as business representatives. With the purpose of providing stakeholders with better information about proposed sites, maps and respective site descriptions were distributed at the community and district level. These were also available via the Internet or on CD-ROMs. A special telephone service was provided to answer related questions. Affected stakeholders were given the opportunity to make written objections to their land being proposed, and a total of 20,000 rejections were filed.

Consultations with stakeholder groups in other countries, such as France, UK, or Finland (see also Welp et al. 2002) have often resulted in excluding proposed sites because of opposition by local people (WWF 2000). These have typically been intensively used agricultural areas and communally owned industrial areas. At the same time, 3,000 new proposals were put forward, 550 of which were ultimately included as newly proposed Sites of Community Interest. These were finally reported to Brussels after district councils had revised all proposals, increasing the total area of proposed sites to 500.000 ha. This included 6.7[1] of the Bavarian state area (BayStMLU 2000c). Of these 6.7%, 64% was forest area with only 36% open land (WBV, personal communication; of which 62% is state-owned forest, 23% private, and 15% community forest), indicating the preference to designate areas owned by the state with the purpose of avoiding conflicts with private landowners (cf. BayStMLU 2000d).

When comparing Bavaria to other German federal states, it becomes evident that conservation efforts could be improved. For example, Brandenburg has designated 11.3% of its federal territory and Thuringia 10.0%. States such as Schleswig-Holstein (8.0%) and Hessen (9.9%) have designated slightly fewer sites than Bavaria (BfN 2006).

Comparing these figures on the total area of proposed protected sites in Germany with other European countries, the difference in the total area of selected sites becomes more obvious (see Table 9.1). However, additional sites under the Habitats Directive have already been included in the latest

[1] In 2006: 9.2% (BfN 2006)

national lists (in Den Haag 2002 and in Potsdam 2003) by countries such as Germany (183), Spain (57), and France (65), and it was expected that these numbers will rise as a result of some bio-geographical seminars to be held in the near future (EU Commission 2003). In this context, in January 2004 a bilateral meeting took place between Germany and the European Commission whose goal was to ascertain whether the Bundesländer's statements of intent to supplement the list of existing protected areas with new ones were complete. In most cases the EU Commission assessed the proposals as adequate, but in a few, additional demands were made. Based on the results of this meeting, in January 2005 the Bundesländer's proposition for the new sites was forwarded to Brussels (BfN 2005).

Table 9.1 The percentages of designated national territory under the Birds Directive (SPAs) and of proposed national territory under the Habitats Directive (SICs) in various European countries (EU Commission 2006)

Member state	% of national territory designated as Special Protected Areas (SPAs)	% of national territory proposed as Sites of Community Importance (SICs)
Denmark	5.9%	7.4%
Spain	18.2%	22.6%
Greece	10.1%	16.4%
United Kingdom	5.8%	6.5%
Germany	8.9%	9.8%
France	2.7%	6.9%

9.8 Public participation – a success or failure?

Communication amongst stakeholder groups is vital for establishing trust and spreading information. Experience has shown, though, that involving people in the planning and implementation process frequently does not take place or takes place at late stage, as was the case in Bavaria. German laws contain weaknesses with respect to these consultation procedures, although the possibility to make written petitions does exist. However, taking decisions without the consultation of people results in recalcitrance (Stoll-Kleemann 2001a-c) and has been criticised by the Advisory Council of Experts on the Environment (SRU 1996).

Furthermore, numerous practices have resulted in resentment and uncertainties amongst landowners and land users about how to interpret regulations. These include the inadequate provision of information, documents written in language incomprehensible to 'non-experts', and ambiguities in the interpretation of how to implement the directives amongst “experts” due to unclear formulations by the EU Commission regarding procedures by each of the member states. There have also been delays in or inappropriate timing of measures whose goal was to inform stakeholder groups, such as talks, seminars, etc.

The granting of a relatively short three-month time period implied that a continuous dialogue, exchange of opinions, concerns, information, or 'true' communication was not actively being sought; the timeframe was simply too short for all the necessary or desired activities to occur. This was also criticised in Bavaria (LBV 2001) despite a very confident take-up by landowner groupings, which considered the procedure very positive because they had been given the opportunity to oppose site designations and reduce the originally proposed area of Bavaria’s territory for SCIs from 12% to 7.5%.

Numerous people, though, resented the sudden rush of the initiated dialogue (*Bund Naturschutz*, personal communication). Representatives of environmental NGOs regarded the content, administration, and commitment of some nature conservation officials during the dialogue procedure as being of inadequate. Criticisms of a ‘pseudo democracy’ and a fragmented selection of stakeholder consultations became loud, referring to the fact that not all objections or proposals were paid attention to. This is in line with the classification by Pretty et al. (1999), where participation by consultation is viewed as unsatisfactory when the consulting agent has no real obligation to integrate the outcome (i.e. opinion) into the decision-making. This is corroborated by the fact that acceptance of site designations increased considerably in cases where discussions and information events with local players took place. For example, the *Bund Naturschutz* initiated informational talks near Freising-Munich to brief farmers about the implications of Natura 2000 and how they would be affected, resulting in many of the farmers’ fears being assuaged.

Data-protection laws created another important obstacle in the process. These made it difficult to contact landowners, which in turn impeded informing local players appropriately about which property was being planned for inclusion in Natura 2000. This problem was partly solved by organising public meetings to inform potentially affected people; it proved impossible to obtain individual addresses due to the above-mentioned privacy regulations.

9.9 What can we learn from the Bavarian case?

Despite the various deficiencies, at least a first positive step towards including stakeholders in policy-making has been made. Overall, however, not all of the requisite conditions for protected area management under Natura 2000 were created by the participatory procedures. This indicates the need for a novel, innovative approach that builds on social self-esteem through expanded participatory involvement and stronger emphasis on sustainable rural livelihoods rather than purely building on compensatory measures that leave out the social dimension. Better communication between stakeholders and positive results, in turn, will be influenced by trust, cooperation, and appropriate property rights laws. It is therefore essential to create an atmosphere of mutual trust amongst all stakeholders involved to contribute to the conservation of biodiversity in Germany.

In order to create the basis for more acceptance, various suggestions by all Bavarian stakeholder representatives interviewed during this research project have been collected and are summarised below. These could pave the future way for a more inclusionary approach that results in milder reactions when it comes to implementing nature conservation objectives.

More participation of all stakeholders at an early stage, particularly in the form of 'roundtables' where involved parties are given the opportunity to sit down together and discuss before and during the planning and implementation process is needed. More transparency would also be a significant asset, including adequate information for the public, proper planning and preparation of the consultation process, respecting process results, and having a good public relations programme for stakeholders (i.e. keeping them informed about further steps). As mentioned previously, independent participation through self-mobilisation by the stakeholder groups themselves would also considerably increase the acceptance and success of any new decisions and/or regulations.

Any approach ought to rely on voluntary participation or commitment and compensatory payments rather than obligatory measures. This was an concept strongly defended by landowners, who vigorously oppose inflexible obligatory measures issued by the state. As a model, it was suggested to consult landowners first and then make contracts, as well as provide compensation payments to guarantee that the sites will be managed according to specific conservation criteria.

Better environmental education and raising awareness, particularly amongst the general public, were suggested as important instruments (Stoll-Kleemann 2001c). Currently, nature conservation does not receive high priority and is suffering from a lack of interest amongst the general public. Intensifying efforts to portray nature conservation as something of importance to everybody and as something that does not always have to exclude economic interests would certainly make a valuable contribution to improving the weak relationship between “nature and human beings”.

Improving the relationship between nature conservation professionals and laypersons, i.e. involve stakeholders more and at an earlier stage and respect each other’s views, opinions and interests, was viewed as essential. Nature conservation officials were conscious of the necessity to integrate the factor “people” into conservation approaches; a conference held by the EU Commission on Natura 2000 and People (EU Commission 1998) also emphasised the need to build partnerships with stakeholders right from the beginning (see also Stoll-Kleemann and O’Riordan 2002).

The goal is to inform the public and stakeholders adequately and ‘advertise’ the advantages of living in a protected area, in order to change reservations and feelings of scepticism into a sense of pride. The WWF (2000), for instance, point out that particularly by placing a new value on many remote areas, Natura 2000 offers the potential for implementing innovative and sustainable development strategies as a way to avoid the ongoing loss of biodiversity.

The demonstration of more political will and support by governments for the implementation of instruments such as the Birds and Habitats Directives and to use politicians as 'good examples' to raise the profile of nature conservation issues as something worth pursuing was also regarded as fundamental if success on a large scale is to be achieved.

More opportunities for capacity building and training in communication should be made available, as a lack of these skills can negatively influence the outcome of any consultation effort. Also needed are more funding to provide financial incentives for landowners to include sites in the Natura 2000 network and more staff to carry out all implementation requirements. Landowners were sceptical whether tasks could be accomplished without an increase in funding to provide compensation payments. Despite the availability of funding schemes like the EAAGF (European Agricultural Guidance and Guarantee Fund) or other schemes co-financed by the EU, little use of this financial support has been made so far.

In other words, in order to achieve wider public acceptance and engagement, participatory tools ought to be included more often and more widely as one of the instruments employed to establish good communication and better cooperation amongst various stakeholder groups. The Bavarian case study clearly demonstrates that a lot of work has to be done in this respect.

The first step, namely the substantiation of the existence of problems and – to a degree – the identification and localisation of their origins, has been done, though. Based on this, problem-oriented solutions can now be tailored, taking lessons learnt in the past into consideration. However, only if nature conservation per se rises on the political agenda and the public show a higher interest in nature conservation issues, can participatory measures really show their powerful effect.

The principles of public participation and the right of everyone to stake a claim in decision-making has often been stated, most importantly in the 1992 Rio Declaration as well as the 1998 Aarhus Convention. The problem down to the present, though, is that while these important principles exist on paper, they are often ignored in practice. Natura 2000 has provided a good opportunity to test them in practice; the experience accumulated during the entire designation process will serve as a foundation on which to build. Not only current European Union countries can and will benefit from this experience; it also provides a great deal of opportunity for the new-accession states, who will be able to avoid the mistakes made by the old EU members and profit from past experience when engaging in future site designations for the Natura 2000 network.

References

BayStMLU (Bayerisches Staatsministerium für Landesentwicklung und Umweltfragen) (2000a). Abschluß FFH-Dialogverfahren in Bayern; Überaus große Beteiligung mit 20 000 Einwendungen und 3000 neuen Gebietsvorschlägen. Press release, 31 May. BayStMLU, München

BayStMLU (2000b). Zweite Tranche ‚auf bayerischem Weg' ins Dialogverfahren; erste und zweite Tranche sehen rund 7% der Landesfläche für Natura 2000 vor. Press release, 27 January. BayStMLU, München

BayStMLU (2000c). Interview von Staatsminister Dr. Werner Schnappauf zur FFH-Richtlinie (Bayerische Gemeindezeitung , Nr.1, 20.1.2000). BayStMLU, München

BayStMLU (2000d). Die Umsetzung der FFH- und der Vogelschutzrichtlinie der Europäischen Union: Das Dialogverfahren – der bayerische Weg. BayStMLU, München

BayStMLU (2000e). Gespräch Umweltminister Schnappauf mit EU-Kommissarin Wallström zur FFH-Richtlinie in Brüssel; Verständnis und Zustimmung der EU für bayerischen Weg des Dialogs liegt vor. Press release, 7 February. BayStMLU, München

BBV (Bayerischer Bauernverband) (1999). Gemeinsame Erklärung zur Umsetzung der Flora-Fauna-Habitat-Richtlinie und der Vogelschutzrichtlinie der Europäischen Union in Bayern vom 20. Dezember 1999

BBV (2000). Hinweise des Bayerischen Bauernverbandes zum „Dialogverfahren Natura 2000". BBV, München

BBV (2001). Deutscher Bauerntag in Münster: Rede Gerd Sonnleitner. BBV, München

BfN (Bundesanstalt für Naturschutz) (2000). Nature Data. BfN, Bonn

BfN (Bundesamt für Naturschutz) (2005). URL: http://www.bfn.de/0316_bilateral.html [July 2006]

BfN (Bundesamt für Naturschutz) (2006). Abteilung Biotopschutz und Landschaftsökologie FG I 2.2. Stand: 17.02.2006

BN (Bund Naturschutz) (1999). Netz des Lebens. Vorschläge des Bundes Naturschutz zum europäischen Biotpoverbund (FFH-Gebietsliste) in Bayern. Bund Naturschutz Forschung, Nr.3. Bund Naturschutz, Nürnberg

BN (2001). Stellungnahme zum Dialog über die Weiterentwicklung der Agrarpolitik am 12.2.2001. In: BayStMLF, München

Dieterich, F. (1999). Ausweisung von Natura 2000-Gebieten in Deutschland und daraus folgende Konsequenzen. In: Bundesministerium für Umwelt, Jugend und Familie (1999). Natura 2000: Eine Chance für den Naturschutz Europas. Tagungsband 14. Schriftenreihe des BMUJF, Wien

European Commission (1998). Natura 2000 and people: a partnership. Proceedings of a conference held in Bath, UK. Organised by: UK presidency

of the European Council and the Unit for Nature Protection, Coastal Zones and Tourism of the EU Commission
European Commission (2001). Natura 2000. Newsletter, No.14. European Commission, GD ENV, Brussels
European Commission (2003). Natura 2000. Newsletter, No.16. European Commission, GD ENV, Brussels
European Commission (2006). Natura 2000. Newsletter, No.20. European Commission, GD ENV, Brussels
LBV (Landesbund für Vogelschutz in Bayern) (2001). Umsetzung der FFH-Richtlinie in Deutschland unzureichend. LBV begrüßt Schlussantrag in der Klage gegen Deutschland. Press information, 3 May. LBV, Hilpoltstein
McNeely, J. (1992). The biodiversity crisis: challenges for research and management. In: Sundland, O.T., Hindar, K., Brown, A.H.D. (Eds). (1992). Conservation of biodiversity for sustainable development. Scandinavian University Press, Oslo
Pretty, J., Hine, R., Deighton, K. (1999). Participatory Appraisal for Community Assessment: Principles and Methods. Centre for Environment and Society, University of Essex. URL: http://www2.essex.ac.uk/ces/ResearchProgrammes/pa&caoverview.htm
Stoll-Kleemann, S. (2001a). Understanding and reconciling opposition to protected areas management in Europe: the German experience. Environment 43 (5), 32-43
Stoll-Kleemann, S. (2001b). Opposition to the designation of protected areas in Germany. Journal of Environmental Planning and Management 44 (1), 109-128
Stoll-Kleemann, S. (2001c). Barriers to the implementation of nature conservation in Germany: a model explaining opposition to protected area designation and management. Journal of Environmental Psychology 21 (4), 369-385
Stoll-Kleemann, S., O'Riordan, T. (2002). From participation to partnership in biodiversity protection: experience from Germany and South Africa. Society and Natural Resources 15 (2), 157-173
SRU (Der Rat von Sachverständigen für Umweltfragen) (1996). Konzepte einer dauerhaft-umweltgerechten Nutzung ländlicher Räume. Sondergutachten. Metzler-Poeschel, Stuttgart
SRU (2000). Umweltgutachten 2000: Schritte ins nächste Jahrtausend.
SRU (2001). Stellungnahme zum Entwurf eines Gesetzes zur Neuregelung des Bundesnaturschutzgesetzes (Stand 2. Februar 2001)
VBG (Verband der bayerischen Grundbesitzer) (2001). FFH-Richtlinie trifft Eigentümer unmittelbar. Press release, 4 April. Verband der bayerischen Grundbesitzer, München
Welp, M., Hamidovic, D., Buchori, D., Ardhian, D. (2002). The uncertain role of biodiversitiy management in emerging democracies. In: O'Riordan, T., Stoll-Klemmann, S. (eds). Biodiversity, Sustainability and Human Communieties. Cambridge University Press, Cambridge, 260 - 291
WWF-Österreich (1999). Umweltpolitik 1999. Natura 2000. Chancen und Hemmnisse. WWF, Wien

WWF (2000). EU Habitats Directive: WWF European shadow list. WWF, Brussels

WWF-Europe (2001). A race to protect Europe's natural heritage. European snapshot report on the status of implementation of the Habitats Directive. WWF European Policy Office, Brussels

10 Experiences with Stakeholder Dialogues in Natural Resources Management in Ecuador

Two Case Studies from German Development Cooperation Projects

Ecuador is rich in biodiversity, at the same time the pressure on natural resource use is high. The first case study focuses on participation in the Machalilla national park, one of Ecuador's 11 national parks while the second looks at community forest management in Esmeraldas.

The projects are funded by two German development agencies, both of which have their own traditions with respect to participation. The DED (Deutscher Entwicklungsdienst/ German Development Service) is supporting the participatory processes in the Machalilla National Park while the GTZ (Gesellschaft für Technische Zusammenarbeit) facilitates the dialogues between different actors in Esmeraldas.

The regions differ considerably in terms of natural conditions, social structures and economy. Nevertheless, a comparison between the participatory approaches and methods applied by DED and GTZ in these regions is of interest. Both studies focus on success and failure factors of stakeholder dialogues.

10.1 Participation in the Machalilla National Park, Ecuador

Michael Sturm[1], Jorge Samaniego Rivera[2]

1 Agentur Sturm, Flensburg, Germany
2 Parque Nacional Machalilla, Ecuador

10.1.1 Introduction

The Machalilla National Park (MNP), approximately 55.000 hectares in area, was established in 1979. It is located in the southwest of the Ecuadorian coastal province *Manabí*. The MNP includes two mainland areas, two islands, and a two-mile-wide central reserve along the respective coasts (Inefan, GEF 1998: 3, 8-9) (cf. Fig. 10.1).

The reasons for the creation of the national park include the protection of parts of the only tropical dry coastal forest in South America and finds of old indigenous cultures (Samaniego 1997: 3). The *Manabí* coastal region is also a breeding area for humpback whales (Castro et al. 1999, Scheidat 2001) and turtles (Barragan and Yumiseva 2003).

The main reason for the German Development Service (DED) to become involved in the national park was to guarantee the protection of the local natural resources and to improve the living conditions of the local population. Other considerations for DED were the necessity to develop and implement a management plan for the MNP as well as to create means of local participation during the implementation. DED experts have been active in villages in or near the area of the future NPM since 1978, and for a period of time, some of them campaigned alongside village communities against the "overnight" establishment of the national park, which took place in 1979.

Property ownership in the area is frequently based on local customary law, which has complicated the achievement of the national park's goals: in 1997, people from the villages and settlements within the MNP alone owned or laid claim to more than 23,400 ha of the national park's territory (Vaca Bucheli 1997: 11). This corresponds to around 43% of the total area. Also farmers, traders, lumberjacks, and cattle owners who live adjacent to the park have made claims or hold estates in the MNP (Vaca Bucheli 1997: 9-10).

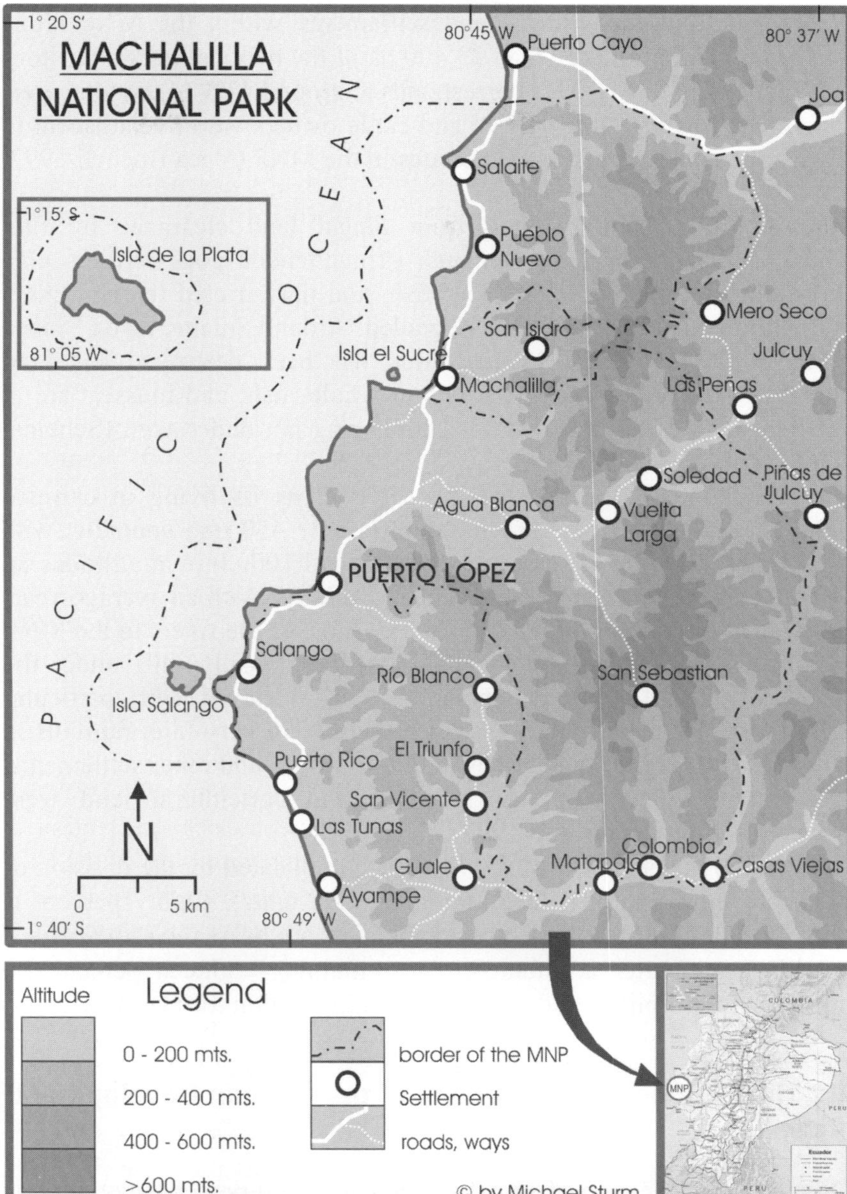

Fig. 10.1 Map of the south-western part of the Ecuadorian coastal province Manabí

Property ownership in the area is frequently based on local customary law, which has complicated the achievement of the national park's goals: in

1997, people from the villages and settlements within the MNP alone owned or laid claim to more than 23,400 ha of the national park's territory (Vaca Bucheli 1997: 11). This corresponds to around 43% of the total area. Also farmers, traders, lumberjacks, and cattle owners who live adjacent to the park have made claims or hold estates in the MNP (Vaca Bucheli 1997: 9-10).

Other conflicts have resulted from illegal land clearance by fire, poaching, and inappropriate uses such as unattended goat herding, clay brick production, overexploitation of fish, and unrestricted tourism. As a result, natural resources have been degraded: Around Salaite, Agua Blanca and Machalilla (cf. Fig.10.1) vegetation has been destroyed and land erosion has occurred. The populations of whale, fish, and mussels are at risk if regulation is not enacted and monitoring not undertaken (Scheidat 2001).

The population in the area of the MNP earns its living in extreme climatic conditions: average precipitation is only 450 mm annually, with approximately 160 mm at sea level up to around 1000 mm at altitudes up to 840 m. This falls primarily from January to May. With an average year-round temperature of approximately 24 ° C, most of the rivers in the MNP do not flow continually. The local climate periodically falls under the influence of the El Niño phenomenon. This occurred with particular severity most recently in 1997-98, when there were very high rainfalls in short periods that led to inundations in Puerto López and other settlements, bridges, streets, electric power supplies, and agricultural land were destroyed (cf. Mero and Sturm 1997).

Coordinated management of the MNP is complicated by the division of its area into three cantons (local administrative units). Eighty percent is allocated to Puerto López, in whose capital - which bears the same name - the park's headquarter is situated. The remaining eighteen percent are within the cantón Jipijapa and two percent in the cantón Montecristi.

10.1.2 The Participatory Approach of the German Development Service

DED views "modification of social relationships and power structures" as a goal of participation in development cooperation (Löbsack 1998: 3). Participation - in the sense of not only taking part but also true involvement - promotes the target groups' identifying themselves with the development tasks as well as the goals these tasks set out to achieve. The Organisation refers to this as "Project Ownership", which implies diversity in the forms and levels of participation.

As a consequence of negative experiences with development endeavors that were not needs-oriented, participation and involvement in decision-making by those directly affected and their early integration into the planning and implementation of the projects has been recognised as a fundamental requirement for enhanced effectiveness of cooperation in development efforts. In practice, though, there is a broad gap between the aspirations of development cooperation and the actual involvement of the target groups in development activities. (ibid: 4). Compare Chapter 10.1.5 "Conclusions, Transferability and Lessons learned".

The following describes the concrete example of how this has played out in the activities in the Machalilla National Park.

The goal of the projects described below was to enable maximal activism and involvement in regard to the topic areas despite the unfavorable prevailing conditions that surrounded the implementation of participatory methods. Many project concepts or aspects of them were the result of time-consuming efforts by team members, meaning they were produced in small groups that usually included both male and female local colleagues. Interested institutions or individuals were usually found to implement them. The aim was to insure that target groups, e.g. those interested in environmental education, would be reached both in terms of the content and their emotional interest because the main goals were raising acceptance of the National Park and its management plan, improvement of knowledge about both, and developing active cooperation in putting the management plan - or at least portions of it - into practice.

10.1.3 Participation in the Machalilla National Park (MNP), Ecuador

Problems and Conflicts

Participatory methods were almost negligible components of the National Park's practical management There was no "proactive" participation in the sense of regularly scheduled "round tables" or similar get-togethers in which all involved stakeholders met to deal with current or potential problems and conflicts. The hierarchy of the National Park team operated in a strictly authoritarian manner and during the period described herein was geared toward the Head of the National Park at the time.

Already the first DED-expert recorded in his final report the lack of and interest in participatory structures in the National park management team by the then park director (Zenger 1994: 4-5). In her project status report of June-October 1995, another DED expert criticised the fact that: "... the

Park Director continues to provide little transparency concerning new developments and results in the area of his tasks that are of great importance to us development workers and our counterparts" (Sigle 1996a). The DED program planning documents of 1996 - 1998 added (p. 13): "...it is not a modern administration that sets the tone, but rather an extremely hierarchical organisation with no downwards delegation of responsibility, which corresponds to the country's current stage of development (DED Ecuador 1996)." Lincango (1997: 7), too, expressed dissatisfaction about the "lack of the actors' regular participation" and drew attention to the low acceptance, knowledge and poor implementation of the management plan in general and especially with regard to tourist activities in the MNP.

Participatory activities that have been undertaken frequently give impression of being merely "for show". They have proven, to a degree, beyond the comprehension and abilities of the inexperienced participants, and though the resulting proposals indeed exist in documents, they have seldom been put into practice.

For example, during the ongoing rework of the old management plan of 1986 that took place between 1996 and 1998, there were several workshops for the identification of the MNP's particular problems to which NGOs and GOs invited representatives of the municipalities and fishermen, and in which the latter actually participated (INEFAN, GEF 1998: 16). For the identification of the macro-problems of the MNP, a five-day workshop entitled "Destruction of the natural and cultural resources in the Machalilla National Park and its buffer zone" was convened in November 1997 in Puerto López. The results comprised detailed analyses of the problems in the form of complicated diagrams which, however, were only rendered comprehensible to a small group of specialists (ibid.: 16). It is to be feared that the goal of the MNP - the sustainable protection of natural and cultural resources - cannot be achieved if local organisations are not actively and adequately integrated into the Park's management (Vaca Bucheli 1997: 51). Samaniego (1997), too, in his article about environmental training in the new management plan, specifically recommends the participation of communal authorities and other interest groups in the management planning of the Park and its buffer zone (Samaniego 1997: 7-8).

Until today, employees express displeasure at the lack of discussion regarding the projects proposed by NGOs in Quito and the fact that these projects do not conform to the management plan. It is not merely in theory that the endeavours have no participatory character; in practice individual projects have been rejected by the affected local population because they felt they had been left out of the process of formulating the activities.

Although participation is recognised as an important conceptual element of the management plan, which is in force since 1998, the document contains neither financing for this purpose nor a practical concept for participatory processes (Gat-PNM 2003: 3). The Park's management has expressed that they are open for dialogues with the communes, but the latter have come to expect (unrealistic) immediate, direct benefits due to their previous experiences with the administration's strongly paternalistic attitude (ibid: 3). Up to September 2002, none of the park attendants, or even the National Park Director at the time, had ever taken part in a training program in participatory methods that might have been applicable for addressing this shortcoming. A new National Park Director assumed office in 2004. This may bring along a change to better implement of participatory approaches.

DED's activity in the MNP since 1992 has not been able to secure a rapid change to participation or ways of collaboration either between the National Park director and the Park team on the one hand, or between the Park and the local population on the other. Rather, it has been - and continues to be - necessary to focus on correcting past, energy robbing errors that have been made both in regard to the National Park team, but primarily in respect to the communities, as well as to regain the confidence of target groups to join the efforts toward appropriate use of the natural resources.

Who are the actors involved?

The statistics about the number of people living in the MNP are inconsistent. The numbers range from 734, according to Vaca Bucheli (1997: 16) to 1,600, according to Fundación Natura (1996: 27). Macias Parraga (1997: 4, 25) states approximately 15,500 inhabitants are found in the Cantón Puerto López, which is part of the MNP and its buffer zone. Of these 10,572 live in the Cantón capital Puerto López (cf. Fig 10.1.). However, the Puerto López Municipality's website (www.puertolopez.com/hojas/puertolopez. htm), counts 18,900 inhabitants in the Cantón Puerto López, of which just over 7,000 are categorised as urban. Drought, the climate phenomenon El-Niño, as well as the foundation of the MNP have caused both immigration and emigration in the Park region.

The people (about 45 percent) who live in the national park are mainly fishermen (Macias Parraga 1997: 5-6). The catch is sold at local and regional markets. The same is true for agricultural products (about 18 percent). In the larger settlements, about 20 percent of the people earn their living from small-scale trade. The remainder of the labor market is involved in the service sector: construction, transport, public administration, social services, and tourism (ibid: 5-6).

Tourism is becoming increasingly important, and between 1999 and 2001, 78,000 people visited the National Park. Entry costs up to $20 (US) for foreign tourists and up to $5 for Ecuadorians. This information is important because decentralisation of the National Park administration has become an official goal since 2002: the management of the MNP is scheduled to be put in the hands of a yet-to-be-established "mancomunidad", which is an association created for this purpose and is to include communal representatives, ministries, the National Park administration, and local interest groups. This will make it possible for the National Park, as well as the other institutions involved, to have responsibility for the disposition of the Park's income, which heretofore has been channeled to Quito.

Important local actors are the *Municipio* of the Cantón Puerto López, led by the mayor and the heads of the various departments, such as the Department for Hygiene and the *Departamento Municipal de Desarrollo Turístico, Saneamiento y Manejo Ambiental* (DETSAM). DETSAM was established in 1995 as a cooperative effort between the *Municipio*, the National Park administration, and DED. It is situated at the *Municipio* to facilitate joint planning that concerns the MNP and the Cantón. In the first few years of its existence, DETSAM had only one employee, who worked as a counterpart with a DED expert (the two of whom have collaborated in the authorship of this article). Today it has two departments, the *Dirección Municipal de Turismo* and the *Unidad Municipal de Gestion Ambiental.* There are eighteen employees, twelve of whom in the tourist Department.

There are, of course, other local actors. One is *Cercapez*, a theater group in Puerto López that has conceived several plays about environmental awareness and protection and performed them in the communes. Another is *Telaraña*, a working group composed of members of the theater group *Sercapez* in Puerto López and other individuals. They have developed radio spots for environmental education, which have been broadcast on Ritmo Azul, a local station. Both worked on behalf of the MNP administration and the DED expert team. There is also the *Organización de Servidores Turísticas*, a regional tourism organisation that supports environmental education activities of the DED and organised an "International Environment Day". Yet another is *la Unión de Comunas de Zona Sur de Manabí*, an alliance of the communes in the southern part of the *Manabí* province.

There have also been important actors at the national level. The *Insituto Nacional Ecuadoriano de areas Naturales y Vida Silvestre* (INEFAN) was the national institution responsible for the management of protected areas in Ecuador and a government agency. Since 2000, the administration of these protected areas has fallen under the authority of the *Ministerio de Turismo y Ambiente*, which is currently undergoing the process of decentralisation. In the middle of the 1990's, the *Fundación Natura* funded approximately 50 percent of the infrastructure of the MNP, for example projects to provide alternative income sources for villages in the MNP as well as vehicles. This NGO was a partner of The Nature Conservancy and later of GEF (see below) in the preparation of the management plan that came into effect in 1998. Since 2002, *Fundación Natura* has no longer been active in the MNP. The *Centro de Apoyo para la Vivienda Popular* (CAVIP) is an agency of the Ecuadorian National Bank for the improvement of the living conditions of marginal urban settlements. The *Centro de Datos para la Conservación* (CDC) is a scientific institution in Quito that has conducted studies into the fauna and flora of the MNP for the management plan as well as undertaking other activities. *Ecociencia*, an NGO in Quito, supported the MNP in the development of the infrastructure for tourism. Together with the Italian organisation CISP (see below)), the Programa *Manejo de Recursos Costeros* (PMRC), a governmental institution for the protection of marine resources and the coasts, developed plans for sustainable use of the coastal area, including fishing, the protection of mangroves, and the problems of waste management.

On the international level, the German Development Service (DED) is one actor. Since 1992 four DED experts have been assigned to MNP, and DED has financed small projects for the sustainable use of natural resources and securing alternative income sources for the people in the villages of the MNP and in his buffer zone. The organisation has also provided equipment, e.g. vehicles, and has financed the counterparts of the DED experts as well as small projects. The Global Environmental Facility (GEF), a World Bank agency, financed the preparation of the new MNP management plan that appeared in 1998. It has also provided equipment such as vehicles, computers, etc., technical specialists, and sponsored activities like workshops, studies, and surveys. The *Comitato Internazionale per lo sviluppo del popoli* (CISP) is an Italian NGO that supported the PMRC in coastal protection projects and others including a waste-collection project.

10.1.4 Examples of participation in the MNP

Resource management: tree planting, tree nurseries, central waste collection areas

Beginning in 1994, the trees were planted in Puerto López and other settlements in the National Park and its surroundings. This was not only done in forests but also in schoolyards, along streets and paths, and in public places. Both students and adults were usually involved. Prior to the activities, interested teachers were sought who planned the event in collaboration with their classes, local counterparts, and employees of the nurseries providing the trees. Participation, however, was not always voluntary, with schools sometimes requiring their students to take part. All the necessary saplings came from the tree nurseries in Agua Blanca and El Pital. As a rule, the principals, which included the *Municipio*, communes, and the schools shared responsibility for the inputs in that they provided the materials for the bamboo protection fences and constructed them. They also assumed responsibility for the care of the saplings. Unfortunately, many projects failed after a while due to climatic conditions and the lack of care of the trees, i.e. there was inadequate ongoing care and monitoring of the projects.

In establishing the above-mentioned tree nurseries in 1992, DED pursued several motives. On the one hand, they served to provide income from "sustainable" activities, i.e. activities conforming to the preservation aims of the National Park. On the other, they formed the basis for forestation projects foreseen in the first management plan of 1987 and implemented in the operational plans of the INEFAN and DED to create resources (firewood, wood, and fruits) and for preservation, e.g. of land and groundwater. Furthermore, the saplings produced were to be made available for both communal tree plantings and agricultural forest projects.

Village inhabitants were decisively involved both during the planning stages of the tree nurseries' establishment and in undertaking their care and commercial exploitation. DED experts and Ecuadorian counterparts provided training of the tree nursery workers and held workshops for further education in which responsible representatives from the respective communes also participated. Despite production successes in 1995 in El Pital and Río Blanco (Sigle 1995: 3-4), the existence of the tree nurseries was brought into question because of low demand, low prices, and purchase contracts that went unfulfilled. Also, care of the saplings and the collection of high-quality seed were inadequate. There was only sporadic advertising for the tree nurseries' products on the radio and in the magazine *Danielito* (cf. below).

Fig. 10.2 Tree nursery in Agua Blanca. The selling of containerised seedlings was insufficient in the beginning because, amongst other reasons, no marketing was done. Source: Michael Sturm.

Fig. 10.3. Plantation of trees along the main street of Puerto López. Project of the Machalilla National Park in cooperation with DED and local schools. Source: Michael Sturm.

The tree nurseries no longer exist and there are no more tree plantings. Courses in subjects like marketing, bookkeeping, and management were not offered in necessary quality and quantity and the knowledge acquired by participants in them was not put into practice.

Central waste collection

The *Municipio* of Puerto López sought a solution to waste problems in the *Cantón* related to sanitation and hygiene but also to economic and tourism issues. There was no practical experience with waste avoidance strategies, organised refuse disposal, or landfill. Sources of the waste pollution were private dwellings, small enterprises like car repair workshops, as well as restaurants and fishing. Garbage was burned on the property, thrown into the street, or disposed of the two rivers that flow through the city. The river outflows occasionally accumulated behind the beach wall in the city, giving it the appearance of a sewer. The bio-waste from the central market was simply deposited behind the market hall in the center of the city opposite the main entrance to the National Park information office, where it was at least partially eliminated by free-roaming cattle, donkeys and dogs.

During the study period, there were very few villages or projects that viewed the waste as a problem, let alone considered it a potential resource and recycled it. However, in Agua Blanca, people sorted waste into paper, plastic, metal, etc., and items of value were sold. Also, in the tourism project *Hostería Alandaluz* in the southern part of the National Park, bio-waste was composted and the fertilizer produced used in the flower gardens. Nearby, earthworms were raised to create compost from biological waste. And in Puerto Rico and Salango, neighboring villages to the south of Puerto López, waste was collected, the initiator here being the local NGO *Pro Pueblo*.

With representatives of the *Municipio*, the authors undertook several trips to the negative and positive foci of the area in order to show possibilities for solutions of the problem. In the city's eastern hinterland, a 0.7 hectare site for a waste dump was identified, surveyed, and put into operation with technical consultation by the *Centro de Apoyo para la Vivienda Popular* (CAVIP). Funding was made available through the German Embassy and the partner institutions CISP and PMRC. In the spring of 1997, *Municipio*, DETSAM, the National Park administration, DED and *Comité Zonal* prepared the concept for a waste collection system that included a time and route schedule for the refuse trucks, the waste capacity, and personnel costs (Cisp-PMRC 1997). Prior to and during the implementation phase, the public was informed through posters, radio spots and home visits, and the advantages of a controlled garbage disposal were explained at workshops and town meetings. Consideration was given to individual desires and input from the population regarding route or time changes was considered in order to achieve high acceptance of this new concept.

Problems came from other sources. Strong rains cases in May and June 1997 (caused by the phenomenon El Niño) destroyed submerged the waste disposal site and proved that it had not been adequately safeguarded for extreme situations. This resulted from an erroneous engineering-geological construction of the dump. Also, over time many waste receptacles disappeared and put to use for other purposes. Then CISP terminated the funding for two of the refuse trucks it had financed. Renewed efforts at waste disposal have not yet been undertaken.

Environmental education: radio, the National Park magazine Danielito, International Environment Day

Up to the end of 1997, the *Centro de Educación Popular* (CEDEP, a radio station in Quito) developed, revised, and produced 23 radio spots that were broadcast on the regional station *Ritmo Azul.* In Puerto López, these had been preceded by intensive work with the local theater group, *Sercapez*, and interested children, adolescents (from 1997 with the youth club *Club Juvenil Albatros*) and adults. *Ritmo Azul* broadcast the spots, which dealt with the MNP and its resources, problems related to waste, and topics related to tourism. Three spots called attention to the tree nurseries in Agua Blanca and El Pital as suppliers of trees and ornamental plants. Preparation of the spots did not incur any significant costs, but in return for the broadcasts at regular intervals, the street on which the radio station was located was planted with trees from the nurseries. Two members of the theater group and the Director of the National Park worked with the authors of the present article to decide on the topics and execution of the spots. Last but not least, the fact that all the readers were known local people and the ads themselves relied on local customs and characters resulted in a positive response from the listening audience.

In August 1995, the authors came up with another idea: the publication of a National Park magazine. The original concept envisioned a monthly periodical, but ultimately only two editions appeared, one in May of 1996 and the other in December 1996: A third was begun but never completed. Sales were only moderate. Why didn't the idea work?

Fig. 10.4 School in Soledad, a small village in the Machalilla National Park. Children of different ages are taught together because classrooms and teaching material are scarce. Source: Michael Sturm.

Fig. 10.5 Environmental education in Casas Viejas, a village adjacent to the Machalilla National Park. Source: Michael Sturm.

A survey done in Puerto López and the villages had indicated great interest in an information medium about the National Park and village-related news. The initiators viewed the publication as a way to make the National Park more transparent through the delivery of content about the Park agency's protection efforts, ecological topics, and the Park's employees and their activities. The composition of the editorial staff was supposed to guarantee that the journal would not be a one-sided organ dominated by the National Park, but that it would also give village residents the opportunity to articulate their expectations and problems vis-à-vis the NPM. To achieve this, DETSAM and *la Unión de Comunas de la Zona Sur de Manabí* were involved in the effort alongside INEFAN and DED.

A children's competition to come up with a logo for the magazine in which the Cantón Puerto López schools participated was held in connection with the International Environment Day, and the winning entry, by an eleven-year-old, was presented in a festive public ceremony. *Danielito*, the diminutive form of the first name of the individual after whom Cantón Puerto López is named (Daniel López), was selected in order to increase readers' identification with the publication.

In addition to regular features like the plant or animal of the month, there were also items about the communities and villages as well as articles written by organisations. The majority of the pieces were written by Ecuadorians of local or regional origin, and almost all contained black-and-white photos or drawings.

There are a variety of reasons why the magazine project failed. The texts of many articles were too long and complicated, and the sales price was too high. Delivery was delayed primarily because of organisational problems, the main one being that editors often had to wait weeks for promised articles. However, conflicts over areas of authority proved to be the ultimate undoer. For example, the former National Park director assumed editorial responsibility for all articles and their contents, which culminated in censorship of the pieces written by representatives of the communes or development workers. Participation in the form of publishing an article expressing an opinion contrary to that of the director's was not possible. This resulted in a total lack of a sense of "ownership". As a consequence, the *Unión de Comunas del Sur de Manabí* - the only institution not directly connected with the National Park - withdrew from the endeavor.

The International Environment Day, at the beginning of June, was celebrated three times during the period under review. Its climax was a parade through the streets of the canton's capital, Puerto López, by classes of the schools in the city and the surrounding villages, as well as by associations, committees, and other social groups. The girls and boys, whose participation was obligatory, undertook preparation for these strikingly military-like marches weeks in advance. This also held true for some of the competitions, like suggesting slogans in favor for the National Park, which the schools coordinated with the Park administration.

Some local participants, the counterparts, and the DED experts felt that many of the events had a very formal character: those taking part merely executed ideas that had been dictated to them. In contrast, participation in the competition for the development of a logo for all the DETSAM's activities related to the environment, as well as in the magazine *Danielito* was voluntary. Plays with environmental themes by the local theater group Sercapez, coordinated with the DETSAM and the Park administration, were very popular.

In 1997, at the suggestion of the DED experts, the counterparts, and others, more lighthearted events and competitions were introduced into the International Environment Day celebrations, all having themes related to tourism or the environment. A "wrap-up" of the events in *Danielito* would have been optimal, but this could not be done for the above-mentioned reasons. Surveys assessed the changes positively and expressed the wish of many participants for a continuation of the event the following year.

Local development: Community gardens and fishery project

The goal of these projects in the villages of Agua Blanca, Casas Viejas, El Pital and Salaite (cf. Fig. 10.1) was the creation of a wide range of alternative income opportunities in the respective villages and the protection of essential resources. *Fundación Natura* and DED provided financial, technical, and advisory support.

In the case of the community garden project in Agua Blanca, no inquiry regarding needs or local interests or acceptance was made prior to the installation of expensive watering systems. In this village, individuals who raise vegetables for their own consumption have their own private gardens.

Fundación Natura provided start-up financing for the community garden in Casas Viejas in 1993. The group who ran the garden was well organised and produced different types of fruits and vegetables. However, due to poor accessibility to the local market, the group had problems in selling the products.

The community garden in El Pital / Río Blanco received start-up financing from DED. This made it possible to initiate the project, but there was a lack of consensus in the community regarding responsibilities and authority, which led to management problems that had a negative impact on output and profits. The garden was ultimately split up and managed with varying degrees of success by small groups or individual families. (Sigle 1996b: 2).

The planning of all these garden projects was obviously based on faulty or missing knowledge of the conditions, which can be traced back to factors such as inadequate involvement of the target groups or incomplete investigation and evaluation of data before the projects began. These shortcomings, for example, led to existing competition within the village going undetected. All hindrances might have been recognised in time if more intense inquiries in the planning stage had been undertaken. Furthermore the ideas of individuals in the target groups should have been taken notice of. The NGO, however, suffered from time constraints because it faced an early deadline to account for the use of its project budgets, which resulted in its failure to undertake such studies.

Thus it became clear that the two sides had different interests and needs. The project partners viewed DED and *Fundación Natura* as institutions "... who provided opportunities for access to staff and financial resources." They also perhaps served "... the partners (the DED development workers) as projection surfaces for their own ideals, such as solidarity and partnership", as it was generally formulated by Grundmann (1998: 14).

In Salango, local inhabitants were supposed to cease production of bricks for dwelling construction, since it resulted in considerable destruction of vegetation and land. The village of about 50 inhabitants lies within the MNP, which is why the National Park administration tried over a period of years to apply pressure to change this practice. Since that proved unsuccessful, means of alternative earnings were considered. Finally the NGO Fundación Natura financed an outboard motor for a small fishing boat that was to make offshore fishing possible.

The village was quickly divided into several camps, each of which sought to control the use of the motor. A survey of the inhabitants carried out by a DED expert revealed that the project had not even been desired because the village inhabitants actually foresaw the problem. Not wanting, though, to reject the offer and its accompanying financial resources, they voted for the project. The NGO itself had obviously shown little interest in the real needs of the inhabitants in the process of "meeting its quota", and the national park administration merely hoped for a fast solution. Later attempts to reduce the harmful effects of the brick production through training and other means such as a plan to resettle the villagers also failed or were not implemented. Today, due to the tourism, brick production is at an all time high and remains one of the unresolved problems in the MNP.

10.1.5 Conclusions, Transferability, and Lessons Learned

Looking generally at the history of the conflicts involved in the dialogues among the various stakeholders, it must be acknowledged that the resolution methods applied in all projects wound up being reactive - rather than proactive. The main problem was the *Machalilla* National Park itself and the way it was established: “under the cover of darkness”. Most of the distrust vis-à-vis the MNP arose because of this and because of the Park administration’s very restrictive view of the project in the years that followed: all projects, in addition to their technical goals, were also supposed to heal the wounds inflicted in the early years by the Park’s management.

The conception and founding of the MNP should not have been allowed to occur without some participation by the resident population. Its boundaries and land uses should have been identified and defined in advance with their input, and the general conditions necessary for meaningful changes were inadequate both in terms of time and financial resources. Until the beginning of the 1990’s, the population was involved in only a limited way. An improvement only came about with the replacement of the Park director and the beginning of DED’s involvement, but this, too, was constrained by the above-mentioned restrictions and hindrances.

Due to a lack of adequate prior research and involvement of the local population, all of the projects were either condemned to failure, like the community gardens, or experienced only a short existence, such as the tree nurseries and the magazine Danielito. In the environmental projects, there was a lack of continuity, human resources, and content. Finally, the technical training given to staff was inadequate in regard to participatory techniques, and this holds true for the DED development workers’ pre-departure training in Germany, where knowledge of the subject is only imparted rudimentarily. Until 2000 this component of the development workers’ training comprised three months, during which courses of a few days’ duration provided only basic information in areas such as “Goal-oriented Project Planning” and “Participatory Rural Assessment”. Since 2001, based upon perceived need, this preparatory period lasts from one to three months, in which even less time is allocated to participatory techniques or information about them, the emphasis being rather planning methodologies and program monitoring and evaluation.

Applying participatory techniques, however, is actually not always desired by either the funding agencies or the development workers themselves, although as Holthusen and Paulus (1998: 40) and Löbsack (ibid.: 4) observe, it is a fundamental prerequisite for successful project work. Among the reasons for rejection of the approaches involved are inadequate knowledge, time pressure, insecurity, fear of failure, limited competence in regard to the selection of appropriate techniques, and also a lack of acceptance on the part of the target groups. The Ecuadorian project agencies, too, frequently block participatory methods because they could mean a loss of influence and power or due to time constraints.

The above-mentioned time pressure and competition with other activities are an essential factor in the inadequate application of participatory methods: decisions frequently have to be rushed into because national and international donors require the disbursement of project budgets almost in “real time”. This often results in making project costs exorbitantly high and all too often the very inefficient use of inputs. The other side of the coin, though, is that the dragging out of development processes engenders a burn-out of interest in topics and projects. Thus, when those responsible for a project recognise that the participation processes will take a long time, this fact must be made clear and explained to the target audience well in advance. Only in this way can what is initially seen as "lost time" become “time saved” because fewer resources (time, money, personnel) need to be allocated to correct unwanted developments.

Table 10.1 History of conflicts and participation in the Machalilla National Park

Date / Event	Actors / Measures	Goals	Impacts	Stakeholders Affected
1978-1985: First generation of DED experts in the region of what later became the MNP	INEFAN: Fundación Natura, DED, German and Ecuadorian government / cooperation contracts	Resource protection, income generating measures, conflict management	Power conflicts: National park administration's loss of authority; DED experts support park-based communities in demands against park administration - to a degree against the wishes of DED	MNP team, communities
1979: Establishment of the MNP	INEFAN: Decree, new legislation (top-down approach)	Protection of the tropical coastal dry forest, prehistoric artefacts and marine resources	Restrictions and prohibition of land use and some types of migration; led to some violent conflicts with the MNP administration	Residents of villages situated in the MNP and its buffer zone
1986: First director of the MNP replaced	INEFAN: Decree, creation of the first MNP management plan	Easing of the tense situation in the MNP; more effective protection of the natural resources	Start of cooperation between MNP administration and local populations - marked by mutual suspicion; corruption involving members of the MNP team and community leaders	MNP team (administration), people in the communities
1992-94: First DED expert of the second generation in the administration of the Machalilla National Park	DED, INEFAN: DED expert and related financial and material resources made available to MNP	Involvement of local population in protection concepts by means of participatory methods, close cooperation with counterparts (including all DED experts); formulation of new management plan for Park	National park administrator loses authority; involvement of foreign institutions in events surrounding stimulation of target groups' awareness of needs and related increased demands; loss of mutual trust between DED expert and Park administrator	MNP administration, DED expert, target groups (people living in the MNP and its buffer zone)

Date / Event	Actors / Measures	Goals	Impacts	Stakeholders Affected
1994-97: second DED expert takes over from the first and is succeeded by the third	DED and INEFAN: Extension of cooperation until 1997	Assured participation of the MNP team and target groups in development of the MNP and its new management plan; agreement to protect natural and cultural resources and artefacts and to provide environmental education; formulation of new management plan for Park	Ongoing mistrust between DED expert and Park manager; suspension of project followed by new projects and goals; disruption of information flow due to failure to overlap contracts among DED-expert generations and generally deficient communication between Park manager and team	Mainly people in the communities of the MNP and its buffer zone, DED expert
1996: Participation of GEF	GEF, INEFAN, Ecuadorian government: Creation and funding of 35 environmental protection projects in Ecuador including 5 national park management plans (among them the MNP)	Target groups, mainly in workshops and assemblies, brought into process of partially guaranteeing effective park management and resource protection; participation of national experts in data collection survey	Parallel data-collection surveys and studies to support the creation of the management plan; DED loses influence; GEF and INEFAN gain influence; completion of a management plan in 1998	DED expert, MNP administration, target groups

Date / Event	Actors / Measures	Goals	Impacts	Stakeholders Affected
1998-2004: Fourth DED expert takes over from the third	DED, INEFAN: Extension of cooperation until 2004	Continuation of cooperation between DED and MNP; target groups gain influence in deciding content of management plan	Project again suspended; new projects and goals; disruption of information flow due to failure to overlap contracts of third and fourth DED experts	People in the MNP and its buffer zone
2001: Third Park director assumes duties; she is not from the region	Ministry of Environment (Ministerio del Medio Ambiente), successor of INEFAN: Decree	Declaration of nationwide goal to facilitate decentralisation of national parks	MNP transformed into public authorities (communities, tourist associations and regional environmental authorities), management plan ignored, uncontrolled growth of tourism activities (primarily all whale watching)	Natural resources (including humpback whale population) endangered; population of the Cantón Puerto López population and officials offended; park administrator loses authority vis-a-vis community leaders

Date / Event	Actors / Measures	Goals	Impacts	Stakeholders Affected
2002 : Beginning of the process of decentralization of the Park management (for which a special-purpose association is created - mancomunidad)	Ministry of Environment: Responsibility for national park management devolves to the Manabí provincial administration and to the cantons Puerto López, Jipijapa and Montecristi	Delegation of responsibility for issues of tourism, environmental education, and nature conservation to regional and local political authorities	Increasingly problematic decision-making processes accompanied by opacity of structures and dominance by lobby groups	People in favour of protection or sustainable use of natural resources in the MNP and its buffer zone (target groups)
2003: Fourth national park leader takes over	Ministry of Environment: Decree	Delegation of management responsibilities to regional and local political authorities and MMA	Power conflicts; disorientation in the transitional phase	People in the MNP and its buffer zone, cooperation partners, "natural resources"
5 Jan 2005: Newly elected mayors and heads of office assume	Political parties, voters	Political representation of regional inhabitants, realisation of local goals without consideration of predecessors' commitments	Delays or breaks in association building	Projects continued, target groups, natural resources
Since March 2005: Start of GEF 2	GEF, Ecuadorian government: Re-formulation of 1998 management plan goals	Improvement of management plan implementation	Better protection of natural resources, more effective project transformation, potential recurrence of delays in implementing current projects	Projects continued, target groups, natural resources

Between 1988 and 2002, over US$ 3.1 million was made available by national and international NGOs alone for various National Park projects (INEFAN, GEF 1998: 13). In the following years, investments in the National Park and related projects will include US $5 million from the *Programa para la Reducción de Pobreza* (Program for the Reduction of Poverty, PRO-LOCAL), 2 million from GEF II, 6 million from PMRC-BID, and US $5 million from the International Conservation. This created great expectations among the target groups and the national and local partners, and engendered a real "fight for the dollars". In this environment it is difficult to implement projects with limited means but great personal contributions, though the latter refers to active, time-intensive participation rather than money. Backes (1998: 32), who fulfilled several contracts as a development worker for DED, expressed the concern that "Professional diversion of development funds is becoming increasingly frequent through innumerable NGO that operate under the guise of being non-profit organisations and spring up like mushrooms. They are often founded by drop-outs from the national government scene". Furthermore: "In the worst case, participatory measures are abused as legitimisation for action directed from outside." (ibid.: 33). Thus it is not only important to invest an adequate amount of money in the participatory planning and implementation of projects, but also to make funds available for the participatory accompaniment of projects (evaluation), particularly for follow-up activities and the staff education.

Sometimes, however, it is the project partners who are mistrustful of participatory methods. Grundmann (1998), a methodology consultant for DED, identifies reasons she feels are involved in the failure of participation. The concept of participation comes from the North: "However, those we term ‘project partners’ call us ... ‘donor institutions’ … which clearly points out the different perceptions of one and the same relationship” Grundmann (1998: 14). Organisations involved in development cooperation have power because they have highly qualified staff as well as financial resources. There is great danger “... that we deny or romanticise this potential for power and dominance, whereas our partners are only too well aware of it". The partners are not only "victims", but actually "instrumentalize the term ‘partnership’ for their own interests and have strategies how to get at the money and make certain their needs are met" (ibid.: 14).

The development worker depends on recognition that basically results from the success of her or his work on site. Projects that have long-term success give rise to emulation. It is possible that a development worker receives recognition for a participatory approach on site, but for reasons such as promoting a viewpoint regarding the emphasis of a project that is not shared by DED. Thus the development worker must also exert influence within her or his own institution and provide timely information about the goals in order to receive approval for the plans.

Since almost all the projects described had only a brief lifespan, the question arises as to whether the participatory methods applied in the MNP will be repeated. This can clearly be answered with a qualified “yes”. Frequently, "only" preliminary phases were looked at, and these were subject to a period of observation that was too short. Since too little time was available to invest in pre-project research about the target groups (as well as opportunities and requirements), the projects were either misunderstood or were carried out based upon inadequate or even incorrect data, both of which - of course - led to failure.

Even if the investment of more time in the investigative and post-evaluation phases had held out the promise of greater project success, though, it would have required a guarantee of permanence of staffing at all the decisive levels, but at least on site in the National Park. This has not yet proven to be the case. In fact, personnel have changed rapidly: The DED experts’ contracts last only two years, and their counterparts - and in the last four years the Park director - have come and gone much more frequently. Excellent knowledge of participatory methods and how to implement them is indeed a requirement for the success of projects, but favorable conditions regarding staff on the site facilitate their application considerably.

10.2 Community Forest Management in Esmeraldas - Is Constructive Dialogue Possible?

Jörg Linke

Gesellschaft für Technische Zusammenarbeit (GTZ), Ecuador

10.2.1 Introduction to a Community Forest Management Project in Esmeraldas

The strategic goal of the GTZ-supported "Community Forest Management – Esmeraldas" project (MFC-E) is the conservation of the biodiversity of the Chocó bio-region in Ecuador. The specific project goal is to enable forest owners and other actors in the forestry sector to manage their natural resources sustainably.

The north of the Province of Esmeraldas belongs ecologically to the Chocó bio-region and is characterised by its exceptionally varied biodiversity (Sierra et al. 1999, Dinerstein et al. 1995, WWF 1997, Myers et al. 2000, Stattersfield et al. 1998, Wells et al. 1998). It is among the world's top ten "biodiversity hot spots" (Myers et al. 2000). The region is subject to intense utilisation, and the palm oil industry, agricultural industries (bananas, etc.) timber exploiters, gold mining, shrimp breeders, settlers, indigenous and Afro-Ecuadorian communities, and other groups contribute to deforestation, the rate of which is one of the highest in Latin America (Esmeraldas: aprox. 22,500 hectares per year). The area's poverty rate is the highest in Ecuador.

A particular focus of the project is the acquisition of experience in cooperation with specific indigenous and Afro-Ecuadorian communities regarding the sustainable use of natural resources. Especially important is experience related to consultation methods that, through dialogue, appear to hold out the greatest promise for success in terms of real and sustainable improvements in the living conditions of the target groups and the conservation of biodiversity (Schein 1987, 1998, Fatzer 1999). This experience in sustainable resource management is already in demand in other communities and will be shared with them (see also Rice et al. 1993).

The project's objective is to bring the various actors together in order to stop illegal resource utilisation and deforestation, promote the development of indigenous and Afro-Ecuadorian communities, and raise

awareness of regarding sustainable resource use. A main activity in which the project cooperates with other organisations is the clarification of land ownership issues, particularly those pertaining to indigenous and Afro-Ecuadorian communities.
Some of the instruments being employed include:

- technical training in sustainable management (low-impact logging systems, improved quality standards of forest products);
- consultation in organisational development for small and medium-sized producers' groups, communities, the Ministry of Environment, and timber companies (financial planning and financial administration, book keeping, development of articles of association), management quality, forestry certification, etc.);
- conflict consultancy;
- building public-private partnerships.

Numerous local, national, regional and international aid agencies are active in Esmeraldas, but they often work with the same target groups and on the same issues without knowing about each other's efforts or sharing information about successes or failures. One of the project goals is to promote dialogue among these agencies in order to establish knowledge exchange as well as to maximise cost and investment effectiveness through precise reciprocal coordination. It is crucial, though, not to give the appearance of trying to coordinate the various agencies' activities but rather to effectively initiate the dialogue without having a lead role, unless responsibility for this role has been designated to the organisation. It has been demonstrated that jointly organised and actualised *mesas de diálogo* (dialogue tables) successfully serve this function, and a great number of agencies have responded affirmatively to their appeal.[1]

Participation and self-help are introduced as primary organisational principles of gender and poverty-related efforts in all the programs supported by the GTZ. Participation is understood as involvement and partaking of the people, principally the lower economic strata, in the decisions and the benefit of the activities (BMZ 1995, 1996), thus simultaneously supporting the process towards democratisation. Participation happens not only in the project framework but also comprises combined involvement in the creation of the framework conditions at the macro level toward the goal of consensus building (BMZ 1997, 1999).

---

[1] Since 2004 the GTZ has brought its interventions in the "green area" together into one program (GESOREN) in order to systematize, make available, and make use at the national level of the lessons learned from experiences at the regional level (see www.gtzecuador.org/a-coop/frame.html).

The supported actors - the actual target groups - are to be as actively involved as possible in all relevant decisions on project planning and implementation. Only in this way can the acceptance of the endeavours (ownership) be guaranteed. This is an absolute requirement for the sustainability of the project's impacts. Special consideration is to be given to the interests and participation of women (GTZ 1993).

10.2.2 What kind of problems and conflicts existed before the stakeholder dialogue was established?

The population of northern Esmeraldas, mainly inhabited by indigenous and Afro-Ecuadorian communities, is among the most marginalized in the whole country: these groups' educational level is very low and their access to information very limited. Until about ten years ago, they were virtually excluded from all political decision-making processes.

The high illiteracy rate in the region combined with the unavailability of information has prevented these population groups from articulating themselves and their needs in the political system. Compounding the overall negative situation have been the prevailing poverty and concomitant lack of mobility, as well as the region's generally poor accessibility.

In many cases, land tenure has not been adequately documented, i.e. communities who have lived in the area for decades do not have any official title deeds, and timber companies and other actors have frequently taken advantage of this situation. The result is that small and medium-sized landowners or communities have at times been deprived of their land possessions. On other occasions, through corruption, resources have been exploited without the putative owners receiving any corresponding indemnity or a share of the profits. Moreover this exploitation has often completely disregarded environmental standards, resulting in severe damage to the ecosystem or even its destruction.

The continuous loss of traditional structures, migration, and unregulated utilisation of the natural resources has produced numerous conflicts. Within communities these have involved perception of roles and responsibilities, and at higher levels there has been conflict between neighbouring communities, between timber companies and communities, between the Ministry of Environment and the communities, between the timber companies and the Ministry of Environment, etc. Most of these conflicts can be traced back to disparities and mismatches in information levels but also to various, unknown economic or socio-cultural interests or the pervasive corruption.

The extremely marginalized situation of the rural population, the high and diverse utilisation pressure on the natural resources, and the exceptionally high biodiversity in the Chocó bio-region have contributed to the past and present engagement in the region of a wide range of aid organisations including religious groups, NGOs, and international donors. Unfortunately there has usually been inadequate synchronisation and coordination of these efforts, and sizable inputs have been allocated to one and the same target groups and activities without the resulting positive and negative experiences being shared.

10.2.3 Who are the actors?

A variety of groupings and organisations are involved in the area. Among them are the indigenous and Afro-Ecuadorian communities, who possess more than 60% of the regional forests, national timber companies, intermediaries who usually pre-finance the activities of bigger timber companies, the Ministry of Environment at the national, provincial and local level, the palm oil industry, a variety of aid agencies including NGOs, as well as the Catholic Church.
The most relevant of these are described in greater detail below:

NGOs

The national and international non-governmental organisations active in the region are mainly occupied with social, public health activities as well as in the area of natural resource management. Among them are:

- *Fondo Ecuatoriano Populorum Progressio* (FEPP) - consulting services related to land title regulations, organisational development, production systems in agriculture, forestry and grazing, public health, rural credit systems, and community forest management for indigenous and Afro-Ecuadorian communities
- *Fundación Altrópico* - supports the development of the Awá indigenous communities in the area of natural resources management, forestry certification and cooperates with the WWF in Columbia
- *Unidad Coordinadora* (UC) - a coordination association for sustainable forestry.

Other NGOs engaged especially in conservation, environmental education, and biological research include *Jatun Sacha* Foundation, *Fundación Natura* and *Ecociencia*.

Most grass-roots organisations are characterised by very poor organisation and conspicuous paternalism. The latter has become ingrained in the region due to the "distribution of gifts" in the communities by a number of "aid organisations" who "distributed presents" but did little to promote participation in the project implementation and thereby achieved little ownership for the stakeholders involved. Among them have been the Arenales Afro-Ecuadorian community, the Centro Chachi Capulí, the Centro Chachi El Encanto, as well as an association of farmers and cacao producers active in several communities. Furthermore the Federación de Centros Awá, a grouping of the Awá indigenous communities, has played a significant role. Generally, umbrella organisations such as UONNE (Black communities) and FECHHE (indigenous Chachi communities) are political alliances that incorporate the legal establishment of indigenous rights, bilingual education, and public health among their issues.

Federal Ministries and other agencies

The Ministries are very weak in their professional and organisational skills and maintain very little on-site presence. Corruption is widespread, especially in the environmental sector, in the control of timber felling and transport, as well as in the granting of authorisation for forest management plans. The influence of the timber industry is structurally very firmly established.

The government's declared target of decentralisation and the devolution of responsibilities and decision-making authority from the central level to the local level have so far failed in Esmeraldas. This has been due to a lack of stakeholders' qualifications to assume responsibilities on the one hand, and the strength of political and private interests on the other. The situation in the Ministry of Environment on the provincial level and in the Eloy Alfaro (Borbón) Canton must be emphasised. The Ministry of Environment is generally considered to be a weak agency that neither plays an important role nor enjoys much prestige within the Ecuadorian government. The influence of the timber industry, in contrast, is strong on all levels. As a consequence of the frequent political changes, there is a high fluctuation of decision-makers and officials. The governmental agency INDA is in charge of the registration of title deeds, and the Ministry of Agriculture is controlled by the politically powerful palm oil industry.

PRODEPINE (Proyectos de Desarrollo para los Indígenas y Negros Ecuatorianos) is a decentralised federal organisation that finances the infrastructure and productive projects of the indigenous and Afro-

Ecuadorian communities and encourages the autonomy of the indigenous and Afro-Ecuadorian organisations.

Furthermore two universities that have their areas of influence in the north of the country must be mentioned: the *Universidad Técnica Luis Vargas Torres* - Esmeraldas with a Faculty of Agriculture and Forestry and the *Universidad Técnica del Norte - Ibarra* that houses an identical faculty.

Foreign Governmental Development Organisations

The area has seen the involvement of several international cooperation organisations. Among the most important are Germany's GTZ - Forest Policy Consulting Project (PPF), which cooperated closely in the region with the MFC-E until 2001 and is now (2004) active in the capital, Quito. A further important organisation is the German Development Service (DED), which has provided experts in the field of forestry and agro-forestry via organisations such as *Jatun Sacha*, the Federación of the Awá, and the MFC-E. DED furthermore manages a program called Civil Peace Service, which has been active in Ecuador since 2002. There is a cooperation agreement with the *Unidad Coordinadora* and the MFC-E concerning the management of social and environmental conflicts in the north of the Province of Esmeraldas. By the year 2003 USAID had already been intervening in the region for about ten years, giving advice to several indigenous and Afro-Ecuadorian communities through capacity development and technical assistance in the field of management of natural resources. Several other international agencies such as the Spanish cooperation service, Canadian cooperation organisation, European Commission, etc. are still active in natural resource management or in the health sector.

Timber Companies

Private timber companies have political influence that reaches up to the President, but environmental NGOs and trade unions are challenging the present business practices. For a long time, the companies have obtained timber from all province areas illegally and without resistance. Claims of private landowners have been violated as well. This situation has been partially changed through the influence of national and international NGOs and international cooperation. In spite of this, the firms successfully exercise all the means at their disposal to keep governmental oversight of on-site timber to a minimum. Given their poor internal organisation and management, limited effectiveness, and, to a certain degree, their modest technical expertise, most of the companies in this branch cannot be referred to as modern enterprises. Their weakness is particularly manifest in comparison with companies that are active in the international market.

The biggest national company in the forestry sector is Endesa / Botrosa, which belongs to the Durini group. Among other things, Endesa / Botrosa has concluded an agreement with the Centro *Chachi Capulí* community for the sustainable management of 6000 hectares of community forest. The MFC-E acts as a facilitator in negotiations and is recognised by both parties. Furthermore, the company has entered several PPPs (Public-Private Partnerships) with the GTZ in the field of cacao production and commercialisation, forest management, agro-forestry, public health, and education. Setrafor is another company belonging to the Durini group; it is responsible for the on-site timber utilisation. AIMA and COMAFORS are joint organisations of the forest companies. Their main activities include marketing, promotion, and forest management.

The Catholic Church

Another relevant actor in social and environmental matters is the Catholic Church, which offers support for projects in Esmeraldas' productive, social, and public health projects. The Bishop of Esmeraldas in particular is a recognised personality who has committed him to the social and environmental interests of the poor population.

10.2.4 What have the objectives of the stakeholder dialogue been?

The stakeholder dialogues aim at the coordination of activities among various aid organisations. The main concerns of their efforts have been the reduction of the prevailing paternalism and the promotion of the target groups' responsibility for and "ownership" of processes that have been agreed upon. The continuous exchange among the aid organisations, though, is also supposed to contribute to the flow of information in regard to the activities that are carried out (see Río 1992, Principle 10 in Burger and Happel 1997); when this is achieved, these activities are jointly financed and executed.

Another objective pursued through the promotion of the stakeholder dialogue is the availability of information and knowledge concerning land ownership rights, utilisation rights, production systems, forestry, environment legislation, and other issues. Many measures, studies, and surveys have been carried out in the region but without adequately reappraising the "lessons learned" or describing them and making them public. The dialogue must supply, among other things, access to such information and experience. The participation of the marginalized population groups in the political discussion must be classified as one of the most significant aims of the dialogues.

Since the need for capacity building, e.g. in the technical matters, organisational learning, or conflict management is high across the various target groups as well as within the different aid organisations, the dialogue should lead to the identification of common training needs and finding ways to meet them.

Raising awareness of the other actors' interests is considered essential, and this can only be achieved through dialogue. The interventions of the Civil Peace Service in the conflict consultancy area had made a strong contribution toward achieving this goal, especially because they have tried to impart tools and methods for conflict resolution to all actors in this sector. In this regard, the recognition of and respect for the interests of the various actors play a decisive role.

The building of alliances (synergy creation, "win – win") is also only possible by promoting dialogue. In capacity building, this has led to many activities being undertaken in concert and to companies from the private sector declaring their willingness to co-finance development activities (PPP) - from which they also expect to profit - outside their original commitment area.

Promoting dialogue between users and authorities in the environmental field should contribute to enabling the on-site reality to be included in the process of formulating laws. This applies particularly to the reforms in the forestry legislation. Here it became possible to introduce both the experiences of the indigenous and Afro-Ecuadorian communities in forest management and the on-site situation. The coming together of the actors furthermore enabled the Ministry of Environment to recognise and utilise the possibilities for the enforcement of national legislation. In cooperation with the Ministry of Environment, didactic materials and methods have been developed on how the reformed forestry legislation can be enforced in the local context.

At all levels (communities, provincial administration and federal), the dialogue contributes to the transparency of the decisions to be taken and to making financial administration comprehensible. As a result, increasing transparency and reducing corruption also are among the pursued goals.

10.2.5 Description of the communication tools

Starting with a stakeholder analysis in the forestry and environmental sector, a "Participatory Rural Appraisal" (PRA) was carried out with Afro-Ecuadorian and indigenous communities in order to jointly identify potentials and development priorities.

PRA is one of several approaches for rapid design, implementation, monitoring and evaluation of rural development (Molnar 1989). It is a systematic yet semi-structured activity carried out in the field by a multidisciplinary team and designed to acquire quickly new information on and new hypotheses for rural development. PRA helps communities mobilise their human and natural resources to define problems, consider previous successes, evaluate local institutional capacities, prioritise opportunities, and prepare a systematic and site-specific plan of action – a village resource management plan for the community to adopt and implement. PRA is an excellent tool to bring together, on the one hand, development needs defined by community groups and, on the other hand the resources and technical skills of government, donor agencies, and non-governmental organisations. In so doing, it integrates traditional skills and external technical knowledge in the development process (Centre for International Development and Environment of the World Resource Institute 1990).

This tool was incorporated at the beginning of the MFC-E program and contributed to the identification of the principal needs of the communities with which the MFC-E wanted to cooperate and to the joint development of strategies to meet those needs for which the MFC-E has been given authority. Group diversity, group process, social identity, gender, change of individual commitment, etc. are important factors in the use of the tool, stamping it unequivocally with features of social psychology theories.

Participatory planning workshops were organised during the four-month planning period of the project activities to which representatives of all the relevant actors were invited to attend. The objective of these 2-3 day events was to bring the actors together and jointly generate ideas. The activities for the coming four months were set out together. The same applies to the yearly evaluation workshops in which, with the support of all cooperation partners, the results and impacts achieved were evaluated and the indicators for the coming year agreed upon. Both planning tools were perceived positively and appreciated by most of the actors as an opportunity for direct exchanges and dialogue. These instruments enabled the MFC-E to gather active feedback from the partners (feedback loops) and adjust the planning accordingly.

Another tool that has received continuous support by the program is the creation of permanent or temporary thematic platforms and forums. In this context, the foundation of a coordinating body (*Unidad Coordinadora*, U.C.) was supported in which principally NGOs, but also umbrella organisations and Afro-Ecuadorian communities, universities, the Ministry of Environment, and timber companies take part.

Since GTZ has become active in the region, the need for the coordination of interventions has repeatedly become apparent. The project originally tried to support the Ministry of Environment in the assumption of this role, especially because of its being, in most cases, the counterpart organisation of national and international aid agencies. The Ministry, however, proved incapable of doing so. The Unidad Coordinadora has therefore taken over this role with some success. It is noteworthy that the interest in this coordination has continued for years among most of the active public and private sector organisations, although not all of them have been willing to co-finance the operation costs.

In spite of these clearly stereotyped relationships between environmentalism and other social and economic interests - with the expected social tensions and conflicts (Mackie and Hamilton 1993) - the coordination efforts have revealed a numerous deficiencies in the environmental situation of Esmeraldas. Furthermore the coordination efforts have supported activities that have only indirect connection with the environment, such as public health, and basic and further education.

Among these is another forum that has the goal of sustainable use and conservation of natural resources as well as the sustainable development of the population in the watershed area of the Río Ónzole ("Foro Permanente del Río Ónzole"). This, too, has been created with the support of the GTZ project. It is most prominently and primarily committed to the educational problems in the region and secondarily involved with issues directly connected to resource utilisation and conservation. There can be no protection or sustainable use of resources without the satisfaction of basic needs.

Besides applying instruments like PRA and forums, the project has, for example, joined ranks with the Catholic Church and a local newspaper in reacting spontaneously to acute environmental hazards. One of these has been the threat to local forests caused by gold mining.

Another resource upon which the project drew was an analysis of perceptions of networks and cooperation. This study was done in the context of a doctoral thesis and comprised three phases: 1) identification of the representative and individual actors, 2) an historical analysis of the changes in the convictions and alliances between the actors, and 3) the analysis of the current points of view, interactions, and alliances among the actors. The usefulness of the investigation for the MFC-E project lies in the strategic knowledge it presented about the actors and alliances, which in turn led to more efficient consulting. The actors profited from the study due to better communication and the possibilities for the resolution of problems it engendered (see: Sabatier and Jenkins-Smith 1993).

The study took the form of interviews in which a series of questions were posed to the interview partners. The questions included, for example, how they judged the interests and rights of various parties and what were the strengths and weaknesses of the interaction among the actors, especially with regard to information flow, cooperation, and conflict resolution. Through the absolute (e.g. regarding interests and convictions) and relative (e.g. concerning socio-economic and ethnic characteristics) responses, interpretations could be made as to their positions. The interviews also spawned statements about the status of the project's interaction with various network and possible means of improving interaction.

Particularly valuable information was rendered available by means of the "gossip matrix" tool, which encouraged the actors to talk unprejudiced and confidentially about third parties. This methodology is based upon the generation of information about "who said what about whom" and thus establishes a tripartite relationship among the interviewer, the informant, and a third party. Given the opportunity to describe the point of view of a third party, the respondent has little interest in withholding information.

The utilisation of this tool assumes an understanding of an organisation, which is based on systems thinking for management practice. The organisation is seen as a living organism that is constantly questioning and redefining itself. This point of view is clearly in line with theories of organisational learning outlined in Chapter 2.

The analysis is used by the project to question ongoing processes and re-orientate stuck structures and points of view.

The existing land ownership conflicts in the area have led the project to apply to the Civil Peace Service for an expert to act as an advisor. Over the past two years, the consultant, working in cooperation with both the U.C. and the MFC-E, has trained various actors in the forestry sector in conflict management, negotiation, etc. Furthermore, local conflict consultation centres that benefit from the services of well-known people from the vicinity have been established. These have not only been utilised by the communities; NGOs and logging companies active in the region have also resorted to them, and they have contributed to the creation of a more amicable atmosphere for dialogue.

The MFC-E also functions as the facilitator in an agreement between a logging company and an indigenous community concerning utilisation of a community forests. Both have accepted the project as a mediator, and they take advantage of its services. Taking the divergent interests - development versus profit - of the parties into consideration, it attempts to create synergy that guarantees a benefit for both.

Another instrument that promoted participation and dialogue was the project evaluation, carried out in December 2002 as joint reflection process with external moderation. The MFC-E workers were instructed to set their own perspectives aside and to take a distanced view in analysing their activities, efforts, successes, etc. and in setting guidelines for their future activities. One of the roles the moderation assumed was guarding that the collaborators did not fall into the position of defending their interventions.

The exercise proved to be an organisational learning process, and it identified the paternalism that prevails in the region as one of the main obstacles to the promotion of self-help. The experience gained resulted in the development of a new approach to the project's activities: henceforth all collaboration was to be done on the basis of a written agreement in which a clear delineation of duties and responsibilities was documented, as was an agreed-upon timetable.

Some of the tools mentioned above are also applicable in other social, cultural, and political contexts. For example thematic forums and conflict resolution are widely applied in development activities (see Coleman 2002, Minda Batallas 2002). Here the type of communication in the various cultural contexts must be considered.

Public Private Partnership (PPP) is also widely used and can be seen as a tool that contributes to exchange and leads to an appreciation of mutual interests and points of view. It is based on an agreement between a national or international company and a state organisation that, focusing on a development goal, provides benefits to both and thus results in a win-win situation.

The "Political Power and Coalition / Stakeholder Constellation Analysis" (Sabatier and Jenkins-Smith 1993) can also be seen as a transferable element. This approach was applied during various project phases: at the beginning to identify alliances and target groups, during the execution phase to help determine the external changes and effects within the network, and toward the end of the project to assess the future stability of the actors' network. Also, when used at the beginning of a project, it can replace the often very extensive and effort-intensive stakeholder analysis and deliver target-orientated, directly usable information.

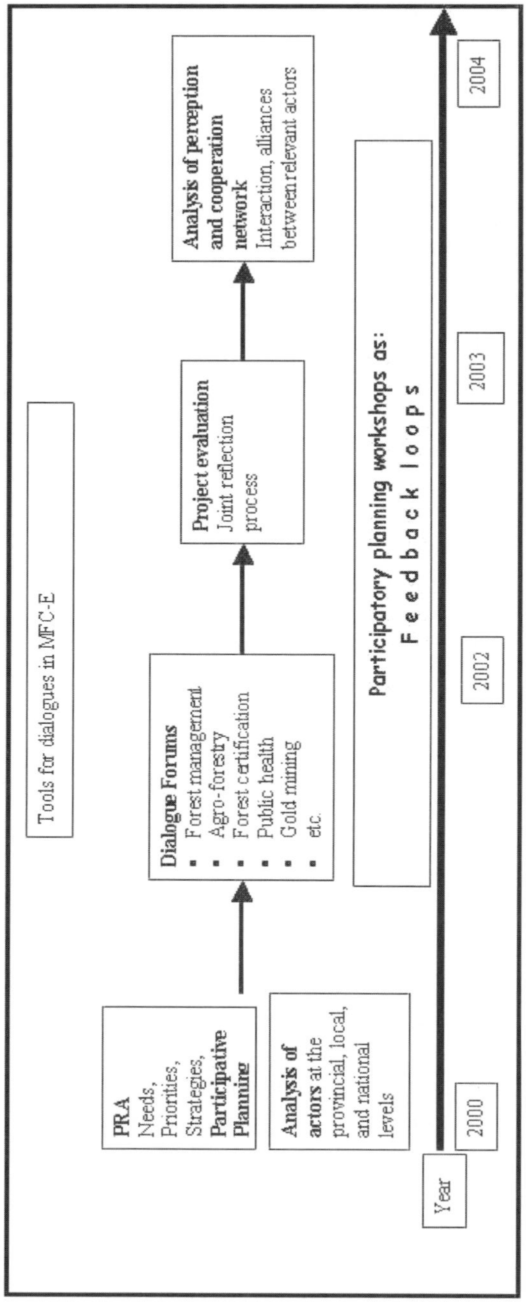

Fig. 10.6 Tools for dialogues in MFC-E. Source: J. Linke

10.2.6 Outcome analysis

One of the factors observed as determining the success of cooperation activities is that the project focuses on putting processes into action and not on being the "implementing agency". Every activity must be preceded by a comprehensive agreement regarding the expectations of all parties involved as well as a definition of goals. This establishes ownership by underlining the fact that the effort is being directed at achieving the objectives of the local actors and not the GTZ's project goals. The course of action to reach the target is jointly analysed and defined.

At this stage it is crucial that the assumption of responsibilities and financing of specific steps taken toward meeting the objectives be clearly agreed upon and documented. Doing so will prevent expectations that remain unfulfilled, prove "erroneous", or are not communicated from leading to misunderstandings - or paternalism - that could hamper reaching the target (see: Burger and Happel, 1997).

The tools presented, like PRA, establish the basis for the consulting intervention of the project, whilst the target groups define the priorities of their needs and strategies for the achievement of the needs are defined jointly. Participative planning precedes the implementation. Continuous evaluations of the achievements and impact lead to appropriate adjustments. The mere acceptance of responsibility (and co-financing) helps actors to consider the process as their own one (ownership).

The project has increasingly focused on its role as consultant and service provider, strengthened by the project evaluation, which emphasises the importance of clear definitions in regard to responsibilities, functions, chronology and financing. Furthermore, when it proved impossible to accomplish a project activity in what was considered an ideal way, those steps that were carried out on the basis of clearly defined arrangements proved successful and sustainable. As a result, the actors see - and have articulated their perception; that, ideally, participation is a necessity for success, and that the project should react as an advisor and facilitator to insure its inclusion in the overall process.

The conditions mentioned for cooperation, such as financial self-participation of the actors in order to achieve their acceptance of responsibility and initiatives, contribute to fight the reigning "paternalism" in the region and to assure the appreciation as well as the usefulness of an activity.

The project consultation includes technical as well as organisational development, capacity building, support in the forging of alliances, supplying of information, and creating the availability of space and possibilities for dialogue. The last-mentioned is considered a success factor as well. Frequently there is neither a coming together nor an exchange of interests due to the lack of mobility or initiative.

The planning tools of the participatory activity planning and yearly evaluation workshops instituted under the feedback configuration demanded major inputs of time, finance, and organisation. However, the mere fact that a representative of the Ministry works together in the same task force with a representative of the community on the analysis of specific activities or results contributes to the dialogue and the mutual understanding of varying points of view.

The mutual scepticism that initially prevailed - rooted in the diverse cultures of communication within the Ministries, logging companies, universities, NGOs, indigenous and Afro-Ecuadorian communities - already decreased significantly after the second planning meeting; the participants got to know each other better and gained confidence. Besides the moderated task force sessions, with prearranged rules (non-reflexive dialogue), the participants always had adequate time during which the organisers set no preconditions for the dialogue. The various stakeholders appreciatively made use of these moments to discuss issues regarding their needs or interests with individual actors (reflexive dialogue).

The models of the unique forums about certain issues (*mesa de diálogo*) or the permanent forums were also welcomed by many actors and perceived as a possibility for information and the exchange of interests. Meanwhile other institutions and organisations also adopt this tool (Bishop, NGOs). It seems intrinsically important to bring the negotiating partners to a uniform knowledge level at the very beginning and to state conditions (moderation) that guarantee equal rights or - said in another way - to prevent the ascendancy of one of the actors during the dialogue, which is a common occurrence in interactions among actors who have disparate educational levels. The question of how seriously one actor takes another other shows how far a balance of interest with equal rights and a dialogue are really possible.

The tools of the shared reflection process (project evaluation, PFK) and the analysis of political power and coalition / stakeholder constellation are considered very successful in promoting the dialogue within the project teams and between project team and partner. Both have contributed to a critical analysis of the intervention approach and strategies, thus making it possible to take appropriate steps or make suitable adjustments.

The corruption that exists at all levels in the timber sector must be named as a factor leading to failure. The financially influential timber companies, which have placed strategic allies in positions at all levels, are able to annul or promote political decisions or even obtain changes in legislation within a very short time. This occurs, e.g. in on-site timber felling monitoring and during timber transport, where the influence of the timber industry led to the failure of a very innovative international model for the delegation of certain responsibilities to a neutral international organisation.

The positions of the timber organisations are not based on scientific or empirical data but rather on mostly short-term economic interests. Thus the confidence that small and mid-sized forest owners and indigenous and Afro-Ecuadorian communities place in national legislation, in the corresponding authorities, and in the big timber companies, is virtually nil. Frequently they have no other choice but to play the game and to swallow their convictions or postpone sustainability aspects of their activities. For this reason it is important to try to gauge carefully the degree to which the participation of the most influential actors in such forums is truly based on willingness to dialogue, or is merely a means to pursue strategic political intentions.

10.2.7 Lessons learned

Natural resources management nowadays can no longer be isolated and executed without consideration of all stakeholders. Based on a corresponding stakeholder analysis, i.e. an analysis of "Political Power and Coalition / Stakeholder Constellation", creative tools should be developed and adapted that promote the exchange of interests and dialogue between them. Solely on the basis of the knowledge of interests, sustainable political decisions can be made and laws developed. Without knowledge of the interested parties' priorities and the development targets of the actors, consultation by aid organisations and project implementation lacks any firm basis. The MFC-E project has mainly succeeded in building these bridges between the conceptual level and legal implementation, incorporating experiences in the on-site cooperation with the target groups (micro level), national and international NGOs, and with the private sector in the discussion process at the macro level. In the opposite direction, through its experience in cooperation with target groups, it has supported the process of translating laws, decrees, and strategies into action.

Therefore managers and political decision-makers should concentrate more on the creation of conditions to make articulation, the increase of knowledge levels, and - last but not least - dialogue possible. These procedures can be implemented possible and should be institutionalised in order to allow weak actors to express themselves and not be overlooked.

The academic training of technicians in the resource management area should also include communication, moderation, and visualisation techniques. This has been the author's experience, not only in his role of principal adviser of the MFC-E project, but also in other cultural contexts.

It also seems important to promote negotiation and conflict resolution skills among managers and political decision-makers in the social and cultural context: "pigeonhole thinking" should be avoided. Another factor that can facilitate dialogue considerably is the return to appreciation of the spiritual and ethical values that still exists in many indigenous communities (Cosmovision of the Chachis) but which are increasingly losing significance in private business in the developed world and on the political stage (see: Coleman 2002). The pervasive corruption can be interpreted as a symptom of this.

If the regular evaluation of the approaches, directions, and successes of the interventions is not done by external evaluators, but - as in the case of the MFC-E - as an internal reflection process that includes collaborators, partners, and target groups, it contributes significantly to agreement on and to internalisation of the intervention approaches and of the "how" within the team. This reflection process furthermore promotes dialogue with the target groups and partners and the comprehension of the "intervention philosophy".

References

Backes, M. (1998). Die Kunst der Gratwanderung. In: DED-Brief No. 4/1998 "Partizipation". DED, Oktoberdruck, Berlin, 30-33

Barragan, M. J., Yumiseva, C. A. (2003). Interacción de Tortugas Marinas y Pesquería Artesanal. Parque Nacional Machalilla-Manabí y su Zona de Influencia. Alianca Jatun Sacha/CDC-Ecuador, Quito

BMZ (1995). Armutsbekämpfung – warum, wozu und vor allem: wie? Bundesministerium für wirtschaftliche Zusammenarbeit.

BMZ (1996). Entwicklungspolitische Konzeption des BMZ. Bundesministerium für wirtschaftliche Zusammenarbeit aktuell 072

BMZ (1997). Konzept für die Förderung der gleichberechtigten Beteiligung von Frauen und Männern am Entwicklungsprozess. Gleichberechtigungskonzept. Bundesministerium für wirtschaftliche Zusammenarbeit aktuell 084

BMZ (1999). Übersektorales Konzept. Partizipative Entwicklungszusammenarbeit. Partizipationskonzept. Bundesministerium für wirtschaftliche Zusammenarbeit aktuell 102

Burger, D., Happel, J. (1997). Das Leitbild nachhaltiger Entwicklung – Handlungsleitende Orientierung der GTZ? Arbeitsgruppe „Orientierungshilfen" im Themenfeld „Nachhaltigkeit im Management natürlicher Ressourcen". Diskussionspapier. Eschborn

Castro, C., González, J. Denkinger, J., Scheidat, M. (1999). Historic presence of marine mammals on the Ecuadorian coast. Proceedings of the thirteenth annual conference of the European Cetacean Society. European Research on Cetaceans 13 (181) (abstract), Valencia, 5.-8. April 1999

Centre for International Development and Environment of the World Resource Institute (1990). Participatory Rural Appraisal Handbook, Natural Resource Management Support Series No. 1. National Environment Secretariat, Egerton University, Clark University

CISP-PMRC (1997). 1er Plan Técnico-Financiero para el manejo de residuos urbanos del Cantón Puerto López. Unidad Técnica para el manejo de reesiduos urbanos del Cantón Puerto López. M.I. Municipio de Puerto López-DHS, Pryecto CISP-PMRC, Servicio Alemán de Cooperacón-DED, Comité Zonal. Puerto López, July-December 1997

Coleman, P. (2002). The way of change. 7 basics für erfolgreiche Veränderungsprozesse im Unternehmen. Wirtschaftsverlag Langen Müller Herbig, München, 239

Davidsen, C. (2004). La organización institucional del manejo forestal comunitario: Análisis de la red de actores con el Manejo Forestal Comunitario – Esmeraldas (MFC-E). Institut für Internationale Forstwirtschaft. Technische Universität Dresden, 54

DED (1998). DED Brief No. 4/1998 "Partizipation". DED, Oktoberdruck, Berlin

DED Ecuador (1996). Programmplanung 1996-1998. Quito, unpublished

Dinerstein, E. et al. (1995). A conservation assessment of the terrestrial ecoregions of Latin America and the Caribbean. World Bank, Washington DC

Fatzer, G. (1999). Organisationsentwicklung für die Zukunft. Ein Handbuch. Edition Humanistische Psychologie EHP

Ferraro, P.J. (2001). Global Habitat Protection: Limitations of Development Interventions and a Role for Conservation Performance Payments. Conservation Biology 15, 990-1000

Ferraro, P.J., Kiss, A. (2002). Direct Payments to Conserve Biodiversity. Science 298, 1718-1719

Fundación Natura (1996). Estudios de alternativas de auto financiamiento para el Parque Nacional Machalilla y el Parque Nacional Podocarpus. Quito

GAT-PNM (2003). "Gestión Participativa" Parque Nacional Machalilla. Puerto López

Grundmann, G. (1998). Partizipation – ein neuer Modebegriff? In: DED-Brief No. 4/1998 "Partizipation". DED, Oktoberdruck, Berlin, 13-15

GTZ (1993). Waldkonzepte: Konzeption und Handlungsfelder des Arbeitsfeldes Waldwirtschaft. GTZ – Abteilung Waldwirtschaft, Waldprodukte und Naturschutz, Eschborn

Holthusen, B., Paulus, I. (1998). Ein Venn-Diagramm ist noch keine Lösung. In: DED-Brief No. 4/1998 "Partizipation". DED, Oktoberdruck, Berlin, 39-41

INEFAN, GEF (1998). Resumen ejecutivo del Plan de Manejo del Parque Nacional Machalilla. Quito

Lincango, M. (1997). Estúdio Turístico para el Plan de Manejo del Parque Nacional Machalilla. INEFAN y Dirección Nacional de Areas Naturales y Vida Silvestre (DNANVS), Quito

Löbsack, L. (1998). Partizipation - und kein Ende. In: DED-Brief No. 4/1998 "Partizipation". DED, Oktoberdruck, Berlin, 3-6

Macias Parraga, R. (1997). Mejoramiento del Sistema de Recolección y Tratamiento de Desechos Sólidos del Cantón Puerto López. Proyecto de Manejo Costero Ecosustentable Integrado de la Zona Sur de Manabí CISP-PMRC. Ecuador, Puerto López, unpublished

Mackie, D.M., Hamilton, D.L. (1993). The influence of affect on stereotyping: The case of illusory correlations. In: Affect, Cognition, and Stereotyping: Interactive Processes in Group Perception. Academic Press, New York, 39-61

Mero, O., Sturm, M. (1997). Mit den Unwägbarkeiten des Klimas leben. In: DED-Brief No. 2/1997 "Wasser". DED, Oktoberdruck, Berlin, 60-61

Minda Batallas, P., Pablo, A. (2002). Identidad y conflicto. La lucha por la tierra en la zona norte de la provincia de Esmeraldas. Abya-Yala, Quito, 118

Molnar, A. (1989). "A review of rapid Appraisal Tools for Use in Natural Resource Management Planning and Project Design and Execution". Draft Paper for UN Food and Agriculture Organization Forest Department and Asia Environment Division, World Bank

Myers, N. et al. (2000). Biodiversity Hotspots for Conservation Priorities. Nature 403, 853-858

Nohlen, D., Nuscheler, F. (1993). Handbuch der Dritten Welt, Band 1: Grundprobleme Theorien Strategien. Dietz Nachf.-Verlag, Bonn

Norton, P., Garcia, M. V. (1992). 5000 años de Ocupación Parque Nacional Machalilla. Centro Cultural Artes, Abya-Yala, Quito

Rice, R., Linke, J., Brunner, A., Suarez, L., Hardner, J. (2003). Conservation incentive agreements: An approach to linking conservation and economic development on indigenous lands in Ecuador. International Conference on rural livelihood, forestry and biodiversity. Bonn, 19.-23.05.2003

Rio (1992). Prinzipien 10. Volle Information und Bürgerbeteiligung bei Umweltentscheidungen

Sabatier, P.A., Jenkins-Smith, H.C. (1993). Policy Change and Learning – An Advocacy Coalition Approach. Westview Press, Boulder/Colorado, USA

Samaniego Rivera, J. (1997). Programa, Subprogramas y Proyectos de Educación y Difusión Ambiental para el Parque Nacional Machalilla y su Área de Amortiguamiento. INEFAN y Dirección Nacional de Areas Naturales y Vida Silvestre (DNANVS), Quito

Scheidat, M. (2001). Abundance, habitat use, behaviour and management of humpback whales (megaptera novaeangliae) in the Machalilla National Park, Ecuador. Dissertation der Mathematisch-Naturwissenschaftlichen Fakultät der Christian-Albrechts-Universität zu Kiel

Scheidat, M., Castro, C., Denkinger, J. González, J., Adelung, D. (2000). A breeding area for humpback whales (Megaptera novaeangliae) off Ecuador. Journal of Cetacean Res. Manage 2 (3), 165-171

Schein, E.H. (1987). Drei Grundmodelle der Beratung. Zürich, Schweiz

Schein, E.H. (1998). Organizational culture and leadership. Jossey-Bass Publishers, San Fransisco, California, 358

Sierra, R., Campos, F., Chamberlin, J. (1999). Areas Prioritarias para la Conservación de la Biodiversidad en el Ecuador Continental. Ministerio del Ambiente, EcoCiencia y WCS, Quito, Ecuador

Sigle, M. (1995). EH-Jahresbericht 1995. Puerto Cayo, 3-4, unpublished

Sigle, M. (1996a). Projektfortschrittsbericht, Berichtszeitraum Juni-Oktober 1995. Puerto Cayo, unpublished

Sigle, M. (1996b). Abschlussbericht. Puerto Cayo, unpublished

Stattersfield, A.J. et al. (1998). Endemic Bird Areas of the World: Priorities for Biodiversity Conservation. Birdlife International, Cambridge, UK

Vaca Bucheli, R. (1997). Diagnóstico Socio Poblacional para el Plan den Manejo del Parque Nacional Machalilla. INEFAN y Dirección Nacional de Areas Naturales y Vida Silvestre (DNANVS), Quito

Velasquez, F. G. (1997). Administración y proyectos para la Reformulación del Plan de Manejo del Parque Nacional Machalilla. INEFAN y Dirección Nacional de Areas Naturales y Vida Silvestre (DNANVS), Quito

Von Loebenstein, K., Trux, A., Welte, T. (1993). Kompensation und Interessenausgleich in der Pufferzonenentwicklung, Band 1: Erfahrungen – Schlussfolgerungen - Empfehlungen. Erstellt im Auftrag der Deutschen Gesellschaft für Technische Zusammenarbeit (GTZ), Bonn.
www.puertolopez.com/hojas/puertolopez.htm

WWF (1997). (World Wildlife Fund). Global 200 Ecoregions (map). WWF, Washington D.C.

Zenger, S. (1994). DED-Abschlussbericht Nationalparkverwaltung, Machalilla. Instituto Ecuadoriano Forestal y de Areas Naturales y Vida Silvestre (INEFAN)

11 Incorporating Local People through Economic Incentives at Lake Mburo National Park, Uganda – Africa Works!

Christiane Averbeck

Free University of Berlin, Germany

Chambers (1991: 515) stated that to "increase the capacity of a community is to increase its ability to do things for itself. It means increased ability and strength, more skills, more confidence, and more efficient organisation. It can be facilitated through action such as community-based projects, but only when all community members become involved from the beginning, to decide upon a community action, to identify hidden resources from within the community, and by developing a sense of ownership and responsibility of communal facilities from the start to the finish".

The wildlife utilisation project around Lake Mburo National Park (LMNP) in Uganda is a good case in point. Right from the beginning all important stakeholders became involved, they decided upon the community action and were helped to identify the hidden resources. They realised that it is possible to do things for themselves and established an efficient organisation in order to realise a community wildlife utilisation programme. Without technical or financial assistance from foreign donors they are running the Rurambira Wildlife Association. But before the community participation process started in the early 1990s, the inhabitants of the Lake Mburo area went through a hard time of exclusion and suppression.

11.1 Non-participatory Conservation History of Lake Mburo National Park

LMNP is situated in southwestern Uganda in the Mbarara District, Nyabushozi County, near the Equator. At the beginning of the twentieth century, Lake Mburo was known to naturalists as one of the premier wildlife areas in East Africa. The Lake Mburo ecosystem included large areas that provided the grazing land of the Banyankole people and their

Ankole cattle. The ethnic group of Banyankole is divided into two subgroups, the Bahima and the Bawiru. Whereas the Bahima traditionally have been pastoralists and never hunted wildlife, the Bawiru have been cultivators and have occasionally hunted. The Lake Mburo area formed part of the Nkore Kingdom, which was controlled by the Omugabe, the king of the Banyankole. The Omugabe controlled the access to the land around Lake Mburo, allowing his people to graze their cattle only in times of drought. By tradition he had to be a Muhima (singular for Bahima), and because hunting has not been part of the tradition of the Bahima, there was no hunting in the Lake Mburo area. In the first decades of the twentieth century, an outbreak of the cattle disease rinderpest decimated the Bahima's herds. Livestock numbers took 20 years to recover, and competition between wildlife and cattle was low during this time (Snelson and Wilson 1994).

In 1935, the area around Lake Mburo was declared a Controlled Hunting Area by the British colonial government. The British colonial government permitted both regulated big-game hunting and traditional human activities.

In the 1940s, a severe outbreak of sleeping sickness and nagana (a form of sleeping sickness found in cattle) carried by tsetse fly (Glossina sp.) forced pastoralists out of the area. However, many of the farmers and fishers remained. Tsetse flies need two basic resources to persist in an area: shade and blood. It was assumed that if all shade and wild animals were removed, then the tsetse fly would be eradicated. The United States funded a drastic tsetse eradication program of spraying, bush burning and cutting, and shooting, which severely reduced game populations.

By the early 1960s, tsetse had been eradicated, once again opening up the area to pastoralists. To protect the remaining wildlife, the newly independent Ugandan Government gazetted the Lake Mburo Game Reserve. All forms of use except controlled hunting were banned, although resident farmers were permitted to stay. Private land ownership was becoming common and communal lands were accroached by the influential and rich people, leaving out the poor who could not afford land, especially the pastoralists (Kafureka 1992). To worsen the situation, the government of Uganda decided that ranching was the best use for this dry and sparsely populated land and alienated a big area for ranching. This pushed the pastoralists on to the margins of the ranches since they were not considered the kind of people to run ranches. Therefore the pastoralists resorted to occupying the Lake Mburo area which was mainly occupied by wildlife (Ayorekire 2000).

In 1983, the Ugandan Government, due to increased encroachment on Lake Mburo Game Reserve and the need to preserve the biodiversity, established LMNP within the boundaries of the original game reserve. All previous forms of land tenure, traditional or otherwise, were effectively terminated. LMNP was established without the consent of local people and involved their forced removal. They were neither consulted nor compensated for the loss of their homes and land. No real attempt was made to explain to them what conservation entailed or what the intention of the government was (Ayorekire 2000).

By 1986, the entire park was again occupied by settlers. The government realised that the option of re-evicting the people would not be a success. In order to resolve the land crisis, a government task force was established and it was decided that 390 km^2 should be degazetted, leaving only 260 km^2 for the park.

Historically the prevailing approach of conservation authorities toward local communities was simply to keep them out of the protected areas. Emphasis was on strict protection, and as a result hunting was banned even outside the park. The LMNP today is directly bordered by farm and ranch land; the park has no buffer zone. Pastoralists and cultivators live on its periphery. Some pastoralists were resettled in the newly established Kanyanyeru Resettlement Scheme, as an option for terminating the land-use conflict. Between 1980 and 2000, a human population growth of 2.7 percent per year in the region adjacent to LMNP reduced the area to a remnant of what had formerly been a much more extensive wildlife area. The reduction in size and the availability of former communal rangelands outside the park caused by privatisation and cultivation of land has confined pastoralists to an area that is too small to support the numbers of cattle needed to sustain their lifestyle and basic livelihood. The result has been overstocking and range deterioration. The pressure on people to develop new and sustainable forms of land use is intense (Averbeck 2001).

11.2 Participatory Conservation History of Lake Mburo National Park

In the late 1980s and early 1990s the authorities' approach to conservation changed. In 1989, realising the extent of the land-use conflict in the Lake Mburo National Park area, the Uganda National Parks, responsible for managing wildlife in the country, established a Community Conservation Programme (CCP), in cooperation with the African Wildlife Foundation (AWF). The theme of the project was 'Neighbours as Partners.' Wildlife managers realised that relationships between rural resource users and conservation agencies were a prerequisite for building sustainable community systems (Barrow et al. 1995). Starting in 1991 and ending in 1999, the project instituted a process of problem solving with stakeholder dialogues. The CCP has worked closely with the people through the local council structure and through the Park Management Advisory Committees (PMAC) which were set up at each parish to act as a link between the local community and the park management authorities (Ayorekire 2000). The CCP mainly put emphasis on formal environmental education, capacity building, and support for community development. However, this approach did not turn out as successful as hoped for (Hulme and Infield 2001, Infield and Namara 2001). It did not stop the local communities from using wildlife in an unsustainable manner inside and outside the park.

In the late 1990s Lake Mburo National Park was not large enough (260km^2) to maintain viable populations of top predators such as lions (Panthera leo), leopards (Panthera purdus) or hyenas (Crocuta crocuta); staffing was insufficient, the Park boundaries were arbitrarily drawn, fishermen were living in the Park, pastoralists were grazing illegally, and hunters were poaching wild animals. The populations of most of the big mammals were already threatened. The pressure was high to continue with new approaches of wildlife management as the old ones were about to fail.

11.3 A New Wildlife Management Policy towards Participation in Uganda

In 1996 a new agency responsible for wildlife in Uganda, Uganda Wildlife Authority (UWA) was created. Uganda adopted, furthermore, a new wildlife statute (The Uganda Wildlife Statute 1996). The statute vests ownership of wildlife in the state but makes provision for people to own wildlife that had been lawfully taken. Part IV of the statute provides for different categories of 'use rights,' such as hunting, farming, ranching, and general extinction of and trading with wildlife products (Averbeck 2002). By implication, the assigning of use rights was intended to motivate communities and individual landowners to conserve wildlife through sustained extractive use. Mechanisms were established to enable local communities to manage their wildlife themselves, rather than having this control imposed from outside (Okua et al. 1997).

11.4 The Lake Mburo Wildlife Utilisation Study

The change of the legal frame opened up the opportunity for consumptive wildlife utilisation in Lake Mburo National Park. On the basis of the approach and the achievements of the USAID funded project, with financial support from the Gesellschaft für Technische Zusammenarbeit (GTZ) GmbH, I conducted a study on wildlife use (Averbeck 2002, 2003). My intention was to assess whether community-based conservation through sustainable use of wildlife could become a viable solution for integrating rural communities and wildlife conservation in the Lake Mburo area. The resource potential of the wildlife population, the objectives and wishes of the target group, the landowners of Nyabushozi, possible impediments and resistance, and socio-cultural problems with adoption of sustainable utilisation by the target group were considered. I assessed and analysed the intricate network of ecological and socio-economic interrelationships between the Park and its surroundings, pointed out the problems that arise for wildlife preservation, and developed a concept for a new approach to ecosystem management and community-based wildlife conservation in Uganda. My research project did not aim at implementing the concept. It was up to Uganda Wildlife Authority to decide whether the results of the study and concept would be a relevant and practical approach to wildlife management for Lake Mburo National Park.

11.5 Participatory Aspects of the Lake Mburo Wildlife Use Study

With the conservation history of Lake Mburo National Park in mind, and in order to prevent mistrust and misunderstanding, the different phases of the research project were conducted in a participatory way. Stakeholder dialogues laid the foundation for mutual planning, transparency, and the cohesion of stakeholders. Furthermore, this approach created ownership and commitment by the Wildlife Authority, the Local Authorities, and the community of Nyabushozi at the same time.

Different factors and phases were important for the communication process.

- *Intercultural communication* was an essential factor. It meant to be aware, to learn about, and partly follow the social-cultural obligations and taboos of stakeholders involved in the process. Without a considerable knowledge of social-cultural patterns and ethnic behaviour the stakeholder dialogue would have failed.

- The *historical* and *political situation* of Uganda and especially Nyabushozi was considered.
- *Understanding and respect* for the situation of the others were essential. Stakeholders were, therefore, given the opportunity to communicate about their strengths, weaknesses, and opportunities not only during the planning phase, but also in the research phase.
- The *aim of the process* was specific and thus, to a certain extent, calculable for the stakeholders involved.
- *Stakeholders exercised patience,* bearing with consequences due to delays.
- And, in all of the following phases *I tried to be flexible*, in order to dynamically adapt to newly arising challenges or opportunities which had an impact on the side of the people.

11.5.1 Planning

Two workshops on "problems and opportunities of impala conservation" by the Senior Warden in Charge of Lake Mburo National Park, started the process in 1994. The aim of the workshops was to bring together different stakeholders, including landowners, wildlife managers, poachers, politicians, representatives of the District, and researchers in order to discuss conservation and opportunities for and problems of sustainable wildlife utilisation in Nyabushozi. As a result of those workshops a pilot impala utilisation scheme in the Lake Mburo area was proposed involving stakeholders. Aim of the pilot project was to determine whether legitimising and formalising controlled off-take of impala might result in a stabilising and gradually increasing impala population. Intentionally, the project was restricted to impala since they were the most common big wild mammal species in the ecosystem.

11.5.2 Introduction

The Lake Mburo Wildlife Use Study started with the introduction of the research project, the researcher, and her team in 1997. Various discussions were held with representatives of the Uganda Wildlife Authority, Lake Mburo National Park, District Authorities and local leaders. Landowners, farmers, and pastoralists in the area were visited, in order to inform them about the goals of the study, and were asked for assistance and support of the ongoing activities.

11.5.3 Wildlife research

In order to gather information on the movements of impala, a total of 233 impalas inside and outside Lake Mburo National Park were marked with numbered ear-tags. Impalas were captured either by immobilisation with a darting gun, or by night capture with spot lights. Interested farmers were invited to participate in catching the animals. Furthermore, they were encouraged to observe tagged animals and report the numbers and locations to the researcher. Although it was not possible to use all the information made available by the farmers, it gave them the feeling of contributing to the assessment of the wildlife population.

11.5.4 Focus group interviews[1]

Altogether, 23 focus group interviews with a total of 334 participants were carried out. Focus group interviews proved as the ideal method for acquiring an orientation on attitudes towards wildlife, hunting, and the opportunities and challenges presented by wildlife utilisation of the communities around LMNP. Target groups for the interviews were people living in the immediate vicinity to LMNP in Nyabushozi County. This area was chosen as it was assumed that its communities were directly affected by and, vice versa, had the capacity to directly affect wildlife populations. Members from each sub-county were interviewed in homogenous focus groups including pastoralists, cultivators, and mixed farmers. Approximately 80% of the inhabitants work in the agricultural sector. While there are still some members of the population that remain pure cattle keepers or cultivators, the general tendency in earning a livelihood is mixed farming (Ministry of Finance, Uganda 1992). In order to avoid

---

1 Focus group interviews in general are used very early in a research project in order to obtain general background information, stimulate new ideas and creative concepts, diagnose the potential for problems with new programmes, and to generate hypotheses rather than to provide solutions for problems. Questions in focus group interviews are relatively unstructured, because they allow respondents to refer to virtually any aspect of the general stimulus identified in the question. Focus group interviews produce mainly quantitative data with a very rich body of information in the respondents own words and context (Stewart and Shamdasani 1990). A questionnaire with mainly open questions was used. Participants of the interviews were encouraged to express their opinions and to discuss them with one another. The questions were not answered individually; the answers rather reflected the consensus of the group (Averbeck 2002).

community leaders with their higher socio-economic status to dominate the interviews, they were interviewed in separate sessions. Personal contacts in one sub-county made it possible to talk to a group of poachers. More interviews with other illegal hunters were planned. However, activities of the Ugandan army in the Lake Mburo area made it impossible to organise further meetings, as poachers feared to be arrested or to be mistaken for rebels.

11.5.5 Collecting legends, phrases and sayings

The first set of questions on legends, phrases and sayings on wildlife were used as an introduction into the topic, as a way of 'warming up' the participants. The stories were compiled and printed in form of a booklet. The booklet was handed out to community members. Although the compilation does not present a systematic collection of the 'oral tradition' on wildlife, it still presents legends and sayings never recorded before. The legends and sayings collected during the study were not systematically analysed, their effects not studied. However, the mere fact that legends are still told indicates that they are still important value patterns.

11.5.6 Feedback to interviews

Representatives of each focus group were invited to a meeting, where the participants were given a first feedback on their statements and ideas developed in the interviews. The results were presented, analysed, and discussed with the participants. Besides, an invited specialist from a Kenyan wildlife utilisation project explained the concept of an existing community-based wildlife utilisation project. Considering the results of the interviews in Nyabushozi and the experiences gained in Kenya, possible alternatives of a wildlife utilisation scheme in the Lake Mburo area were discussed as a realistic option for the area.

11.5.7 Impala cropping

For this study I looked at the opportunities for hunting in the Lake Mburo area, both in the form of sustainable cropping and of sport hunting. With permission of UWA, a trial cropping was conducted in the Rurambira area, east of LMNP. The objective of a trial cropping was to establish base-line data in relation to health risks associated with consumption of wildlife products. Furthermore, the trial cropping provided information on the procedures of cropping, handling, and skinning of impalas, as well as processing and marketing of game meat in Uganda. Wildlife managers, meat processors, researcher, District officials, members of the communities adjacent to LMNP, restaurant owners, and consumers took part in the trial cropping, processing, and marketing of the animals.

Table 11.1 Participatory aspects of the Lake Mburo Wildlife Use Study and Pilot Project

Year	Activity	Organised by	Participants
1994	1st workshop "Problems and opportunities of impala conservation"	UNP/GD	Community members, wildlife managers, poachers, politicians, District representatives, researchers
1995	2nd workshop "Problems and opportunities of impala conservation"	UNP/GD	Community members, wildlife managers, poachers, politicians, District representatives, researchers
1996	Preparatory consultations	IP	District authorities, wildlife managers
1997	Introductory visits	IP	Community members, citizens of Nybushozi
1997-1999	Marking and observation of impala	IP	Community members, citizens of Nybushozi
1998-1999	Focus group interviews	IP	Community members, wildlife managers, poachers, politicians, District representatives, researchers
1998-1999	Collecting fairy-tales, sayings and phrases on human beings and wildlife	IP	Community members, citizens of Nybushozi
1998/ 1999	Feedback meetings on results of focus group interviews; presentation on wildlife utilisation projects in Kenya	IP	Community members, wildlife managers, poachers, politicians, District representatives, researchers

Year	Activity	Organised by	Participants
1998	Cropping of impala	UWA/IP	Community members, wildlife managers, researchers
1999	Jerseys for football team	IP	Schoolchildren of Rurambira
2000	Presentation of study results	IP	Community members, wildlife managers, poachers, politicians, District representatives, researchers
2000	Study tour to Kenya	IP	Community members, wildlife managers, politicians, District representatives, researchers
2000	Sharing of impala cropping revenue	UWA	Community Conservation Unit (LMNP), community members
2001	Sending out study report	IP	Community members, wildlife managers, poachers, politicians, District representatives, researchers
2001	Establishment of Rurambira Wildlife Association (RWA)	RWA	Community members
2001	Authorisation of RWA and Game Trails (U) (GT)	UWA	Community members, Game Trails (U) Ltd.
2001–?	Pilot hunting scheme	RWA, Game Trails (U) Ltd. UWA, Safari Company, Germany	Sport hunter, community members of pilot area, District leaders, local leaders, poachers
2002	Evaluation of the pilot project	UWA	District leaders, community members of Nyabushozi

UNP: Uganda National Parks, GD: Game Department; UWA: Uganda Wildlife Authority, IP: Impala Project, RWA: Rurambira Wildlife Association

11.5.8 Presentation of study results

The main results were presented and discussed with stakeholders in a final meeting after completing the data collection in Nyabushozi.

11.5.9 Study tour to Kenya

In Uganda, due to the legal and political situation, there had been little interest in legal consumptive wildlife utilisation. Attitudes ascertained during the interviews with the local communities in Nyabushozi and informal discussions with staff of UWA revealed lack of knowledge and experience concerning the challenges and opportunities of community wildlife utilisation projects. However, elsewhere in South and East Africa, many people have experienced community-based wildlife programmes. People's culture changes through interactions among its own members and with those from other cultures (Marks 1984, Kellert 1997). Resources not considered important at one point in time may well become so or may become valued for a different reason (Marks 1984). Therefore landowners of Nyabushozi, local politicians, members of staff of UWA, an environmental journalist, meat processing experts from Uganda Meat Technology Centre, and a veterinarian travelled to Kenya. The objective of the study tour was to give the participants an opportunity to interact with members from another culture, the Maasai pastoralists, who are living in a similar situation, in order to broaden the knowledge of the participants and to potentially change their attitude on wildlife utilisation. The visit and discussions exposed the participants to the core problems but also to the opportunities of wildlife utilisation and conservation projects.

At the end of the tour they decided to form an interim steering committee. Task of the committee was to organise follow up meetings in Nyabushozi with other community members to share the experiences made during the study tour.

11.5.10 Sharing of impala cropping revenue

The impala cropping activity was aimed at illustrating the financial opportunities for sustainable utilisation of wild mammals in the Lake Mburo area. A total of 100 impalas were cropped, their meat sold to selected hotels in Kampala, and their skins to the public. In total, 6,000,000 Uganda Shilling (1,000 Uganda Schilling [Ug. Shs] = 1 US$) were handed over to the owners of land on which the impalas were cropped, as part of the revenue generated in 2000. They received 4,000,000 Ug. Shs for skins and 2,000,000 Ug. Shs. for meat. The landowners decided to build a bridge and to reconstruct a school building in their parish.

11.5.11 Sending out the final report of the research project

The final results of the research project were compiled and published in a book by GTZ (Averbeck 2001). More than 50 copies of the report were distributed among the stakeholders.

11.5.12 The Lake Mburo Wildlife Utilisation Pilot Project

With the end of the research project the responsibility for further activities was put on the stakeholders. No funding or organisational assistance through local or foreign donors was available.

Landowners of the Rurambira parish East of Lake Mburo National Park without financial support from outside established the Rurambira Wildlife Association in the beginning of 2001. Landowners started to support law enforcement programs by reporting illegal activities by poachers to the park officials.

In 2001, UWA on pilot basis authorised Game Trails Ltd., a private Ugandan hunting operator, in collaboration with Rurambira Wildlife Association, to implement sport hunting on ranches in the Rurambira parish. Unlike the sport hunting activities that used to take place in Uganda in the 1960s and 1970s, this project is managed by the local communities themselves, with the advantage of maxiumum benefit. A hunting quota was determined by UWA, based on a 2–3% offtake of the total population estimates and other factors such as birth rate, behavioural, and distribution patterns. It was agreed by different stakeholders to share the revenue accrued by giving 65% of the amount to the Rurambira Wildlife Association, 5% to the sub-county, 25% to the Lake Mburo Conservation Area / UWA, and 5% to the Community-Protected Area Institution (CPI). CPI was formed by the key stakeholders as an institution to manage wildlife outside LMNP and to handle issues concerning the pilot project (UWA 2002).

The pilot project was evaluated in 2002 and the Wildlife Authority and the other stakeholders decided its continuation. In 2004, due to the success and the demand of other communities, the project was extended to two other parishes. From 2001 to 2004, the Rurambira Wildlife Association received 50,000 US$ from the project, which they used for building two schools, six teachers' houses, a health centre and a dam to provide water for livestock (New Vision 2004). Furthermore, the project has lead to an increase in the number of wild animals, has reduced disputes between wildlife authorities and pastoralists, and reduced poaching and charcoal burning in the area (New Vision 2004). According to experiences made in the first three years even landowners will receive a percentage (10%) of the total revenue, which reduces UWA's revenue accrued to 15%.

11.6 Lessons Learnt on Participation

Development can be seen as a process of expanding the real freedoms that people enjoy (Sen 1999). The Nobel laureate Sen's (1999: 3) notions on development go beyond neo-liberal ideas: "...freedoms depend also on other freedoms, such as social and economic arrangements as well as political and civil rights (for example, the liberty to participate in public discussion and scrutiny). Development requires the removal of major sources of unfreedom: poverty as well as tyranny...". This theoretical approach of Sen can be confirmed by the practical experiences made in the Lake Mburo area.

After a long period of mistrust between Government and community members in the Lake Mburo area, the responsible authorities realised that in order to gain understanding and acceptance of conservation politics it is necessary to instigate stakeholder dialogues and implement community conservation programs. Reviewing different community conservation projects of that kind, Adams and Hulme (2001: 22) concluded: "The achievement of the concept is not that it has proved that community conservation 'works:' it is that it has created the space for a set of community conservation experiments that take forms and are achieving very different results." The community conservation program has not altered significantly the cost-benefit equation for communities around the park (Infield and Namara 2001). The ultimate achievement of community conservation at LMNP, from a conservation perspective, are the ways in which it has changed the ideas local communities hold about conservation. Its initial contribution has been positive in helping to reduce anti-wildlife values so strongly held by local people in the late 1980s (Hulme and Infield 2001).

Therefore, as a next step the necessity of community members around LMNP deriving tangible and legitimate benefits from the wildlife on their land became apparent. It was assumed that they would have an incentive to protect wildlife from illegal hunting. The consequence was the establishment of a study on wildlife utilisation in the Lake Mburo area.

Although the aim of the study was not to implement a wildlife utilisation project around LMNP and the outcome of the study was open, the participatory approach led to the implementation as it created ownership and commitment by the different stakeholders. Action and change of the situation and research were pursued in the Lake Mburo area at the same time. The prominent approach of the study, therefore, can be classified as a form of 'action research' which according to Dick (1999) "is an emergent process which takes shape as understanding increases; it is an iterative process which converges towards a better understanding of what happens".

Without intending to do so, my role changed from a mere researcher to a facilitator of a process. My presence over five years in the project area, including the time before and after the official duration of the research project, and my prior experiences as development worker with participatory methods contributed to this shift. Over the years, interactions with different stakeholders increased my understanding of the people, their living conditions and culture which became a prerequisite for the development of the project. On the other hand, it took time to create mutual trust between the other stakeholders and me, as local people were mislead by the notion that I was a stooge of Uganda Wildlife Authority trying to implement what had been already decided upon by the Ugandan Government. It would have been impossible to conduct interviews with poachers at the beginning of the project.

It is not possible to single out one factor which contributed to the success of the project. In fact it was an intricate net of factors was important.

- First, the idea of a wildlife utilisation project in the Lake Mburo area was created by the Senior Warden in Charge of LNMP who originated from the region. His practical experiences made him think of new approaches to wildlife management.
- Second, in the planning workshops the voices and interests of the different stakeholders were heard and considered. They participated in "public discussions and scrutiny," as Sen (1999) would put it. The activities of the research project took place with their consent and support. Right from the beginning the ownership was clearly with the Ugandan stakeholders. It was not merely a project of a researcher from Germany with her own interests funded by a German NGO.
- Third, at all levels and times the activities and outcomes of the study were transparent. The introductory visits, focus group interviews, presentation of study results, open discussions, and even the assessment of the wildlife population and the sharing of the impala cropping revenue involved community members and other stakeholders. The final results were not only made available to the technocrats but also to the other stakeholders

- Fourth, the existing attitudes and knowledge were appreciated and utilised. The organisational structure and the management concept of the wildlife utilisation scheme were developed on the basis of the interviews and discussions. Information on wildlife was collected together with community members and staff of UWA.
- Fifth, room for activities was created in the study, which helped all stakeholders to gain experiences in "learning by doing." The assessment of the wildlife population, the trial cropping exercise, and the revenue sharing revealed information on how to implement these important factors of a wildlife utilisation scheme best in the future.
- Sixth, the exposure to other cultures. The study tour was an opportunity for the stakeholders to interact with members from another culture living in a similar situation. Through interaction with other people they gained new information and learned how to face challenges, solve problems, and conduct themselves in a variety of situations. As described in the concept of the experiential learning cycle[2] (Knowles 1996), the participants gained experience, reflected, drew a conclusion, and tried to apply their experiences in their own setting.

Intentionally, participatory methods such as Rapid Rural Appraisal (RRA) and Participatory Rural Appraisal (PRA), the most frequently used planning methods for the management of natural resources, were not applied in this study. Experiences show that participatory methods can be irresponsible exercises if a serious follow up is not guaranteed (Schönhuth and Kievelitz 1994) and should be applied as soon as the project is entering or has reached the planning and implementation phase. However, even the participatory approaches of this study raised not only the commitment but also the expectations of the people of Nyabushozi. It, therefore, was important that the management of Uganda Wildlife Authority was inherently open for such a project and willing to implement it.

A socio-economic evaluation of the first phase of the pilot project indicated that the realisation on part of the residents concerning the economic value deriving from wild animals on private land contributed to a positive change in attitude towards wild animals on private lands (UWA 2002). However, the evaluation revealed that political leaders and

[2] Experiential Learning Cycle (Knowles 1996) is a model with four phases, experiences, reflection, drawing conclusions, and applying lessons learnt. The learning cycle begins with experience. After we experience something we tend to reflect on it. Following a period of reflection, we draw conclusions about die experience. Lastly, we apply the lessons learnt. The model is representing the concept of experience-based training.

administrative officials were not involved sufficiently. Regular reports on the ongoing activities were not circulated among all stakeholders. The evaluation reports illustrate that, even after the beginning of the utilisation project, it is important that all stakeholders are involved and informed in a participatory way. Otherwise one stakeholder left out or feeling neglected might, by withdrawing his support, jeopardise the success of the whole project.

Furthermore, one has to realise that such a wildlife utilisation project does not create a win-win situation for all stakeholders; i.e. the interests of most of the poachers were not considered. Although it was possible to employ some of the leading poachers as guides and porters it did not involve all of them, and as a result of the utilisation project community members started to support law enforcement programs by reporting illegal activities by poachers to the park officials. In addition, the project increased the expectations amongst residents in areas not yet covered by the pilot project.

11.7 Conclusion

Only the integrated approach to the ecological, socio-economic, and socio-cultural interrelationships between the Park and its environment and a process of awareness creation as well as stakeholder dialogues enabled the implementation of a new approach to ecosystem management in Uganda. The experiences show that non-participatory approaches to nature conservation were bound to fail and rather created a situation of mistrust, tension, and even violence. The implementation and evaluation of the pilot project on “user rights” shows that this new management tool is the right way to wildlife conservation in Uganda. As envisaged before, Uganda Wildlife Authority should expand this type of projects to other areas in Uganda as well. It might not prove possible to implement all activities in the same manner. Adaptive management is required according to the situation of the particular project areas. Funds might be limited. However, it must be realised that the involvement of all stakeholders does not involve as much as the consequential costs of failed top-down approaches. Experiences from other countries, too, indicate that a transfer of a participatory approach is not only necessary but probably the only option for effective nature conservation (Duffy 2000, Prins et al. 2000). More than ten years after Chambers’ (1991) statement, “ownership” and “commitment” of stakeholders are recognised by the German Development Cooperation as an essential prerequisite for successful development politics (Klingebiel 2003). It is considered more important than ever.

11.8 Summary

Like many other protected areas in Africa, LMNP in Uganda is in danger of becoming a paper park. For ecological as well as socio-economic reasons, the park's existence and potential to sustain wildlife depend heavily on its surroundings.

In this chapter, I illustrate the non-participatory and participatory conservation history of LMNP and describe the participatory approaches of a research project on community wildlife utilisation. The participatory approach comprising a vivid stakeholder dialogue laid the foundation for mutual planning, transparency and the cohesion of stakeholders. Furthermore, it created ownership and commitment by the Wildlife Authority, the Local Authorities, and the community members of Nyabushozi at the same time. As a result, a community wildlife utilisation pilot project was established adjacent to LMNP helping to conserve the protected area.

Acknowledgements

I would like to thank the people of Nyabushozi, Mbarara District, and the Uganda Wildlife Authority for their cooperation and support. This study was funded by the Gesellschaft für Technische Zusammenarbeit (GTZ) GmbH, Tropical Ecology Support Program. Thanks go to Volker Riehl, Susanne Stoll-Kleemann, and Martin Welp for their comments on this chapter.

References

Adams, W., Hulme, D. (2001). Conservation and communities: Changing narratives, policies and practices in African conservation. pp 9–24. In: Hulme, D., Murphree, M. (eds.) African wildlife and livelihoods: the promise and performance of community conservation. James Currey, Oxford, UK

Averbeck, C. (2001). Integrating local communities and wildlife conservation in Nyabushozi, Uganda: sustainable use as a viable solution. TZ-Verlagsgesellschaft mbH, Rossdorf, Germany

Averbeck, C. (2002). Population Ecology of Impala (Aepyceros melampus) and Community-based Wildlife Conservation in Uganda. Dissertation. TU München, Germany. URL: http://tumb1.biblio.tu-muenchen.de/publ/diss/ww/2002/averbeck.html

Averbeck, C. (2003). A Park Is Not an Island: Linking Different Wildlife Management Strategies in the Area of Lake Mburo National Park, Uganda. pp 283–302. In: Bissonette, J.A., Storch, I. (eds). Landscape Ecology and Resource Management. Linking Theory with Practice. Island Press, Washington

Ayorekire, J. (2000). Conflicting Land-use: Lake Mburo National Park. IFRA – Les Cahiers. 22, 1–22

Barrow, E., Bergin, P., Infield, M., Lembuya, P. (1995). The peoples voice: partnership and community conservation. In: Bissonette, J.A., Krausmann, P.R. (eds). Integrating people and wildlife for a sustainable future. Proceedings of the 1st International Wildlife Management Congress. The Wildlife Society, Bethesda, Maryland, 255–259

Chambers, R. (1991). Shortcut and participatory methods for gaining social information for projects. In: Cernea, M. (ed). Putting people first: Sociological Variables in Rural Development. World Bank, Washington, 515–537

Dick, B. (1999). What is action research? URL: http://www.scu.edu.au/schools/gcm/ar/whatisar.html

Duffy, R. (2000). Killing for Conservation. Wildlife Policy in Zimbabwe. The International African Institute. James Currey, Oxford; Weaver, Harare; Indiana Press, Bloomington and Indianapolis

Hulme, D., Infield, M. (2001). Community conservation, reciprocity and park-people relationships. Lake Mburo National Park, Uganda. In: Hulme, D., Murphree, M. (eds). African wildlife and livelihoods: the promise and performance of community conservation. James Currey, Oxford, UK, 106–130

Infield, M., Namara, A. (2001). Community attitudes and behaviour towards conservation: an assessment of a community conservation programme around Lake Mburo National Park, Uganda. Oryx 35(1), 48–60

Kafureka, B.M. (1992). The dynamics of the land question and its impact on agricultural productivity in Mbarara District. Working Paper No. 25. Kampala, CBR Publications

Kellert, S.R. (1997). The value of life: biological diversity and human society. Island Press, Washington

Klingebiel, S. (2003). "Ownership" ist essenziell. epd-Entwicklungspolitik 22, 24–27

Knowles, M.S. (1996). The adult learner: a neglected species. Gulf Publishing, Houston, Texas

Marks, S.A. (1984). The imperial lion. Human dimension of wildlife management in Central Africa. Westview Press, Boulder, 53–57

Ministry of Finance and Economic Planning (1992). The 1991 population and housing census. Mbarara District. Statistics Department, Kampala, Uganda

Okua, J.M., Buhanga Aralli, E., Opolot, F.H.O., Anywar, J., Karugaba, K., Heath, B.R. (1997). Guidelines for the implementation of wildlife use rights. UWA, Community Conservation for Uganda Wildlife Authority Project. Unpublished Report

Prins, H.H.T., Grootenhuis, J.G., Dolan, T.T. (2000). Wildlife Conservation by Sustainable Use. Kluwer Academic Publishers

Schönhuth, M., Kievelitz, U. (1994). Participatory Learning Approaches. TZ-Verlagsgesellschaft mbh, Rossdorf, Germany

Sen, A. (1999). Development as freedom. Oxford University Press, Oxford

Snelson, D., Wilson, A. (1994). Lake Mburo National Park Guide Book. African Wildlife Foundation, Kampala, Uganda

Stewart, D.W., Shamdasani, P.N. (1990). Focus groups. Theory and Practice. Applied Social Research Methods Series Vol. 20. SAGE Publications, Newbury Park, London, New Delhi

Tenywa, G. (2004). Mburo people benefit from hunting project. New Vision, Uganda, URL: http://www.newvision.co.ug. [July 2004]

The Uganda Wildlife Statute (1996), Uganda Gazette 32 (89), 1–77

Uganda Wildlife Authority (2002). Management of Wildlife Outside Protected Areas. Workshop. Unpublished Report

Part IV
Perspectives

12 Linking Case Studies to the Integrative Theory of Reflexive Dialogues

Susanne Stoll-Kleemann[1], Martin Welp[2]

1 Humboldt University of Berlin, Germany
2 University of Applied Sciences Eberswalde, Germany

12.1 Case studies in view of the Integrative Theory of Reflexive Dialogues

The Integrative Theory of Reflexive Dialogues outlined in Chapter 2 aims to link different ways of representing stakeholders' assessments and to foster the development of analytical and communication tools for management, policy, and research. In this final chapter, we want to test our theory by looking at the experiences with stakeholder dialogues and participatory procedures in natural resources management that were outlined in the case study chapters.

Questions to be discussed include:

- To what degree have truly "reflexive" dialogues taken place?
- Were the rules of the dialogues fixed or negotiable?
- How successful were stakeholder dialogues in building mutual trust and knowledge, in developing a common language, and in securing commitments (time, resources) from all participants?
- What attention did different ways of knowing (scientific knowledge, expert knowledge and lay knowledge) receive in the dialogues?
- How successful were the stakeholder dialogues in combining different knowledge bases?

The analysis of the case study chapters in the light of the Integrative Theory of Reflexive Dialogues will proceed based upon the five key concepts of the theory, namely actors, structures, methods, processes, and outcomes (see Figure 12.1, which is the same as 2.2: "Elements of the Integrative Theory of Reflexive Dialogues").

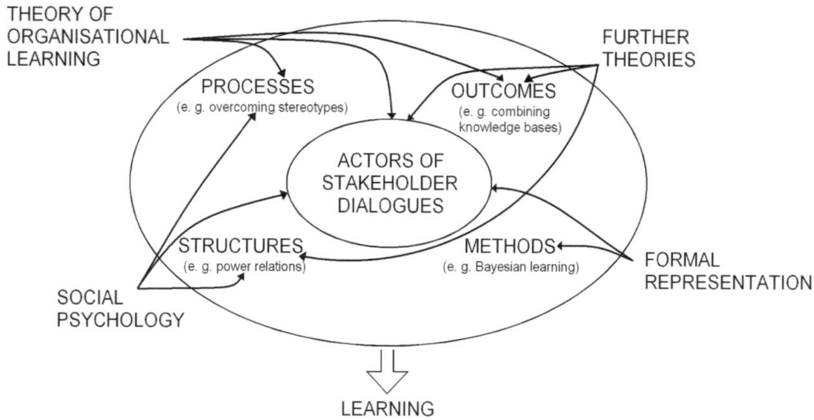

Fig. 12.1 Elements of the Integrative Theory of Reflexive Dialogues.

12.1.1 Actors: who were the stakeholders?

In this section, we ask which actors were involved in the stakeholder dialogues described in the case study chapters and to what degree they represent various target groups and play different roles. This is of importance since it is possible that actors are simultaneously members of very different social groups or organisations. A further question that will be addressed in this section is whether the role of the stakeholders varies depending on the type of dialogue and the attention cycle of an issue. Finally, we ask if the different individual preferences, values, and knowledge bases of stakeholders have been taken into account in the stakeholder dialogues that have been outlined.

The stakeholders in the scientific dialogues described in Chapter 7 (Welp et al. 2006) were representatives of various organisations that have an interest in climate change. A distinction can be made between those having a special interest in climate mitigation, i.e. how to slow down climate change by controlling greenhouse gas emissions or storing carbon, and those having an interest mainly in adaptation to climate change. The latter refers to how organisations and individuals can cope with climate change so that negative impacts are reduced. So far, these two areas have been largely unrelated. Representatives of those having a special interest in mitigation include energy companies, which have to make long-term investments under great uncertainty regarding political regulations in future, and environmental NGOs, which act as pressure groups in climate policy. On the adaptation side, typical stakeholders include those involved in various fields of natural resources management, such as forestry, agriculture, and coastal and water management. Inter-linkages between adaptation and mitigation exist on various levels. Thus an energy company may also have an interest in adaptation with regard to extreme weather events or the melting of the permafrost, which poses a threat to the energy infrastructure in Alaska and Siberia. Similarly, foresters are not only interested in adaptation, such as choosing the right tree species for the future climate. Increasingly the role of wood biomass is discussed as an option for climate mitigation and increasing the share of domestic energy supply and thus reducing dependence on fossil fuels.

The role of the stakeholders in the scientific dialogues varied. It ranged from identifying new research questions to commenting on research and model results. Very close interaction in the form of being involved in the actual research and modelling activities turned out to be very difficult. Being involved occasionally at different phases of the research process was more often the case. The preferences, values, and knowledge bases of various stakeholders were of interest for both the scientists and for other stakeholders. For this reason, having a safe space for dialogues was attractive to companies, climate negotiators, NGOs, and other stakeholder groups.

The actors in the chapter written by Hellström (2006) were the Finnish Ministry of Environment, which set up a working group comprising the scientific community, mainly specialists in ecology and protection biology. The Commission described in Hellström's chapter ("METSO" Commission) had 25 members, representing a broad variety of economic, social, and environmental interests related to forests. This new process was based on interest group representation, including scientists, to whom several roles "inside" the process were assigned. Although officially labelled as experts to the Commission, the experts were not the only scientists to participate in the process. For example, the Finnish Association for Nature Conservation appointed one of the best-known ecological scientists in Finland as its representative. That is, not all scientists involved operated from a neutral position in relation to the interests involved. Moreover, several of the people involved (e.g. the Secretary General and the vice-chairman of the Commission) had scientific educations and careers prior to their positions in administration.

The site selection for Natura 2000 areas in Bavaria, described in Chapter 9 (Eben 2006), was an issue that was of interest for different types of stakeholders. Opposition and critique towards the process and proposed sites were voiced particularly by farmers, private landowners, forest owners, and in some cases by municipalities or political leaders at the local level.

The landowners voiced critique as individuals or via landowner associations such as the Bavarian Farmers' Union, the Bavarian

Landowner and Forest Owner Associations. Governmental actors included Environment and Agricultural Ministries, agencies such as the Bavarian State Agency for Environmental Protection, and the Bavarian Forestry Agency, and furthermore the District Council. Two nature conservation NGOs were involved as well.

The landowners were not involved until a late stage, which was a main point of critique. Having been omitted from the process of selecting the sites, their role was reduced to accepting or rejecting those that had been designated. The local knowledge of landowners and environmental NGOs was not taken into account in the selection process. Thus nature conservation groups criticised that the proposed list was incomplete and did not cover all important areas.

Compensation of landowners was a further point of contention. The expectation of landowners was that any limitations regarding use rights should be compensated accordingly, a view that was not necessarily shared by those designating the areas.

A broad range of stakeholders were involved at the different stages of the German development projects in Ecuador described in Chapter 10 (Sturm and Samaniego Rivera 2006, Linke 2006). Although a distinction can be made among the international level, the national level, and the local level, the close interaction of the three was a special characteristic of the two case studies. The German Development Service (DED), the Gesellschaft für Technische Zusammenarbeit (GTZ), and the Global Environmental Facility (GEF) - as funding agencies who collaborated with national bodies such as relevant ministries and agencies - set the direction for development projects. These agencies put strong emphasis on the involvement of a variety of stakeholders, including the local people. In both Ecuadorian cases, many different groups were involved, among them the NGOs, with their focus on public health, environmental education, or natural resource management. Churches, too, were among the organisations having an interest in such projects.

Local people's involvement in wildlife management in Uganda was the focus of Chapter 11. Averbeck (2006) describes how the local population is not a single homogenous group but rather consists of several subgroups. Among the groups with divergent interests were pastoralists, cultivators, and mixed farmers. The project was interested in the beliefs and perceptions of these stakeholders. The oral tradition and legends were compiled and printed as a booklet. Institution building was one of the core activities of the project as well. The Wildlife Authority, the local authorities, and community members worked together closely to establish a pilot project for wildlife management. The efforts also contributed to institution building, and a new organisation, the Rurambira Wildlife Association, was established during the project period.

To sum up, the main actors that can be identified across the cases are scientists, international organisations, governmental organisations, the media, industry, and non-governmental organisations. Groups or individuals who were excluded from the dialogue process for one reason or another were not explicitly addressed in the case studies. A key finding in the different areas of stakeholder dialogues discussed in this book - science, policy and management - is the importance of making a proper stakeholder analysis at the beginning of dialogue activities. This type of analysis should at least identify the most relevant stakeholder groups and subgroups, predict their likely attitude toward the dialogue and the issues at hand, and gauge the kinds of power they possess. A general observation about the work at hand is that a gender perspective was not strongly emphasised in the case studies. This may have been due to the character and location of the areas selected for scrutiny.

12.1.2 Structures

In the Integrative Theory of Reflexive Dialogues, the element "Structures" encompasses some of the general conditions in which dialogues take place. Public understanding of science is one key component of such structures. In this section, we ask whether unequal distribution of the power relations among the actors involved was visible in the stakeholder dialogues analysed in the case study chapters. Power relations are one of the most important factors influencing the structures of stakeholder dialogues. In the Integrative Theory of Reflexive Dialogues, we assume that in practice, the ideal of a power-free discourse postulated by Habermas will never be achieved. Following this we aim to test whether and if so, to what degree this assumption is reflected in the case studies. That is, we hope to discover, e.g. if at least a certain awareness of asymmetric power relations can be detected or if asymmetries are addressed and corrected by applying, for example, certain rules or communication tools. A final question that is asked is whether rules and principles related to the fairness of the processes are defined and specified by the people involved in the dialogues.

Structural aspects of the science-based stakeholder dialogues were different from the case studies described below insofar as they represent structured communication processes that link scientists with actors who are relevant for the research problem at hand. Since the dialogues are not directly decision oriented, the power relations may be of less importance in comparison to the other kinds of stakeholder dialogues, although they should not be ignored, for example, in relation to power issues that are of particular relevance in the framing of issues.

Public understanding of science is more relevant within science-based stakeholder dialogues than in the other case studies because it can be regarded as a distinct approach to knowledge creation, which actively seeks to incorporate non-scientific knowledge. The dialogue initiatives considered here range from individual projects funded by different sources to the creation of platforms for dialogues (e.g. associations and forums). The stakeholder dialogues, which have focused on issues such as ecosystem service provision and European vulnerability, local forest management responses to global change, and the European Climate Forum explore how the generation of knowledge can benefit from the exchange of arguments between science and the wider society. All science-based stakeholder dialogues were perceived as fair. Problematic was that some researchers were not ready to engage in dialogues with stakeholders. Among the reasons given were that they found the uncertainty of scientific results too challenging to communicate adequately and that it was feared that this might deviate project time and resources away from "real research".

The Finnish stakeholder dialogue would have been a typical example of an instrumental utilisation of science (understood as direct influence in decision-making) if it had led to a decision on the future of the protection of forests in Southern Finland. Information needs concerning forest protection in Southern Finland were identified, scientific information was gathered, and to a minor extent also produced. While at the beginning there was considerable concern about the political utilisation of science, e.g. using science to support particular policies, by the end, the sad opposite proved to be the case. Because the information was only interpreted in an ecological framework, the working group was never given the mandate to finalise the instrumental use of science by deciding on the choice of a solution. Instead, this issue was left to an explicitly politically dominated policy process that was to follow. Concerning the question of how far rules and principles related to fairness were followed, the author of this chapter states that they (the rules and principles) were distracted by the lack of trust among the parties involved (Hellström 2006).

In Bavaria (Germany), strong opposition towards the implementation of European nature conservation law (NATURA 2000) was the starting point for the stakeholder dialogues and public participation exercises. A three-month public consultation procedure ('Dialogverfahren') was initiated by the Bavarian Ministry of Environment. The general conditions in which these dialogues took place were good insofar that in various districts, public meetings, organised by the Bavarian Environment Ministry and the Bavarian State Agency for Environmental Protection, were held beforehand. The stakeholders were informed quite fully about the proposed nature conservation sites through the use of maps and respective site descriptions. A special telephone service was offered to answer related questions. A weakness that emerges when analysing the general conditions of these stakeholder dialogues is that they took place too late in the process. Furthermore, the limited time period of three months implied that a continuous dialogue could not occur, as any exchange of opinions, concerns, or information was restricted to a short period of time. In general, the public understanding of science was low in Bavaria, e.g. there was very little understanding of why nature protection is also vital to maintain and improve local livelihoods and not only as a value in itself.

Concerning power relations, rules, and principles related to fairness, the situation in the Bavarian case study is not too bad, e.g. landowners were powerful enough to reduce the percentage of protected sites, but it could be improved by the addition of more transparency in the process. This, for example, would imply planning and preparing well the consultation process, respecting process results, and maintaining public relations with stakeholders (i.e. keeping them informed about further steps). The political will regarding power sharing is still low, but at least nature conservation officials' awareness of the necessity to integrate the factor "people" into conservation approaches is rising.

The goal of the stakeholder dialogues in the Ecuadorian National Park Machalilla was to enable the involvement of local actors in order to raise acceptance of the National Park and its management plan. A further goal was to foster active cooperation among stakeholders in the implementation of the park's management plan. The general conditions and power relations of the stakeholder dialogues in the National Park were bad due to a long history of mistrust between several actors and agencies involved in the National Park on the one hand and the local communities on the other. One problem was that the stakeholder dialogues were mainly reactive rather than proactive. Power relations and rules were described as unequal, and principles related to fairness were not sufficiently considered, e.g. the Park administration always acted in a restrictive way. Furthermore, the Ecuadorian project agencies frequently blocked stakeholder dialogues because of the possibility of losing influence and power or due to time constraints.

The objective of the second Ecuadorian project, the GTZ-supported "Community Forest Management – Esmeraldas" project (MFC-E) was to bring the various actors together in order to stop illegal resource utilisation and deforestation and to promote the development of indigenous and Afro-Ecuadorian communities. A further project goal was to promote dialogue among the large number of agencies in order to foster the exchange of knowledge and the participation of the marginalized groups in the political discussion. The general conditions for the stakeholder dialogues were good insofar that power relations and rules and principles related to fairness were considered at all levels: communities, provincial administration, and federal. The stakeholder dialogue contributed to the transparency of decisions and made financial administration comprehensible. In order to achieve more balanced power relations, it was important to bring the negotiating partners to a uniform knowledge level and to agree on rules and principles (and provide professional moderation), which guarantee equal rights and prevent the superiority of one of the actors during the dialogue. A problem related to power relations was that the most influential actors (such as representatives of the timber sector) not necessarily participated in the stakeholder dialogues because of their own strategic political intentions.

The general conditions and power relations in the Ugandan stakeholder dialogues were very good, especially compared to the previous situation when the inhabitants of the Lake Mburo National Park area had to endure a difficult period of exclusion and suppression. The conservation history of Lake Mburo National Park made it important to conduct stakeholder dialogues as the foundation for mutual planning, cohesion of stakeholders, and creation of ownership. The goal of the stakeholder dialogues was to bring together various groups, including landowners, wildlife managers, poachers, politicians, representatives of the District, and researchers, in order to discuss conservation issues, in particular opportunities for and problems of sustainable wildlife utilisation in the area.

The general conditions, power relations, and rules and principles related to fairness can be described as brilliant for the following reasons:

- the idea of the project originated from the region;
- in the planning workshops, the voices and interests of different stakeholders were heard and right from the beginning, the ownership was with the Ugandan stakeholders;
- at all levels and times, the activities and outcomes of the study were transparent.

The introductory visits, focus group interviews, presentation of study results, open discussions, and even the assessment of the wildlife population, as well as the sharing of the impala cropping revenue, took place involving community members and other stakeholders. The final results were not only made available to the technocrats but also to other stakeholders. However, some weaknesses concerning the structures remain, e.g. that political leaders and administrative officers were not involved sufficiently, and that regular reports on the ongoing activities were not circulated among all stakeholders.

12.1.3 Processes

Processes in our conceptual framework refer to meta-communication, learning and different modes of communication, and stereotyping. In this section we reflect on how the processes in the stakeholder dialogues analysed in the case study chapters took place in detail. For example, were the rules of the process agreed upon by all actors? Can they be described as consensus-seeking processes (like policy dialogues) or are they rather processes that tolerate radically different views (science-based stakeholder dialogues)?

The main analytical indicator of this section is whether learning processes on different levels (i.e. the individual level, the group level, or the organisational level) can be identified. In particular, we want to test if social and organisational learning (as outlined by Senge 1998, see also Chapter 2 Welp and Stoll-Kleemann 2006 as well as Chapter 6 Maarleveld et al. 2006) took place. This is defined as a cycle of discovering problems or issues, issue framing, public attention to a new issue, debating possible solutions, and creating instruments, policies, and management structures to cope with problems (see, e.g. Kolb's learning cycle in Chapter 6).

The processes of the science-based stakeholder dialogues, which have focused on bringing together researchers, companies and NGOs, can be described as successful learning and communication processes: the participants learnt where the areas of agreement and disagreement were and what issues needed to be further researched and debated. The strengths of the learning processes of the science-based stakeholder dialogues were the richness of perspectives and multitude of arguments. The processes were characterised by identifying socially relevant research questions, providing a critical 'reality check' for the research conducted, incorporating ethical and value considerations in assessments, and accessing stakeholders' knowledge. The science-based stakeholder dialogues furthermore showed that the social psychological theories outlined in Chapter 2 are highly relevant since group processes, in-groups and out-groups, and prejudices often play an important role, especially in the phase in which relationships between scientists and stakeholders are consolidated (trust building). Being aware of such dynamics may help to design meetings and events in a manner in which personal relationships can emerge naturally.

Process issues of the Finnish stakeholder dialogue were challenging because of its wide scope and large number of participants (see above 12.1.1 "actors"). However perhaps even more challenges were related to the novel nature of the work, differences related to information production and use, trust, commitment, funding, and innovativeness. The process of the stakeholder dialogue suffered from a substantial lack of trust, e.g. it was not possible to divide the group into smaller working groups because everyone wanted to be present in every meeting in order to be able to safeguard his or her interests at every point. Finally, although the establishment of a common knowledge base did not succeed in all aspects, giving enough time for discussion helped in clarifying some of the concepts used, in learning to communicate with each other, and in building enough seeds of trust and commitment in order to be able to continue with other working methods.

In the Bavarian case, a very confident take-up by landowner groups can be posited. Since they had been given the opportunity to oppose to site designations and reduce the originally proposed area significantly, they considered the procedure as very positive. Many people resented the sudden rush of the dialogue that had been initiated, and representatives of environmental NGOs observed that the content, the administration, and the commitment of some nature conservation officials during the dialogue procedure were of low quality. Criticisms of a 'pseudo democracy' and a fragmented selection of stakeholder consultations became loud, referring to the fact that attention was not paid to all objections or proposals. This is in line with the classification by Pretty et al. (1999), where participation by consultation is unsatisfactory, as the consulting agent is under no obligation to integrate the outcome (i.e. opinion) into the decision-making. However, acceptance of site designations increased considerably in cases where discussions and information events with local players took place. For example, the Bund Naturschutz initiated informational talks near Freising/ Munich to apprise farmers about the implications of Natura 2000 and how the scheme would affect them. Consequently, it became possible to reduce many of the farmers' fears.

The process of the stakeholder dialogues in the Ecuadorian Machalilla National Park suffered from time pressure and competition with other activities. Time pressure was in some cases the reason for bad communication: too little time was available to invest in pre-project research about the target groups, as well as about opportunities and requirements. Subsequently projects either were misunderstood or were carried out based upon inadequate or even incorrect data.

The stakeholder process in the second Ecuadorian case study profited from an analysis of perceptions of networks and cooperation because it presented strategic knowledge about the actors and alliances (the so-called "gossip matrix"). This led to more efficient communication and consultations in problem solving. Concerning the process, Linke draws attention to "the mutual scepticism that initially prevailed - rooted in the diverse cultures of communication within the Ministries, logging companies, universities, NGOs, indigenous and Afro-Ecuadorian communities". This scepticism decreased significantly after the second planning meeting; the participants got to know each other better and gained confidence. Besides the moderated task force sessions with prearranged rules (non-reflexive dialogue), the participants always had adequate time, during which the organisers set no preconditions for the dialogue. The various stakeholders appreciatively made use of these moments to discuss issues regarding their needs or interests with individual actors (reflexive dialogue).

Averbeck describes the Ugandan stakeholder dialogues as a process of problem solving. She points out that the initiators of the stakeholder dialogues worked closely with the people through local council structures and through Park Management Advisory Committees that were set up to act as a link between the local community and the park management.

Various factors and phases were important for the success of the (communication) process of these stakeholder dialogues. These procedural success factors included considerable knowledge of socio-cultural patterns and ethnic behaviour of those involved in the stakeholder dialogues, consideration of the specific historical and political situation of that specific area of Uganda, and understanding and respect for the situation of the groups. The moderator of the stakeholder dialogues furthermore tried to adapt to newly arising challenges or opportunities that had an impact on the site or the people.

During the whole process, the existing attitudes and different knowledge bases were appreciated and utilised. The organisational structure and the management concept of the wildlife utilisation scheme were developed on the basis of interviews and discussions. An extraordinary part of the process – which is at the same time an outcome as well (see below) – is the sharing of impala cropping revenue. The process is still ongoing insofar as the responsibility for further activities is now in the hands of the stakeholders, and no funding or organisational assistance is available through local or foreign donors. One example of the ongoing process beyond the stakeholder dialogues in the Ugandan Lake Mburo National Park is the establishment of a Wildlife Association consisting of landowners from the area east of Lake Mburo National Park without external financial support. Furthermore, landowners started to support law enforcement programs by reporting illegal activities of poachers to the park officials.

12.1.4 Methods

In this section we ask which methods were used in the stakeholder dialogues in the case study chapters, whether they matched with the objectives of the dialogue, and if the experiences with the use of different methods were seen as successful or not. In Chapter 6, Maarleveld describes how various GIS-based and visualisation tools can enhance learning in stakeholder dialogues and public participation. In particular, place-based management decisions can benefit greatly from the use of such tools. In other kinds of dialogues, such as science-based or policy dialogues, visualisation of model results or alternative policy strategies can spark dialogues and help to identify areas of disagreement and consensus.

Typical methods used in science-based stakeholder dialogues on climate change, as described in Chapter 7, included joint studies, different kinds of events enabling direct communication, and dialogues such as conferences, workshops and teleconferences, as well as methods of collecting information and insights (questionnaires and interviews). A distinction can be made between communication tools and analytical tools - both of which help to formalise actors' mental models and assessments - and communication tools for facilitating dialogues. Some of the most interesting analytical tools include Bayesian learning, multi-criteria analysis, and the Q Methodology. These were not applied in the scientific dialogues described in the chapter, the interest being rather in the further development of such tools for later use.

The methods applied in the policy case from Finland (Chapter 8) included hearings, a committee, and several working groups with stakeholders from different backgrounds and with varying interests. The issue at hand, protection of biodiversity in forest ecosystems in Southern Finland, was conflict laden, and the relationships among several stakeholders had been negatively influenced in previous disagreements over nature conservation. Thus the facilitators had to be sensitive with respect to these open and latent points of contention.

The controversy in the Bavarian case was also about nature conservation although on a different level than in the Finnish situation. The implementation of EU policy, in this case the Habitats Directive, was done to a great extent without any participatory procedures. Eventually a consultation procedure was launched by the Environmental Ministry of

Bavaria, but only in face of heated conflicts – far too late, as Eben observes. This shows that not only is the selection of participatory methods essential but also the timing of their use. The methods applied in this consultation process, which was a novelty in this form in Bavaria, included public meetings and dissemination of relevant information about the proposed sites in printed versions, on CD-ROMs, and via the Internet. Furthermore, a telephone hotline was launched to meet the information needs of various stakeholders. As tools for conflict management, those applied can be regarded as necessary components but insufficient if used alone.

In the Machalilla National Park in Ecuador, the main problem with the participatory methods applied was similar to the Bavarian case in that they were introduced too late. No efforts were made during the phase of the park's establishment to involve local stakeholder groups in the process. Thus, many viewed subsequent efforts with great scepticism. Two problems contributed to the problems with participatory methods: the lack of awareness and the inadequacy of the training of the national park staff.

In the other example from Ecuador, the focus was on community forest management in Esmeraldas and in particular in incorporating the knowledge and experience of indigenous and Afro-Ecuadorian communities. A further aim of the dialogues was to foster cooperation between different agencies that often work on the same issue or target group, but which do not inform each other or know enough about each other's work. The methods applied, Participatory Rural Appraisal (PRA) and a sequence of participatory planning workshops, seem to have fit the objective of the project well.

The methods used in the Ugandan case of community based wildlife management included workshops, focus-group interviews, and collecting fairy tales. The use of a combination of different methods is a positive feature of this project. The stakeholder dialogues laid the basis for joint planning and implementation. An interest in the fairy tales, which related to wildlife, indicated respect for the local culture. Such recognition is likely to enhance trust building and thus make future dialogues easier to conduct.

12.1.5 Outcomes

In this section, we ask how the outcomes of the stakeholder dialogues analysed in the case studies are described and whether they can be deemed successes or not, e.g. if (joint) problem solving and network formation of the actors and/or groups involved took place. Another question is whether stakeholder dialogues have contributed to attitude and behaviour change, i.e. because people were confronted with new information and experiences. Two more issues are whether constructive conflict management took place and if a consensus view emerged in the process of the stakeholder dialogues.

One of the outcomes of the science-based stakeholder dialogues is that the dialogues helped to understand stakeholders' concerns, problems, restrictions, and uncertainties in the research fields investigated. Through the science-based stakeholder dialogues, scientists were sometimes led to question the basic assumptions, methodologies, and indicators used in their scenarios and models, as well as the meaningfulness of the models themselves and of their results (including temporal and geographical scales in particular) for stakeholders.

Stakeholders' suggestions were addressed when time and resources allowed (e.g. new case studies, new focuses for Ph.D. theses, adapted indicators); otherwise their suggestions were used to draw a future research agenda, which fed ongoing research proposals. Stakeholders made clear requests regarding the scientific credibility of the selected approaches and for the need of transparent aggregation and integration of the results. The science-based stakeholder dialogues themselves have been a powerful driver for more inter- and transdisciplinarity and a continuous help in the processes of focusing and prioritising research efforts and resources to better address stakeholders' needs. Good research increasingly takes place in small interdisciplinary teams in which the individual scientists meet regularly for collective thinking. Scientific stakeholder dialogues are an extension of this practice and an effort to bring together not only different academic disciplines but also the various knowledge realms outside the scientific domain.

The idea of the Finnish stakeholder dialogue was that compiling ecological information would facilitate further decision-making in a subsequent multi-stakeholder process. However, this did not prove to be the case. Although the contents or conclusions of the process were only questioned to a minor degree, there was reluctance among many stakeholders to utilise the findings of the working group because its members were involved neither in the process nor in drawing up the conclusions. In terms of content, the outcomes of the Finnish working group reported the need for better protection of herb-rich forests, mineral-soil sites with abundant decayed wood, and spruce mires in southern Finland. The recommendations were made on ecological-scientific grounds only; they did not consider the potential social or economic impacts. Although a large amount of information was compiled by the groups of specialists assigned to evaluate the status and needs of forest protection in Southern Finland, the Finnish stakeholder dialogues were accompanied by problems of availability and credibility of information from beginning to end. Thus, problems related to a lack of trust and commitment were not solved until the stakeholder dialogues were over. Instead of producing, disseminating, or sharing information related to the substance of the work, the scientists offered procedural support to the process.

Concerning the outcome in the Bavarian stakeholder dialogues, it can be stated that despite various deficiencies, at least a first positive step towards including stakeholders in policy-making was made. Overall, however, the necessary pre-requisite conditions for protected area management as stipulated by Natura 2000 were not established by these participatory procedures. This situation points out the need for a new, innovative approach that builds on social self-esteem through expanded participatory involvement and stronger emphasis on rural sustainable livelihoods, rather than being based purely on compensatory measures that ignore wider social dimensions.

Sturm, the author of the first Ecuadorian case study, regards the outcomes of the participatory processes in the Machalilla National Park as disappointing. According to him, the reasons for this were lack of continuity, human resources, and content, as well as inadequate training of staff in the methods of participatory techniques. In terms of environmental education the project was successful. Many different dissemination channels were used including newspapers, radio, and theatre. Involving young people in environment-related activities was a further accomplishment of the project.

Linke, the author of the second Ecuadorian case study, points out that the stakeholder dialogues in his project were perceived positively and appreciated by most of the actors as an opportunity for direct exchange and dialogue. The outcomes were active feedback from the partners (feedback loops) and the adjustment of the ongoing planning process. The coming together of the actors furthermore enabled the Ministry of Environment to recognise and utilise the possibilities for the enforcement of national legislation. In cooperation with the Ministry of the Environment, didactic materials and methods were developed on how the reformed forestry legislation could be enforced on the local level. A factor that had negative impacts on the otherwise successful outcomes is the widespread corruption levels in the forestry sector. The financially influential timber companies, which have placed strategic allies in positions at all levels, were able to promote or annul political decisions or even quickly machinate changes in existing legislation.

The outcomes of the participatory approach of the Ugandan case study comprised a vivid stakeholder dialogue that laid the foundation for collaborative planning, transparency, and the cohesion of stakeholders. Furthermore, it created ownership and commitment amongst nearly all relevant stakeholders. As a result, a community pilot project for wildlife utilisation was established adjacent to the Lake Mburo National Park (LMNP) that helped to conserve the protected area. Furthermore, in the Ugandan stakeholder dialogues, emphasis was put on environmental education, capacity building, and support for community development. However the ultimate achievement of community conservation at LMNP from a conservation perspective was the ways in which it changed local communities' perception about conservation. The initial contribution has been positive in helping to reduce the anti-wildlife values that were so strongly held by local people in the late 1980s.

Although the aim of the study was not to implement a wildlife utilisation project around LMNP and the outcome of the study was open, the participatory approach led to the implementation as it created ownership and commitment by the different stakeholders. In 2004, due to the success and the demands of other communities, the project was extended to two other parishes. From 2001 to 2004, the Rurambira Wildlife Association received US$ 50,000 through the project, which was used to build two schools, six teachers' houses, a health centre, and a dam for providing water for livestock. The project also led to the increase in the number of wild animals, which has mitigated disputes between wildlife authorities and pastoralists and reduced poaching and charcoal burning in the area. The realisation of economic value from wild animals on private land by residents contributed to a positive change in attitude towards wild animals on these properties.

Problematic outcomes of the Ugandan effort include the fact that the wildlife utilisation project did not create a win-win situation that was equal for all stakeholders; namely, the interests of most of the poachers were not considered. Although it was possible to employ some of the poachers' leaders as guides and porters, the project did not engage all of them. Moreover, community members started to support law enforcement programs by reporting illegal activities by poachers to park officials. Also, the project increased the expectations amongst residents in areas, which the pilot project has not yet succeeded in meeting.

12.2 Analysis of the case studies using the book's other theoretical approaches

In this section, we will expand the theoretical reflections to encompass the other theoretical approaches outlined in the first part of this book – in particular those related to development thinking and tool development. In development cooperation, an independent body of literature exists that focuses on specific issues of democratisation, the role of donor agencies with respect to local development, and the power of various groups in development work. Concerning tool development, Chapter 5 by Scheffran and Chapter 6 by Maarleveld et al. are especially relevant. Scheffran outlines seven approaches in the area of stakeholder modelling and simulation, including game theory, agent-based modelling, and dynamic systems theory (see 12.2.2). Maarleveld et al. describe how geo-visualisation tools are important for spatial planning and analysis, as well as for assessment of place-based information.

12.2.1 Participation and development

The underlying theoretical foundations of Berghöfer and Berghöfer's chapter on "Participation in Development Thinking" are relevant to the case studies in developing countries in particular: the two in Ecuador and the one in Uganda. We have already discussed two of the proposed four questions: "Who participates?" (actors) and "How does the process take place?" (processes). Thus in the following, we focus first on the axes "participation in what dimensions and for what purpose?" The first two questions, however, also deserve some attention, as they are discussed here in the specific context of development.

Participation as defined by Berghöfer and Berghöfer can have the following dimensions: economic, political, social, and at the project level. The Ugandan case included three of these. First, the economic dimension: through impala cropping, people were able to generate income. Second, participation was a process of social learning and mobilisation involving all important stakeholder groups. Third, on the project level, local knowledge was of high importance. The project was primarily a research endeavour, and thus the use of interviews, group discussions, and other participatory research methods suited the objectives well. Furthermore, these supported joint action of community-based wildlife management.

In the two Ecuadorian cases, the dimensions of participation were quite different from each other. In the Machalilla National Park, the main focus was at the project level. It can be argued that the intention of the project was to broaden this scope and to include economic and political dimensions as well. However due to various difficulties and the short time period of the rather disjointed activities, the economic and political dimensions of participation remained weak. Social learning and mobilisation took place to some extent but not beyond the concrete project activities. In contrast, community forest management in the Esmeraldas participation was strong on the political and social levels: building alliances among different actors and addressing the need for better integration between the local, provincial, and national levels were major objectives of the dialogues activities. The direct economic dimension was not the primary focus in this project.

Berghöfer and Berghöfer identify aid efficiency, mutual learning, empowerment, and inclusion of the marginalized voices as main purposes of participation. In the Ugandan case study, one can state that 'participation' really was understood in the sense of 'empowerment' of the poor in a natural resource context. Very marginalized voices, such as poachers', were taken into account in the assessment of the situation.

In the first Ecuadorian case, the Machalilla National Park, the German Development Service (DED) was the actor that got involved in the park's management. The organisation views "modification of social relationships and power structures" as a goal of participation development cooperation. This objective was only partly met through the participatory activities. The other Ecuadorian case, namely the Esmeraldas Community Forest Management Project, was evaluated by the participants themselves. In this project, the mutual scepticism that initially prevailed - rooted in different cultures and ways of communication - gradually vanished. This can be regarded as a good basis for learning and empowerment.

Furthermore, Berghöfer and Berghöfer emphasise that the question of who participates is a problematic issue in the development context. They identify three thematic areas that have received attention in recent literature on development cooperation: the dangers of localism, the myths about community, and the identification of stakeholders. In the Ugandan case, the danger of localism was of primary concerns. The new wildlife management policy and the new agency responsible for wildlife management in Uganda strongly supported the management by local communities. Different categories of use rights, such as hunting, farming, ranching, and the use of wildlife products were handed over to local people, motivating them to use their wildlife resources in a sustainable way. With respect to the second critical issue, communities were never regarded as monoliths. Rather, different sub-groups, such as pastoralists, cultivators, mixed farmers, and poachers, were identified and approached. This was done in ways that suited each group respectively. The identification of stakeholders was carefully done based on respect and taking into account possible problems in intercultural communication. Thus, mistrust towards outsiders' engagement in wildlife-management issues was overcome.

With respect to the process of participation, Berghöfer and Berghöfer identify four critical issues: the propriety of rules, costs and benefits of participation, the role of the facilitator, and procedural justice. In the Ugandan case, the participatory process was acceptable with the existing rules of the communities, even though the legal and political situation in the country was very uncertain. Because of the respect demonstrated for local customs and beliefs, no serious conflict in this respect emerged. The costs and benefits of the participatory procedures were in balance in this process. Since the project was framed as a research process and was done with limited resources, local people assumed responsibility for the management of the wildlife. A major role of the facilitator was her being present to help the public discuss and scrutinise ideas that were developed at the local level. Procedural justice was secured so that information, insights, as well as perceptions were collected by interviewing and discussing with people. As a result, not only strong interest groups had the opportunity to express themselves; a broad range of individuals were also consulted.

In the two Ecuadorian cases, the experiences with stakeholder dialogues were very different. While the outcome of the dialogue exercises was less satisfactory in the Machalilla National Park, the community forest management approach adopted in Esmeraldas was in many respects successful. The participatory activities in the Machalilla National Park lacked true impact since an empowerment of local communities was not high on the agenda. Various stakeholder groups were identified and different activities were carried out with the respective groups in mind. In Esmeraldas, the dangers of localism were obvious as, for example, the forest sector was influential on all levels of policy-making. Thus aims of the stakeholder dialogues included improving interagency cooperation across different levels and increasing the availability of information concerning land ownership rights, utilisation rights, legislation, and other key issues influenced by administrative levels above the local one.

With respect to the processes, the examples from the Machalilla National Park were of very different kinds, ranging from activities with theatre groups to the establishment of tree nurseries. The activities were action oriented and depended on the enthusiasm of the groups involved, and their benefits must be assessed as limited due to the intermittent nature of the efforts. To what extent the rules of the dialogues and procedural justice were an issue in the two cases was not extensively described in the case studies. Self-evaluation in the Esmeraldas case can be regarded as a sign of a high degree of reflexivity.

In the Machalilla National Park, the role of the facilitator was to establish trust (which turned out to be extremely difficult due to mistakes made at the time of the park's establishment) and to initiate projects. The situation in Esmeraldas was different: with respect to the role of the facilitator, the GTZ never sought to be the implementing agency. Rather, each activity was preceded by a comprehensive agreement regarding the expectations of all parties as well as a definition of goals.

12.2.2 Stakeholder Dialogues and Tool Development

In the following, we will have a closer look at the interlinkages between theory and tools. Chapter 5, by Scheffran, gives an overview of modelling and computational tools that can be used in stakeholder dialogues and public participation. Chapter 6, by Maarleveld et al., gives an overview of GIS and visualisation tools. Participatory natural-resource management can greatly benefit from analytical tools that contribute to the understanding of stakeholders' behaviour by depicting it, help to structure and analyse complex issues and preferences, or can be used for fostering dialogues. Furthermore, visualisation tools help stakeholders to better understand the problem situation and create a shared understanding. In the evaluation of stakeholder dialogues (see Oels's overview of approaches to evaluation in Chapter 4), the selection of appropriate tools is an essential question. Oels emphasises that the selected tools have to fit the objectives of the dialogues. They should be easy to apply and support dialogues, and not take limited time from stakeholder interaction.

Scheffran distinguishes among seven approaches under the rubric stakeholder modelling and simulation: computer simulation and dynamic systems theory, decision analysis, game theory, including dynamic games, agent-based modelling, qualitative reasoning, and viability theory. All remain essentially in the scientific domain, and their application in natural resources management and policy-making has so far been rare. The same holds true for tools for environmental conflict resolution, mediation, negotiation support, and group decision-making. The user-friendliness of Decision Support Systems (DDS) has greatly improved in recent years. More intuitive user interfaces combined with possibilities to visualise outcomes as graphs or maps have made DDS interesting for facilitators of dialogues in natural resources management.

Simulations and agent-based models have become increasingly attractive for coupling and embedding social interaction into environmental models. Such tools can be helpful for a better understanding of how complex systems work and what role each actor in this system has. Scheffran provides an overview of such agent-based models in different areas of natural resource management, including water, land-use, agriculture, and recreation.

Models that suit various stages of stakeholder assessment and management thus exist (see Figure 5.1 in Scheffran's chapter), but their broader use in real life settings requires training and education. A person who wants to use such tools requires an understanding of the underlying assumptions and the limitations and areas of application of each tool, and stakeholders may be sceptical about the methods' net benefit or the results such tools deliver. Although it may not be necessary to know each mathematical algorithm in detail, the facilitator should be able to explain the theory that lies behind each tool.

In summary, modelling and computational tools have not yet been widely used in stakeholder dialogues. They have usually been restricted to use in research projects or in pilot projects with a strong research component. Except for the science-based stakeholder dialogues described in Chapter 7, methods of this type were not used in the case studies of this book. In the stakeholder dialogues on climate change, computer models were not developed by the stakeholders but rather by scientists. However, the structure of the models as well as results of model runs were presented and discussed in varying degrees of detail.

People with no profound mathematical training, including facilitators and participants, may be reluctant to learn and apply computer-based tools. On the other side, there are new pieces of software, including modelling tools (such as Vensim) and approaches based on Bayesian learning, which have a relatively short learning curve. Using these methods, of course, requires time and effort, and the benefits of such approaches have to be weighed against the costs of using them. When the issues at hand are very complex and uncertainties are high, the use of modelling and computational tools becomes attractive for people working in the field of natural resources management, which is why GIS tools have been widely adopted in this field. Watershed management (see Maarlevelds' et al. Philippines case study), forest management, coastal management, biodiversity management, and many others are realms where GIS tools can improve planning procedures and help to incorporate different knowledge bases. Extending the use of such tools to show how climate change may impact natural resources management exemplifies a potential future uses of computer-based tools and models.

12.3 Conclusions

Stakeholder dialogues in natural resources management are becoming an important part of processes that require consultation and learning among a broad range of actors. In some areas such as corporate management and development cooperation, it has become a well-established practice or a routine operation, while in other areas such as research and policy-making, stakeholder dialogues are still the exception. Various degrees of stakeholder dialogues can be identified and a minimal level of often unplanned, ad hoc dialogues takes place in many processes (e.g. via the press). In the present work, we have focused on dialogues in natural resources management on the conceptual level that have been properly planned and professionally carried out.

Numerous methodological improvements have taken place in recent years, and the practitioner facilitating stakeholder dialogues can choose from a toolbox with various communication and analytical tools. The use of analytical tools, however, has often been limited to stakeholder dialogues with a strong research component. The use of analytical tools such as computer-based models, multi-criteria analysis, and Bayesian belief networks requires some training and interest in mathematical ways of presenting positions, interests, and preferences. The educational background and profile of a facilitator typically manifests no strong focus on modelling or mathematical representation. Rather the focus is on moderation techniques, peoples' skills, and other qualities often referred to as soft skills, including empathy, the ability to listen to people, respect for others and their opinions, and a measured style of communication. These are key skills for an individual engaged in stakeholder dialogues. In the face of increasing complexity of natural resource management challenges, the training of facilitators should, however, be widened to encompass analytical tools and the use of scientific insights. Teams whose members complement each other with respect to the above-mentioned communication and analytical skills are well suited to facilitate various kinds of stakeholder dialogues in natural resources management as well as in other areas.

In literature on natural resources management, a distinction is made between public participation and conflict management. Although they overlap in some of the methods and approaches used, they are driven by different values: public participation by empowerment and democratic ideals – conflict management by effectiveness in environmental decision-making and accelerating processes. How, then, do stakeholder dialogues relate to these two concepts? The following distinction between conflict management and stakeholder dialogues can be made: conflict management aims at consensus building and avoiding the escalation of conflicts. Stakeholder dialogues, in contrast, aim at exploring a range of options while identifying areas in which stakeholders agree and others in which they do not. In comparison to public participation, stakeholder dialogues are more selective with regard to participants. While, basically, anyone should be given the opportunity to be involved in public participation, stakeholders represent key groups that influence or can be impacted by a policy, project, or idea.

The case studies demonstrate that public participation, stakeholder dialogues, and conflict management in natural resources management can promote learning by all parties. Empirical evidence from case studies shows that success factors (What facilitates learning?) include early engagement, defining the objectives jointly, and agreeing on the rules of the procedure. Failure factors (What hinders learning?) include among others the use of inappropriate communication tools in conflict situations and distrust among those participating in the dialogues.

Three general conclusions can be drawn from the case studies:

- stakeholder dialogues and public participation in natural resources management are worth the effort, but there are implementation barriers that need to be eliminated;
- in the field of tool development and application, a synthesis of analytical and communication tools needs more attention; and
- in future research and dialogues practice, systematic and comparative evaluation needs to be emphasised.

In management, policy-making, and research, stakeholder dialogues are likely to play an increasingly important role. The Integrative Theory of Reflexive Dialogues and the case studies in this book have sought to help individuals involved in dialogues or in tool development to improve their practice and to call attention to issues that are critical factors for the success or failure of their efforts.

A challenge remains in identifying how the different levels - management, policy-making and research - can be united into stakeholder-oriented, sustainable governance of natural resources.

References

Averbeck, C. (2006). Incorporating Local People through Economic Incentives at Lake Mburo National Park, Uganda – Africa works! In: Stoll-Kleemann, S., Welp, M. (eds). Stakeholder Dialogues in Natural Resources Management. Springer-Verlag, Heidelberg

Berghöfer, A., Berghöfer, U. (2006). ‚Participation' in Development Thinking – Coming to Grips with a Truism and its Critiques. In: Stoll-Kleemann, S., Welp, M. (eds). Stakeholder Dialogues in Natural Resources Management. Springer-Verlag, Heidelberg

Eben, M. (2006). Public Participation during Site-Selections for Natura 2000 in Germany: The Bavarian Case. In: Stoll-Kleemann, S., Welp, M. (eds). Stakeholder Dialogues in Natural Resources Management. Springer-Verlag, Heidelberg

Hellström, E. (2006). Science in Support of the Forest Biodiversity Programme for Southern Finland. In: Stoll-Kleemann, S., Welp, M. (eds). Stakeholder Dialogues in Natural Resources Management. Springer-Verlag, Heidelberg

Kolb, D. (1984). Experiental Learning: Experiences as the source of learning and development. Englewood Cliffs: Prentice-Hall

Linke, J. (2006). Community Forest Management in Esmeraldas - Is Constructive Dialogue Possible? In: Stoll-Kleemann, S., Welp, M. (eds). Stakeholder Dialogues in Natural Resources Management. Springer-Verlag, Heidelberg

Maarleveld, M., van de Velde, R., van Uum, J., Pleisier, I. (2006). To See or not to See, that is the Question: Geo-information Visualization Tools as a Means to Facilitate Stakeholder Dialogues in Land and Water Management Planning. In: Stoll-Kleemann, S., Welp, M. (eds). Stakeholder Dialogues in Natural Resources Management. Springer-Verlag, Heidelberg

Oels, A. (2006). Evaluating Stakeholder Dialogues. In: Stoll-Kleemann, S., Welp, M. (eds). Stakeholder Dialogues in Natural Resources Management. Springer-Verlag, Heidelberg

Pretty, J., Hine, R., Deighton, K. (1999). Participatory Appraisal for Community Assessment: Principles and Methods. Centre for Environment and Society, University of Essex. URL: http://www2.essex.ac.uk/ces//ResearchProgrammes/pa&caoverview.htm

Renn, O. (2006). Participatory Processes for Natural Resource Management. In: Stoll-Kleemann, S., Welp, M. (eds). Stakeholder Dialogues in Natural Resources Management. Springer-Verlag, Heidelberg

Scheffran, J. (2006). Tools for Stakeholder Assessment and Interaction. In: Stoll-Kleemann, S., Welp, M. (eds). Stakeholder Dialogues in Natural Resources Management. Springer-Verlag, Heidelberg

Senge, P. (1998). The fifth discipline. The art and practice of the learning organisations. Century Business, London

Sturm, M., Samaniego Rivera, J. (2006). Participation in the Machalilla National Park, Ecuador. In: Stoll-Kleemann, S., Welp, M. (eds). Stakeholder Dialogues in Natural Resources Management. Springer-Verlag, Heidelberg

Welp, M., Stoll-Kleemann, S. (2006). Integrative Theory of Reflexive Dialogues. In: Stoll-Kleemann, S., Welp, M. (eds). Stakeholder Dialogues in Natural Resources Management. Springer-Verlag, Heidelberg

Welp, M., de la Vega-Leinert, A.C., Stoll-Kleemann, S., Fürstenau, C. (2006). Science-based stakeholder dialogues in climate change research. In: Stoll-Kleemann, S., Welp, M. (eds). Stakeholder Dialogues in Natural Resources Management. Springer-Verlag, Heidelberg

Epilogue

Tim O´Riordan

University of East Anglia, Norwich, United Kingdom

Spreading the Ripples

Providing opportunities for a wide array of interested parties to be satisfied by an integrated management decision is the stone that causes the ripples. Where the ripples move depends on the depth and obstructiveness of the water body, in this case the powers that shape how decisions are made. Just because there are ripples does not mean that they reach the shore. Offering participation can mean surprisingly little: those who must be involved are there already. Those who shout loudest make it their business to be there, however inconvenient. Those who may not realise their ultimate interests could well be affected by a decision outcome may only enter when advised or encouraged or enabled to do so. And there will always be many who simply do not want to be part, who cannot know in any reasonable way that their interests are relevant to the decision outcome, yet who may experience costs or suffering by being absent.

In short, stakeholder dialogue is usually a discussion amongst known and informed people to the party. The ones who also count may not be there either because they have sufficient power and influence to bend this outcome in their direction, so can be absent simply because of their virtual presence in the decision setting. They still command power even when not present. Others who are absent are those who are often included by a combination of ignorance, alienation or distraction by other demands on their distracted lives. Stakeholder dialogue may be mirror on social power relations and on the institutional design of decision-making.

Before exploring the scope for reinterpreting stakeholder dialogue, it is necessary to set the context in terms of fresh approaches to national resources management and integrated assessments. The Millennium Ecosystem Assessment set the scene for interpreting national resources in the context of ecosystem services. No normal resource resides outside its placenta of ecosystem functioning that makes up the web of life. There is growing realisation that these functions are hugely valuable for human existence and economy, but that they are also seriously endangered by inappropriate and incomplete integrated assessments of reduced natural resource management. Hence, not only are there potentials for combinations of ecosystems functioning failure, with unknown implications for human livelihood and well-being, let alone ecological viability. There is also no sound scientific basis for conducting the kind of genuinely integrated, integrated assessments that should be able to grapple with such outcomes.

This valuable collection of essays provides a basis for reflecting further on the theory and role of stakeholder dialogue. To begin with the notion of stakeholder may be misleading since key individuals or groups who may be absent do not declare their "stake". We may need to reflect on another set of names for participants. Power participants influence outcomes by virtue of their political bargaining strength or their economic ascendancy. Owners of resources and land have legal property rights that give them a variant of power. They do not need to be present to exert their stake. So they may not enter "dialogue". Illegal and corrupt interests may have huge bargaining power over regulators, non-governmental organisations and politicians. Again, they are neither present nor talking (and certainly not listening), yet they influence the "stake". Politicians influenced by the demands of established lobbies, or specialised entrants, may also not enter any dialogue, for they will have their own agendas. Citizens whose well-being may be influenced by a natural resources decision in say 20 years time, because of associated failure of ecosystem services, will fail to declare or even know of their interests, and hence not declare a "stake" or participate.

So the very notion of stakeholder needs careful attention in both theory and practice. Those who control the agenda need not be present: they are "second dimension" political interests. They are insidious because they infiltrate coalitions, enter into behind the scenes deals and adjust the biases in regulation, patterns of environmental quality or of natural resources sustainability.

The "third dimension" interests may simply not know what their long-term well-being might be, and how it might be affected by a set of natural resources decisions that could well affect their livelihoods and happiness. This is because we are only beginning to realise the wide range of arguments relating to alteration of ecosystem functioning that could result for a change in land use or climate change related factors. For example, the steady toxification of soil due to prolonged intensive agriculture, or because of "rained out" nutrients landing on catchments from air pollution would result in prolonged and indeterminate human health effects in two generations time. We do not know, but there is evidence for soil toxicologists that prolonged deterioration to ecosystems corrupted by toxic additions, may result in food and water contamination that, as yet, has not been removed.

Meddling with ecosystem services can result in prolonged and pervasive long-term consequences for which communities involved are fully prepared. It is almost impossible to incorporate them in any stakeholder dialogue. They are wrapped in ignorance, and distracted by other demands on their attention to make room for dialogue. Even the heady application of the precautionary principle may not bring them in. Much as the precautionary principle has to recommend it for alerting future consequences via uncertain pathways, when the uncertainty is two generations away at least and the pathways worthy of a braided river, there is no feasible way of alerting their attention and encouraging effective involvement.

All this is telling us that the theory of stakeholder dialogues is skewed in two contrasting directions. On the one hand are the second dimension power absentees who manage the short term in their interests and set in train a huge array of possible damages to critical ecosystem functions. On the other there are the third dimension absentees whose "real future" interests are possibly affected by the steady breakdown of ecosystem services, yet who cannot sensibly get involved at a suitable point on the decision channelling.

The result is a heavily distorted picture of stakeholder dialogue that by no means guarantees either long-term sustainable outcomes or overall human well-being in resistant natural processes. Yet surely such an eminently desirable outcome is at least part of the purpose of stakeholder dialogues.

Hence it will be necessary to rechart the character of "interests inclusion" for future natural resource management and integrated assessments. There is no ready answer to how this can be addressed, but here are some thoughts on possible ways forward.

- *Establish mechanisms for exploring long term consequences for ecosystem functioning arising out of all national resource decision making.*

This could be done by a series of community-scientist-planner meeting arrangements designed to explore the likelihood of certain clusters of outcomes arising from particular natural resources, such as water use, coastal redesign in the face of sea level rise, soil care, and whole landscape sustainable stewardship. In essence, the aim would be to establish a setting for exploring a range of outcomes, and set these against the highest standards of sustaining nature and all identifiable social interests.

– *Imaging scenarios with break points.*

These scenarios of possible ecosystem functioning futures need to be shaped by a range of citizen's groups and science-regulatory interests operating a very free flowing manner. Such "floating" groups would deliberately target schools, young people (future residents) and those who may not immediately perceive their interests. The future images need to be realistic, challenging, fully supported by hypothesis and other conditions, and presented in such a way that the various future states are shared by the participant clusters acting in real "dialogue" of open ended creative learning.

– *Exposing the power relations.*

The scenario groups need to be enabled to become award of the layers of power that surround all natural resources decisions. This can be achieved by an equivalent set of "stories" of how power and interest coexist in natural resources management, who wields it and why, and what mechanisms are possible to incorporate new power relations into the setting. Such a precedent will rely on direct engagement by politicians and by knowledgeable insiders. Here is where the regulators and donors may play a role. It may be necessary to require the decision pathway to reduce such "power scenarios" as part of the political possibly legal framing of the ultimate decisions. In addition, such "power scenarios" may need to be talked through by special training and awareness raising sessions.

All of this may appear heavy handed. But in the context of possible long term damage to critical ecosystem functions, such a procedure may become a vital compliment of integrated assessment.

– *Sequential monitoring.*

If the Millennium Ecosystem Assessment has any meaning, it is that biodiversity is losing out and that ecosystem nurturing of species, habitats and ultimately, humans, is diminishing. These important additions of stakeholder interest and dialogue are necessary if there is to be any serious assurance of functioning national processes in two generations' time. Hence regular monitoring, regular correction of initial decisions, regular dialogue amongst the two sets of scenario groups will be necessary if natural resources management is to be truly sustainable. Frankly there is far too much at stake now to shift the ripples so they actually reach a shore that is recreated by their energies.

Index

Printing: Krips bv, Meppel
Binding: Stürtz, Würzburg